T0304623

UNIVERSAL ALGEBRA and APPLICATIONS in THEORETICAL COMPUTER SCIENCE

UNIVERSAL ALGEBRA and APPLICATIONS in THEORETICAL COMPUTER SCIENCE

Klaus Denecke
Shelly L. Wismath

CRC Press
Taylor & Francis Group
Boca Raton London New York

CRC Press is an imprint of the
Taylor & Francis Group, an **informa** business

A CHAPMAN & HALL BOOK

Chapman & Hall/CRC
Taylor & Francis Group
6000 Broken Sound Parkway NW, Suite 300
Boca Raton, FL 33487-2742

ISBN-13: 978-1-58488-254-1 (hbk)

Visit the Taylor & Francis Web site at
http://www.taylorandfrancis.com

and the CRC Press Web site at
http://www.crcpress.com

Library of Congress Card Number 2001054803

Library of Congress Cataloging-in-Publication Data

Denecke, Klaus.
   Universal algebra and applications in theoretical computer science / Klaus Denecke, S.L. Wismath.
      p. cm.
   Includes bibliographical references and index.
   ISBN 1-58488-254-9 (alk. paper)
   1. Algebra, Universal. 2. Computer science—Mathematics. I. Wismath, Shelly L. II. Title.

QA251 .D385 2001
512—dc21   2001054803

Introduction

An algebra is a structure consisting of one or more sets of objects and one or more operations on the objects. Such structures occur in all areas of Mathematics and Science: sets of numbers with operations such as addition and multiplication, sets of relations or functions with the operation of composition, sets of matrices with operations of addition and multiplication, sets of propositions with operations of conjunction, disjunction, and negation, finite automata with a set of inputs, a set of outputs and a set of states, with state transition and output functions, and so on. In each case we are interested in the properties of the operations involved, what their arities are and what laws or axioms they satisfy. In particular, any algebra has a type associated with it, which specifies the number of operations involved and the arity of each one. Most students of abstract algebra begin their study with groups, which are sets with one binary operation which satisfies four specific axioms or properties. The example of groups shows that not only single algebras but classes of similar algebras having certain properties in common are interesting and important.

General or universal algebra is the study of algebras and classes of algebras of arbitrary types. There are two main approaches to the study of such abstract algebras. In the first approach, we look for constructions on algebras which produce new algebras of the same type. For instance, we can look for subsets of the original set which inherit the same operations and properties, and for mappings between sets which preserve the operational structure. There are three main constructions available for producing new algebras from given ones: the construction of subalgebras, homomorphic images and product algebras. These concepts occur in specific algebraic theories such as group theory, ring or field theory, but they can be formulated in a more general way as well. This allows us to prove theorems such as the usual Homomorphism or Isomorphism Theorems for groups in a more abstract setting which covers all types of algebras at once.

The formation of subalgebras, homomorphic images and products gives us three operators on classes of algebras, and we can look for classes of algebras which are closed under these operators. Such classes of algebras are called varieties.

The second approach to the study of abstract algebras involves the study of terms and identities. Here we classify algebras according to the identities or axioms they satisfy. Given any set of identities, of a particular type, we can form the class of all algebras of that type which satisfy the given identities. Such a class is called an equational class. This point of view is quite different from the classical structure theoretical approach, and combines logic, model theory and abstract algebra.

In the first six chapters of this book we develop these two approaches to general algebra. We begin in Chapter 1 with a detailed list of examples of various algebras, arising from a number of areas of Mathematics and Computer Science. The subalgebra, homomorphic image, product and quotient constructions are described in Chapters 1, 3 and 4. In Chapter 5 we introduce terms and identities, to be used in the second approach. This treatment culminates in Chapter 6 with a complete proof of Birkhoff's Theorem, which relates our two approaches: this theorem says that any variety is an equational class, and vice versa, so that the two approaches lead us to exactly the same classes of algebras.

These first six chapters provide a solid foundation in the core material of universal algebra. With a leisurely pace, careful exposition, and lots of examples and exercises, this section is designed as an introduction suitable for beginning graduate students or researchers from other areas.

A unique feature of this book is the use of Galois-connections as a main theme. The concept of a Galois-connection, with the related topics of closure operators, closure systems and lattices of closed sets, is introduced early on, in Chapter 2, and used throughout the book. The classical Galois theory, the relation between groups of automorphisms and fixed points, is touched on in Chapter 3. The main purpose of abstract Galois theory is to develop the interplay between two different sets or classes of objects, on the basis of a binary relation between the sets or classes. This allows us to tackle a problem involving one kind of object by using the theory of the second kind of objects.

We make use of our early introduction of Galois-connections and lattices of closed sets in two main examples. The first of these is the connection Id-Mod, between sets of identities and classes of algebras, which in Chapter 6 gives us the complete lattice of all varieties of a given type. The other main example is the connection Pol-Inv, between sets of relations and sets of operations on a fixed base set. From this connection we obtain the complete lattice of all clones of operations on a base set. Both of these examples are followed up in Chapter 14.

The study of clones is another important feature of this work. Clones (closed sets of functions) occur in universal algebra as clones of term operations or of polynomial operations of an algebra, in automata theory as combinations of elementary switching circuits, and in logic as sets of truth-value functions generated by elementary truth-value functions.

Chapters 7 and 8 deal with applications of general algebra to theoretical Computer Science. The terms and free algebras from Chapters 5 and 6 are used in Chapter 7 to study term rewriting systems, including the important properties of confluence and termination. Having shown in Chapter 1 that finite automata may be regarded as heterogeneous or multi-based algebras, in Chapter 8 we develop the theory of automata and Turing machines as algebraic machines. Finite automata recognize languages made up of words which are in fact terms of the free monoid; so they are a special case of more general machines which recognize sets of terms from the free algebra of any fixed type. Such general machines are called tree-recognizers, since terms are often referred to as trees. In Chapter 8 we also use Turing machines to look at decidability and algorithmic problems in general algebra. Thus these two chapters give some concrete application of the more theoretical material of the first six chapters.

Chapters 9 to 12 cover more advanced topics of universal algebra, including Mal'cev conditions, tame congruence theory and commutators. An important feature is the study of clones of operations, primality and functional completeness, which are also important in Computer Science. We study finite algebras and the varieties they generate, and give an algebraic proof of Post's characterization of the lattice of all clones on a two-element set.

The last three chapters, Chapters 13, 14 and 15, tie together the main themes of Galois-connections, clones and varieties, and algebraic machines. In Chapter 13 we return to the study of complete lattices of closed sets, obtained from Galois-connections. We describe several methods for obtaining complete sublattices of such complete lattices, and in Chapter 14 illustrate these methods on our two main examples, the lattice of all varieties of a given type and the lattice of all clones of operations on a fixed set. This leads to complete sublattices of M-solid varieties (using M-hyperidentities to obtain new closure operators) and G-clones and H-clones. We then return to applications to theoretical Computer Science, by looking at hypersubstitutions as tree-recognizers. Hypersubstitutions involve replacing not only the leaves of a tree by elements but also the nodes by term operations. The main result here is the proof of the equivalence of this "parallel" replacement with the linear approach. Hypersubstitutions can also be applied to the tree transformations and tree transducers of Chapter 8, and to the syntactical and semantical hyperunification problems.

An even more general approach to algebraic structures and to structural thinking is the category-theoretical one. In this approach, clones and equational theories are combined into the concept of an algebraic theory, and some parts of the theory become clearer. We believe however that for a beginning student, the universal algebraic approach is a necessary first step in reaching such higher levels of abstraction.

The material of this book is based on lectures given by the authors at the University of Potsdam (Germany), Chiangmai University and KhonKaen University (Thailand) and the University of Balgoevgrad (Bulgaria), and in research seminars with our students. The authors are grateful for the critical input of a number of students.

The work of Shelly Wismath on this book was supported by a research leave from the University of Lethbridge (January to July 2001) and by funding from the Natural Sciences and Engineering Research Council of Canada. The hospitality of the Institute of Mathematics of the University of Potsdam, and particularly the General Algebra Research Group, during the year July 2000 to June 2001 is also gratefully acknowledged. Special thanks to Stephen and Alice for accompanying me to Germany for a year.

Contents

Introduction   **v**

1 Basic Concepts   **1**
    1.1 Algebras   1
    1.2 Examples   4
    1.3 Subalgebras   13
    1.4 Congruence Relations and Quotients   21
    1.5 Exercises   27

2 Galois Connections and Closures   **31**
    2.1 Closure Operators   32
    2.2 Galois Connections   37
    2.3 Concept Analysis   42
    2.4 Exercises   44

3 Homomorphisms and Isomorphisms   **47**
    3.1 The Homomorphism Theorem   49
    3.2 The Isomorphism Theorems   58
    3.3 Exercises   61

4 Direct and Subdirect Products   **63**
    4.1 Direct Products   63
    4.2 Subdirect Products   68
    4.3 Exercises   72

5 Terms, Trees, and Polynomials   **75**
    5.1 Terms and Trees   76
    5.2 Term Operations   82
    5.3 Polynomials and Polynomial Operations   85

5.4   Exercises 88

6   Identities and Varieties   91
   6.1   The Galois Connection (Id, Mod) 91
   6.2   Fully Invariant Congruence Relations 95
   6.3   The Algebraic Consequence Relation 97
   6.4   Relatively Free Algebras 98
   6.5   Varieties 101
   6.6   The Lattice of All Varieties 109
   6.7   Finite Axiomatizability 110
   6.8   Exercises 113

7   Term Rewriting Systems   115
   7.1   Confluence 116
   7.2   Reduction Systems 123
   7.3   Term Rewriting 129
   7.4   Termination of Term Rewriting Systems 141
   7.5   Exercises 145

8   Algebraic Machines   147
   8.1   Regular Languages 148
   8.2   Finite Automata 150
   8.3   Algebraic Operations on Finite Automata 159
   8.4   Tree Recognizers 165
   8.5   Regular Tree Grammars 169
   8.6   Operations on Tree Languages 174
   8.7   Minimal Tree Recognizers 176
   8.8   Tree Transducers 182
   8.9   Turing Machines 185
   8.10   Undecidable Problems 187
   8.11   Exercises 191

9   Mal'cev-Type Conditions   193
   9.1   Congruence Permutability 193
   9.2   Congruence Distributivity 195
   9.3   Arithmetical Varieties 201
   9.4 n-Modularity and n-Permutability 203
   9.5   Congruence Regular Varieties 205
   9.6   Two-Element Algebras 206

9.7 Exercises 213

10 Clones and Completeness   **215**
10.1 Clones as Algebraic Structures 215
10.2 Operations and Relations 217
10.3 The Lattice of All Boolean Clones 219
10.4 The Functional Completeness Problem 228
10.5 Primal Algebras 231
10.6 Different Generalizations of Primality 240
10.7 Preprimal Algebras 245
10.8 Exercises 249

11 Tame Congruence Theory   **251**
11.1 Minimal Algebras 251
11.2 Tame Congruence Relations 262
11.3 Permutation Algebras 269
11.4 The Types of Minimal Algebras 276
11.5 Mal'cev Conditions and Omitting Types 281
11.6 Residually Small Varieties 286
11.7 Exercises 287

12 Term Condition and Commutator   **289**
12.1 The Term Condition 289
12.2 The Commutator 293
12.3 Exercises 299

13 Complete Sublattices   **301**
13.1 Conjugate Pairs of Closure Operators 301
13.2 Galois Closed Subrelations 308
13.3 Closure Operators on Complete Lattices 316
13.4 Exercises 323

14 G-Clones and M-Solid Varieties   **325**
14.1 G-Clones 325
14.2 H-clones 331
14.3 M-Solid Varieties 334
14.4 Intervals in the Lattice $\mathcal{L}(\tau)$ 342
14.5 Exercises 345

15 Hypersubstitutions and Machines   **347**
  15.1 The Hyperunification Problem 347
  15.2 Hyper Tree Recognizers 349
  15.3 Tree Transformations 357
  15.4 Exercises 361

  Bibliography   **363**

  Index   **373**

Chapter 1

Basic Concepts

In this first chapter we shall introduce the two basic concepts of Universal Algebra, operations and algebras, and provide a number of examples of algebras. One of these examples, lattices, is useful in a theoretical way as well: any algebra of any type is accompanied by some lattices, making the theory of lattices important in the study of all algebras. In Section 3 we examine the concept of a subalgebra, and the generation of subalgebras. In Section 4 we look at congruence relations and quotient algebras. Congruence relations on algebras generalize the well-known notion of the congruence modulo n defined on the ring of all integers, and quotient algebras generalize the construction of quotient or residue rings modulo n.

1.1 Algebras

As a branch of Mathematics, Algebra is the study of algebraic structures, or sets of objects with operations defined on them. These sets and operations often arise in other fields of Mathematics and in applications. Let us consider the following example:

Example 1.1.1 Let $D(\Phi)$ be the set of all mappings of symmetry of a rectangle Φ. As usual, a mapping of symmetry of a rectangle in a plane is understood to be a motion of the plane under which the rectangle is invariant. We will denote by g_1 and g_2 the two axes of symmetry of the rectangle. The set $D(\Phi)$ of all mappings of symmetry of the rectangle Φ consists exactly of the four mappings e, s_1, s_2 and z_0, where

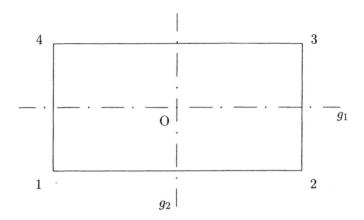

e is the identity mapping,

s_1 is the reflection of Φ through the line g_1,

s_2 is the reflection of Φ through the line g_2, and

z_0 is the rotation around the center point O by π radians.

If we label the corners of the rectangle by 1, 2, 3 and 4, then the four mappings of symmetry can be described by the following permutations, in the usual cyclic notation:

e by (1), s_1 by $(14)(23)$,

s_2 by $(12)(34)$, z_0 by $(13)(24)$.

Then the set of all mappings of symmetry of the rectangle Φ is given by

$$D(\Phi) = \{(1), (14)(23), (12)(34), (13)(24)\}.$$

The composition of any two mappings of symmetry of the rectangle Φ is again such a mapping, so we have a *binary operation*

$$\circ : D(\Phi) \times D(\Phi) \to D(\Phi)$$

defined on our set $D(\Phi)$. Thus we consider the pair $(D(\Phi); \circ)$, consisting of the set of mappings with this binary operation. This is what is called an algebra: a base set of objects, together with a set of one or more operations which are defined on this base set. In this example the set of operations contains only one element, the binary operation of composition of mappings.

Notice that we may use the same example to produce a different algebra, as follows. We can use as our base set the set $\{1, 2, 3, 4\}$ of the four corners of the rectangle, and as our set of operations the set

$$\{(1), (14)(23), (12)(34), (13)(24)\}$$

of four unary operations. This gives us the algebra

$$(\{1, 2, 3, 4\}; \{(1), (14)(23), (12)(34), (13)(24)\}),$$

which is completely different from the first one but which can also be used to describe the geometric situation.

These examples show that in defining an algebra, we must specify in addition to the base set of objects the set of operations: both how many operations there are, and what their arities are. We begin by defining formally the concept of an operation on a set.

Definition 1.1.2 Let A be a set, and let $n \geq 1$ be a natural number. A function $f : A^n \to A$ is called an *n-ary operation* defined on A, and is said to have *arity* n. We let $O^n(A)$ be the set of all n-ary operations defined on A and let $O(A) := \bigcup_{n=1}^{\infty} O^n(A)$ be the set of all finitary operations defined on A.

Remark 1.1.3 1. Any n-ary operation f on A can be regarded as an $(n+1)$-ary relation defined on A, called the *graph* of f. This relation is defined by $\{(a_1, \ldots, a_{n+1}) \in A^{n+1} : f(a_1, \ldots, a_n) = a_{n+1}\}$.

2. Definition 1.1.2 can be extended in the following way to the special case that $n = 0$, for a nullary operation. We define $A^0 := \{\emptyset\}$. A nullary operation is defined as a function $f : \{\emptyset\} \to A$. This means that a nullary operation on A is uniquely determined by the element $f(\emptyset) \in A$. For every element $a \in A$ there is exactly one mapping $f_a : \{\emptyset\} \to A$ with $f_a(\emptyset) = a$. Therefore a nullary operation may be thought of as selecting an element from the set A. If $A = \emptyset$ then there are no nullary operations on A.

3. If A is the two element set $\{0, 1\}$, operations on A are called *Boolean operations*. If we associate the truth-values false with 0 and true with 1, then the Boolean operations are just the truth-value functions of the Classical

Propositional Logic. For instance the functions of negation, conjunction, disjunction, implication, and equivalence are given by the truth-value tables

$\neg$		$\wedge$	0	1		$\vee$	0	1		$\Rightarrow$	0	1		$\Leftrightarrow$	0	1
0	1	0	0	0		0	0	1		0	1	1		0	1	0
1	0	1	0	1		1	1	1		1	0	1		1	0	1

respectively.

Using the concept of an operation we can now define an algebra. As we have seen, an algebra should be a pair consisting of a base set of objects and a set of operations defined on the base set. The set of operations is usually indexed by some index set I, so we will define an indexed algebra here; the more general setting of non-indexed algebras will be considered later, in Chapter 10.

Definition 1.1.4 Let A be a non-empty set. Let I be some non-empty index set, and let $(f_i^A)_{i \in I}$ be a function which assigns to every element of I an n_i-ary operation f_i^A defined on A. Then the pair $\mathcal{A} = (A; (f_i^A)_{i \in I})$ is called an *(indexed) algebra* (indexed by the set I). The set A is called the *base* or *carrier set* or *universe* of $\mathcal{A}$, and $(f_i^A)_{i \in I}$ is called the *sequence of fundamental operations* of $\mathcal{A}$. For each $i \in I$ the natural number n_i is called the *arity* of f_i^A. The sequence $\tau := (n_i)_{i \in I}$ of all the arities is called the *type* of the algebra $\mathcal{A}$. We use the name $Alg(\tau)$ for the class of all algebras of a given type τ.

Notice that in our definition we do not allow the base set A of an algebra to be the empty set. It is possible to define an empty algebra with the empty set as a base, but many theorems about algebras then have to include a separate case to discuss what happens when the algebra is empty. Thus we have chosen to exclude this case from our discussion.

1.2   Examples

In this section we illustrate our basic definition of an algebra, by presenting a number of examples, many of which we will also use later. We note that an operation f^A on a set A will often be denoted simply by f, omitting the superscript. Also, operations of arities zero, one and two are often said to be *nullary*, *unary* and *binary* respectively.

Example 1.2.1 As we saw in the previous section, we may describe the set of symmetry mappings of a rectangle by two different algebras:

$$\mathcal{A}_1 = (\{1, 2, 3, 4\}; (1), (14)(23), (12)(34), (13)(24)),$$

of type $\tau_1 = (1, 1, 1, 1)$, or

$$\mathcal{A}_2 = (\{(1), (14)(23), (12)(34), (13)(24)\}; \circ),$$

of type $\tau_2 = (2)$. Note that these are both examples of finite algebras, where the base set is a finite set.

Example 1.2.2 A *unar* is an algebra $\mathcal{U} = (U; g^U)$ of type $\tau = (1)$, with one unary operation.

Example 1.2.3 An algebra $(G; \cdot)$ of type $\tau = (2)$, with one binary operation, is called a *groupoid*. Here the single binary operation f is denoted by $\cdot$ or even just by juxtaposition, so we write $x \cdot y$ or just xy instead of $f(x, y)$.

A groupoid $(G; \cdot)$ is called *abelian* or *commutative* if it also satisfies

(G0) $\forall x, y \in G \ (x \cdot y = y \cdot x)$ (commutative law).

Example 1.2.4 A groupoid $(G; \cdot)$ is called a *semigroup* if the binary operation $\cdot$ is associative; that is, if G satisfies

(G1) $\forall x, y, z \in G \ (x \cdot (y \cdot z) = (x \cdot y) \cdot z)$ (associative law).

Example 1.2.5 An algebra $\mathcal{M} = (M; \cdot, e)$ of type $(2, 0)$ is called a *monoid*, if the associative law (G1) and

(G2′) $\forall x \in G \ (x \cdot e = e \cdot x = x)$ (identity law)
are satisfied.

Example 1.2.6 A *group* is an algebra $\mathcal{G} = (G; \cdot)$ of type (2), which satisfies the axioms (defining identities) (G1) and

(G2) $\forall a, b \in G \ \exists x, y \in G \ (a \cdot x = b \text{ and } y \cdot a = b)$.
                                                              (invertibility)

A group can also be regarded as an algebra $\mathcal{G} = (G; \cdot, ^{-1}, e)$ of type $(2, 1, 0)$, where the laws (G1), (G2′) and

(G2″) $\forall x \in G \ (x \cdot x^{-1} = x^{-1} \cdot x = e)$ (inverse law)

are satisfied.

Example 1.2.7 An algebra $\mathcal{Q} = (Q; \cdot)$ of type (2) is called a *quasigroup*, if $\cdot$ is a uniquely invertible, but not necessarily associative binary operation on the set Q.

A quasigroup can also be characterized by the property that for all $a \in Q$ the following mappings of Q onto Q are bijections:

$$x \mapsto a \cdot x, \quad \text{left multiplication with } a,$$

$$x \mapsto x \cdot a, \quad \text{right multiplication with } a.$$

Quasigroups can also be defined as algebras of type $(2, 2, 2)$ with the operations $\cdot$, $\backslash$ and $/$, where the following axioms are satisfied:

(Q1) $\forall x, y \in Q \ (x \backslash (x \cdot y) = y)$,
(Q2) $\forall x, y \in Q \ ((x \cdot y)/y = x)$,
(Q3) $\forall x, y \in Q \ (x \cdot (x \backslash y) = y)$,   and
(Q4) $\forall x, y \in Q \ ((x/y) \cdot y = x)$.

Example 1.2.8 An algebra $\mathcal{R} = (R; +, \cdot, -, 0)$ of type $(2, 2, 1, 0)$ is called a *ring* if $(R; +, -, 0)$ is an abelian (commutative) group (with addition as the binary operation and 0 as an identity element) and $(R; \cdot)$ is a semigroup, and also the two *distributive laws*

(D1) $\forall x, y, z \in R \ (x \cdot (y + z) = x \cdot y + x \cdot z)$ and
(D2) $\forall x, y, z \in R \ ((x + y) \cdot z = x \cdot z + y \cdot z)$
are satisfied.

A ring $(R; +, \cdot, -, 0, e)$ with an identity element e for multiplication is called a *skew field* if $(R \backslash \{0\}; \cdot)$ is a group. When in addition the operation $\cdot$ is commutative, the ring $\mathcal{R}$ is called a *field*.

Example 1.2.9 An algebra $\mathcal{V} = (V; \wedge, \vee)$ of type $(2, 2)$ is called a *lattice*, if the following equations are satisfied by its two binary operations, which are usually called meet and join:

(V1) $\forall x, y \in V \ (x \lor y \ = \ y \lor x)$,
(V1′) $\forall x, y \in V \ (x \land y \ = \ y \land x)$,
(V2) $\forall x, y, z \in V \ (x \lor (y \lor z) = \ (x \lor y) \lor z)$,
(V2′) $\forall x, y, z \in V \ (x \land (y \land z) = \ (x \land y) \land z)$,
(V3) $\forall x \in V \ (x \lor x \ = \ x)$,
(V3′) $\forall x \in V \ (x \land x \ = \ x)$ (idempotency),
(V4) $\forall x, y \in V \ (x \lor (x \land y) \ = \ x)$,
(V4′) $\forall x, y \in V \ (x \land (x \lor y) \ = \ x)$ (absorption laws).

If in addition the lattice satisfies the following *distributive laws*,

(V5) $\forall x, y, z \in V \ (x \land (y \lor z) = \ (x \land y) \lor (x \land z))$,
(V5′) $\forall x, y, z \in V \ (x \lor (y \land z) = \ (x \lor y) \land (x \lor z))$,

then the lattice is said to be *distributive*. We remark that (V3) and (V3′) follow from (V4) and (V4′) respectively.

A lattice is called *modular* if it satisfies the *modular law*

(V6) $\forall x, y, z \in V \ (z \leq x \Rightarrow x \land (y \lor z) = (x \land y) \lor z)$.

Lattices are important both as examples of a kind of algebra, and also in the study of all other kinds of algebras too, since any algebra turns out to have some lattices associated with it. We shall study such associated lattices in the coming sections. For now, we show how lattices are connected to partially ordered sets. A binary relation $\leq$ on a set A is called a partial order on A if it is reflexive, anti-symmetric and transitive. There is a close connection between lattices and partially ordered sets, in the sense that each determines the other, as the following theorem shows.

Theorem 1.2.10 *Let $(L; \leq)$ be a partially ordered set in which for all $x, y \in L$ both the infimum $\bigwedge \{x, y\}$ and the supremum $\bigvee \{x, y\}$ exist. Then the binary infimum and supremum operations make $(L; \land, \lor)$ a lattice. Conversely, every lattice defines a partially ordered set in which for all x, y the infimum $\bigwedge \{x, y\}$ and the supremum $\bigvee \{x, y\}$ exist.*

Proof: Let $(L; \leq)$ be a partially ordered set, in which for any two elements $x, y \in L$ the infimum $\bigwedge \{x, y\}$ and the supremum $\bigvee \{x, y\}$ exist. If we put

$x \wedge y = \bigwedge\{x,y\}$ and $x \vee y = \bigvee\{x,y\}$, then the required identities (V1) - (V4) and (V1') - (V4') are easy to verify.

If conversely $(L; \wedge, \vee)$ is a lattice, then we define

$$x \leq y : \Leftrightarrow x \wedge y = x,$$

and show that this gives a partial order relation on L. Reflexivity follows from (V3'), $x \wedge x = x$; for antisymmetry we see that

$$(x \leq y) \wedge (y \leq x) \Leftrightarrow (x \wedge y = x) \wedge (y \wedge x = y),$$

and so from commutativity we get

$$x = x \wedge y = y \wedge x = y;$$

and transitivity follows from $(x \wedge y = x) \wedge (y \wedge z = y) \Rightarrow x = x \wedge y = x \wedge (y \wedge z) = (x \wedge y) \wedge z = x \wedge z \Leftrightarrow x \leq z$. It is easy to check that $\bigwedge\{x,y\} = x \wedge y$ and $\bigvee\{x,y\} = x \vee y$ are satisfied. ∎

Example 1.2.11 A lattice (or partially ordered set) L in which for all sets $B \subseteq L$ the infimum $\bigwedge B$ and the supremum $\bigvee B$ exist is called a *complete lattice*. Obviously, any finite lattice is complete.

A *bounded lattice* $(V; \wedge, \vee, 0, 1)$ is an algebra of type $(2, 2, 0, 0)$ which is a lattice, with two additional nullary operations 0 and 1, which satisfy

(V6) $\forall x \in V \ (x \wedge 0 = 0)$ and
(V7) $\forall x \in V \ (x \vee 1 = 1)$.

These axioms tell us that in the partial order determined by the lattice, 0 acts as the least element and 1 as the greatest element.

Example 1.2.12 An algebra $\mathcal{S} = (S; \cdot)$ of type (2) is called a *semilattice*, if the operation $\cdot$ is an associative, commutative, and idempotent (satisfies $x \cdot x = x$) binary operation on S. This means that a semilattice is a particular kind of semigroup.

Example 1.2.13 An algebra $\mathcal{B} = (B; \wedge, \vee, \neg, 0, 1)$ of type $(2, 2, 1, 0, 0)$ is called a *Boolean algebra*, if $(B; \wedge, \vee, 0, 1)$ is a bounded distributive lattice with an additional unary operation called complementation, which also satisfies the two identities

(B1) $\forall x \in B\ (x \wedge \neg x = 0)$ and
(B2) $\forall x \in B\ (x \vee \neg x = 1)$ (complement laws).

As an example of a Boolean algebra, we recall the set of all Boolean functions on the two element set $\{0,1\}$ from Remark 1.1.3, 3. This gives us the algebra $(\{0,1\}; \wedge, \vee, \neg, 0, 1)$, usually denoted by the name 2_B. Another example of a Boolean algebra is the power set of a set A, with the operations of intersection, union, complementation, empty set and A.

Example 1.2.14 For applications in Logic and the theory of switching circuits, algebras whose carrier sets are sets of operations on a base set play an important role. We recall from Definition 1.1.2 that for a non-empty base set A, we denote by $O(A)$ the set of all finitary operations on A (with nullary operations regarded as constant unary operations).

We can make the set $O(A)$ into an algebra in several ways, by defining various operations on $O(A)$. We first define the following composition operation on $O(A)$. If $f^A \in O^n(A)$ and $g_1^A, \ldots, g_n^A \in O^m(A)$, we obtain a new operation $f^A(g_1^A, \ldots, g_n^A)$ in $O^m(A)$, by setting $f^A(g_1^A, \ldots, g_n^A)(a_1, \ldots, a_m)$ to be $f^A(g_1^A(a_1, \ldots, a_m), \ldots, g_n^A(a_1, \ldots, a_m))$, for all $a_1, \ldots, a_m \in A$. This gives an operation

$$O^n(A) \times (O^m(A))^n \to O^m(A),$$

which is called *composition* or *superposition*. Clearly, the whole set $O(A)$ is closed under arbitrary compositions.

The set $O(A)$ also contains certain special elements called the projections. For each $n \geq 1$ and each $1 \leq j \leq n$, the n-ary function e_j^n defined on A by $e_j^n(a_1, \ldots, a_n) = a_j$ is called the j-th projection mapping of arity n. Any subset of $O(A)$ which is closed under composition and contains all these projection mappings is called a *clone* on A. Of course $O(A)$ itself has these properties, and it is called the *full clone* on A.

There is another way to define an algebraic structure on the set $O(A)$, using the following operations $*$, ξ, τ and Δ on $O(A)$:

$$* : \ O^n(A) \times O^m(A) \to O^{m+n-1}(A) \ \text{ defined by } (f, g) \mapsto f * g$$

with
$$(f * g)(x_1, \ldots, x_m, x_{m+1}, \ldots, x_{m+n-1}) : =$$
$$f(g(x_1, \ldots, x_m), x_{m+1}, \ldots, x_{m+n-1}),$$
for all $x_1, \ldots, x_{m+n-1} \in A$;

$$\xi : \quad O^n(A) \to O^n(A) \quad \text{defined by} \quad f \mapsto \xi(f)$$

with

$$\xi(f)(x_1, \ldots, x_n) := f(x_2, \ldots, x_n, x_1);$$

$$\tau : \quad O^n(A) \to O^n(A) \quad \text{defined by} \quad f \mapsto \tau(f)$$

with

$$\tau(f)(x_1, \ldots, x_n) := f(x_2, x_1, x_3, \ldots, x_n);$$

$$\Delta : \quad O^n(A) \to O^{n-1}(A) \quad \text{defined by} \quad f \mapsto \Delta(f)$$

with

$$\Delta(f)(x_1, \ldots, x_{n-1}) := f(x_1, x_1, x_2, \ldots, x_{n-1})$$

for every $n > 1$ and for all $x_1, \ldots, x_n \in A$; and

$$\xi(f) = \tau(f) = \Delta(f) = f, \quad \text{if} \quad n = 1.$$

We also use the nullary operation e_1^2 which picks out the binary projection on the first coordinate; as defined above, $e_1^2(a_1, a_2) = a_1$, for all $a_1, a_2 \in A$. With these five operations, we obtain an algebra $(O(A); *, \xi, \tau, \Delta, e_1^2)$ of type $(2, 1, 1, 1, 0)$. This algebra is called the *full iterative algebra* on A, or sometimes also a *clone* on A. We shall study clones in more detail in Chapter 10.

Example 1.2.15 Let $\mathcal{R} = (R; +, \cdot, -, 0)$ be a ring such that R is infinite. In this example we consider an infinitary type of algebra, one with an infinite number of operation symbols, one for each element of the ring $\mathcal{R}$. An algebra $(M; +, -, 0, R)$ of type $(2, 1, 0, (1)_{r \in R})$ is called an $\mathcal{R}$-*module* (or a *module over* $\mathcal{R}$) if $(M; +, -, 0)$ is an abelian group and the following identities are satisfied, for all r and s in R and x and y in M:

(M1) $r(x + y) = r(x) + r(y),$
(M2) $(r + s)(x) = r(x) + s(x),$

(M3) $(r \cdot s)(x) = r(s(x))$.

If $\mathcal{R}$ also has an identity element 1, then the following additional identity must also hold in an $\mathcal{R}$-module:

(M4) $1(x) = x$.

In the special case that $\mathcal{R}$ is a field, an $\mathcal{R}$-module is usually called an $\mathcal{R}$-*vector space*. The elements of the module are then called *vectors*, while the field elements are called *scalars*.

Thus modules and vector spaces are algebras with infinitely many operations. There is another way to describe modules and vector spaces algebraically, which avoids this infinitary type, and which is more often used. But it involves a modification of our definition of an algebra. We have so far considered only what are called *homogeneous* or *one-based* algebras, where we have only one base set of objects and all operations are defined on this set. One can also define what are called *heterogeneous* or *multi-sorted* or *multi-based* algebras, where there are two or more base sets of objects and operations are allowed on more than one kind of object. We can use this approach here, by allowing the set V of all vectors and the universe of the field $\mathcal{R}$ as two different base sets in such a heterogeneous algebra. Then we have two operations, $+ : V \times V \to V$ for the usual vector addition and $\cdot : R \times V \to V$ for the scalar multiplication of a scalar from R times a vector from V. Then $(V, R; , +, \cdot)$ is an example of a heterogeneous algebra. Although we will focus on homogeneous or one-based algebras in this book, we will consider the multi-based algebras described in Example 1.2.16 below in more detail in Chapter 8, and we remark that the algebraic theory for multi-based algebras can be developed in a completely analogous way.

Example 1.2.16 This example shows how multi-based algebras may be used in theoretical Computer Science. An *automaton without output*, or an *acceptor* or a *recognizer*, is a multi-based algebra $\mathcal{H} = (Z, X; \delta)$, where Z and X are non-empty sets called the sets of states and inputs, respectively, and where $\delta : Z \times X \to Z$ is called the state transition function. An *automaton (with output)* is a quintuple $\mathcal{A} = (Z, X, B; \delta, \lambda)$, where $(Z, X; \delta)$ is an acceptor, the non-empty set B is called the set of outputs, and where $\lambda : Z \times X \to B$ is called the output function. For any state $z \in Z$ and any input $x \in X$, the value $\delta(z,x)$ is the state which results when the input x is read in when the machine is in the state z. The element $\lambda(z,x)$ is the output element which is produced by input x if the automaton is in state z. If all the sets Z, X and B are finite, then the automaton is said to be finite; otherwise it is infinite. When δ and λ are functions, and so have exactly one image for each state-input pair (z,x), the automaton is called deterministic; in the non-deterministic case we allow δ and λ to take on more than one value, or to be undefined, for a given input pair. Sometimes a certain state $z_0 \in Z$ is selected as an initial state, and in this case we write $(Z, X; \delta, z_0)$ or $(Z, X, B; \delta, \lambda, z_0)$ for the automaton. In the finite case, an automaton may be fully described by means of tables for δ and λ, as shown below.

δ	x_1	$\cdots$	x_n
z_1	$\delta(z_1, x_1)$	$\cdots$	$\delta(z_1, x_n)$
.	.	.	.
.	.	.	.
.	.	.	.
z_k	$\delta(z_k, x_1)$	$\cdots$	$\delta(z_k, x_n)$

λ	x_1	$\cdots$	x_n
z_1	$\lambda(z_1, x_1)$	$\cdots$	$\lambda(z_1, x_n)$
.	.	.	.
.	.	.	.
.	.	.	.
z_k	$\lambda(z_k, x_1)$	$\cdots$	$\lambda(z_k, x_n)$

We shall study finite automata in detail in Chapter 8. We conclude here with the remark that a finite automaton can also be described by a directed graph. The vertices of the graph correspond to states, and there is an edge labelled by x going from vertex z to vertex y when $\delta(z,x) = y$. We can also label the vertex by both x and $\lambda(z,x)$ when the output is $\lambda(z,x)$. For example, the graph below

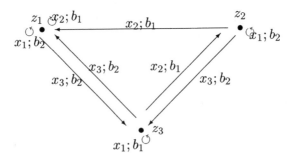

corresponds to the automaton with $Z = \{z_1, z_2, z_3\}$, $X = \{x_1, x_2, x_3\}$, and $B = \{b_1, b_2\}$, with the tables shown below for δ and λ.

δ	x_1	x_2	x_3
z_1	z_1	z_1	z_3
z_2	z_2	z_1	z_3
z_3	z_3	z_2	z_1

λ	x_1	x_2	x_3
z_1	b_2	b_1	b_2
z_2	b_2	b_1	b_2
z_3	b_1	b_1	b_2

1.3 Subalgebras

The set of all mappings of symmetry of an equilateral triangle Φ,

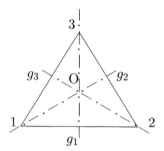

described by permutations of the set $\{1,\ 2,\ 3\}$, is the set

$$D(\Phi) \ = \ \{(1),\ (123),\ (132),\ (12),\ (13),\ (23)\}.$$

Here (12), (13) and (23) correspond to reflections at the lines g_1, g_3, and g_2, respectively; and (1), (123) and (132) correspond to rotations around the point O by 0, 120 and 240 degrees, respectively. We notice that $D(\Phi)$ forms a group with respect to the composition of these mappings; that is, with respect to the multiplication of the corresponding permutations.

Since the composition of two rotations around the point O is again a rotation around O, our operation preserves the subset of $D(\Phi)$ consisting of all rotations around O, and we say that this subset is closed under our operation. This closure can also be seen if we consider the multiplication table of the corresponding permutations:

$\circ$	(1)	(123)	(132)
(1)	(1)	(123)	(132)
(123)	(123)	(132)	(1)
(132)	(132)	(1)	(123)

However, we can see that the composition of reflections through different lines gives a permutation which is not a reflection; so the set of reflections does not have this closure property. To describe this property algebraically, we use the concept of a subalgebra.

Definition 1.3.1 Let $\mathcal{B} = (B; (f_i^B)_{i \in I})$ be an algebra of type τ. Then an algebra $\mathcal{A}$ is called a *subalgebra* of $\mathcal{B}$, written as $\mathcal{A} \subseteq \mathcal{B}$, if the following conditions are satisfied:

(i) $\mathcal{A} = (A; (f_i^A)_{i \in I})$ is an algebra of type τ;

(ii) $A \subseteq B$;

(iii) $\forall i \in I$, the graph of f_i^A is a subset of the graph of f_i^B.

Remark 1.3.2 1. Condition (iii) of the Definition refers to the graph of an operation, as defined in Remark 1.1.3. This condition means that the graph of f_i^A is the *restriction* of the graph of f_i^B to $A^{n_i} \subseteq B^{n_i}$. We write $(f_i^A = f_i^B \mid A^{n_i})$, for all $i \in I$, using $f_i^B \mid A^{n_i}$, or just $f_i^B \mid A$, to denote the restriction of f_i^B to A^{n_i} .

2. If f_i^A is a nullary operation, then
$$f_i^A = (A^0 \times B) \cap f_i^B = (\{\emptyset\} \times B) \cap f_i^B$$
$$= \{(\emptyset, f_i^B(\emptyset))\} = f_i^B.$$
That is, a nullary operation f_i must designate the same element in each subalgebra $\mathcal{A}$ of $\mathcal{B}$.

Whether one algebra is a subalgebra of another algebra can be checked by the following criterion:

Lemma 1.3.3 *(Subalgebra Criterion) Let $\mathcal{B} = (B; (f_i^B)_{i \in I})$ be an algebra of type τ and let $A \subseteq B$ be a subset of B for which $f_i^A = f_i^B \mid A$ for all $i \in I$. Then $\mathcal{A} = (A, (f_i^A)_{i \in I})$ is a subalgebra of $\mathcal{B} = (B, (f_i^B)_{i \in I})$ iff A is closed with respect to all the operations f_i^B for $i \in I$; that is, if $f_i^B(A^{n_i}) \subseteq A$ for all $i \in I$.*

Proof: Assume that $\mathcal{A} \subseteq \mathcal{B}$. Then $f_i^A = f_i^B \mid A$ is an operation in A for every $i \in I$, and so for all $(a_1, \ldots, a_{n_i}) \in A^{n_i}$ we have

$$f_i^B(a_1, \ldots, a_{n_i}) = f_i^A(a_1, \ldots, a_{n_i}) \in A.$$

Thus the application of any operation f_i^B of B to elements of A always gives elements of A.

If conversely A is closed with respect to f_i^B for all $i \in I$, then we have $f_i^B \mid A^{n_i} \subseteq A^{n_i} \times A$. This makes each $f_i^B \mid A^{n_i}$ an operation on A^{n_i} in A, and all the conditions of Definition 1.3.1 are satisfied. ∎

Corollary 1.3.4 *Let A, B and C be algebras of type τ. Then:*

(i) $(A \subseteq B) \wedge (B \subseteq C) \Rightarrow A \subseteq C;$

(ii) $(A \subseteq B \subseteq C) \wedge (A \subseteq C) \wedge (B \subseteq C) \Rightarrow A \subseteq B.$

Proof: (i) Clearly, for all $i \in I$ we have $f_i^A \subseteq f_i^B \subseteq f_i^C$ for the graphs, and thus $f_i^A \subseteq f_i^C$.

(ii) Applying Criterion 1.3.3 to an arbitrary n_i- tuple $(a_1, \ldots, a_{n_i}) \in A^{n_i}$, we get $f_i^B(a_1, \ldots, a_{n_i}) = f_i^C(a_1, \ldots, a_{n_i}) = f_i^A(a_1, \ldots, a_{n_i}) \in A$. This shows that A is closed with respect to all operations of B. ∎

Now suppose that $B = (B; (f_i^B)_{i \in I})$ is an algebra of type τ and that $\{A_j \mid j \in J\}$ is a family of subalgebras of B. We define a set $A := \bigcap_{j \in J} A_j$ and operations $f_i^A := f_i^B \mid A$ for all $i \in I$. We will verify that when the intersection set A is non-empty, the algebra $A = (A; (f_i^A)_{i \in I})$ is indeed an algebra of type τ, to be called the *intersection* of the algebras A_j, and denoted by $A := \bigcap_J A_j$.

Corollary 1.3.5 *The non-empty intersection A of a non-empty family $\{A_j \mid j \in J\}$ of subalgebras of an algebra B of type τ is a subalgebra of B.*

Proof: We will show that A is closed under all f_i^B, for all $i \in I$. Consider an element $(a_1, \ldots, a_{n_i}) \in A^{n_i}$, so that $(a_1, \ldots, a_{n_i}) \in A_j^{n_i}$ for all $j \in J$. Therefore we have $f_i^B(a_1, \ldots, a_{n_i}) \in A_j$ for all $j \in J$ and $i \in I$, since $A_j \subseteq B$ for all $j \in J$. Hence $f_i^B(a_1, \ldots, a_{n_i})$ also belongs to the intersection $A = \bigcap_{j \in J} A_j$, and $A \subseteq B$. ∎

Now we consider an algebra $\mathcal{B}$ of type τ and a non-empty subset X of its carrier set B. There is at least one subalgebra of $\mathcal{B}$ whose carrier set contains X, since $\mathcal{B}$ itself is one such algebra. This means we can look for the intersection of all such subalgebras of $\mathcal{B}$. We define

$$\langle X \rangle_{\mathcal{B}} : \;\; = \;\; \cap \{ \mathcal{A} \mid \mathcal{A} \subseteq \mathcal{B} \text{ and } X \subseteq A, \}$$

and call this subalgebra $\langle X \rangle_{\mathcal{B}}$ of $\mathcal{B}$ the *subalgebra of $\mathcal{B}$ generated by X*, and X a *generating system* of this algebra.

Notice that $\langle X \rangle_{\mathcal{B}}$ is the least (with respect to subset inclusion) subalgebra of $\mathcal{B}$ to contain the set X. In particular, X might generate the whole algebra $\mathcal{B}$, if $\langle X \rangle_{\mathcal{B}} \; = \; \mathcal{B}$.

Example 1.3.6 As an example we will consider the algebra

$$\mathcal{Z}_6 \;\; = \;\; (\{[0]_6, [1]_6, [2]_6, [3]_6, [4]_6, [5]_6, \}; \; +, \; -, [0]_6),$$

where the elements of the base set are the equivalence classes of the integers modulo 6 and $+, -$ denote addition and subtraction of equivalence classes. This algebra is a group, with $[0]_6$ as an identity element. For each one-element subset X of Z_6 we calculate the subalgebra generated by X:

$$\langle \{[0]_6\} \rangle_{\mathcal{Z}_6} \;\; = \;\; (\{[0]_6\}; \; +, -, [0]_6), \;\; \langle \{[1]_6\} \rangle_{\mathcal{Z}_6} \;\; = \;\; \langle \{[5]_6\} \rangle_{\mathcal{Z}_6} \;\; = \;\; \mathcal{Z}_6,$$

$$\langle \{[3]_6\} \rangle_{\mathcal{Z}_6} \;\; = \;\; (\{[0]_6, [3]_6\}; \; +, -, [0]_6), \;\; \langle \{[2]_6\} \rangle_{\mathcal{Z}_6} \;\; = \;\; \langle \{[4]_6\} \rangle_{\mathcal{Z}_6} \;\; =$$

$$(\{[0]_6, [2]_6, [4]_6\}; \; +, -, [0]_6), \;\; \langle \{[2]_6, [3]_6\} \rangle_{\mathcal{Z}_6} \;\; = \;\; \langle \{[4]_6, [3]_6\} \rangle_{\mathcal{Z}_6} \;\; = \;\; \mathcal{Z}_6.$$

Our next theorem shows that this process of subalgebra generation satisfies three very important properties, called the *closure properties*. This makes the subalgebra generation process an example of a closure operator, which we shall consider in more detail in Chapter 2.

Theorem 1.3.7 *Let $\mathcal{B}$ be an algebra. For all subsets X and Y of B, the following closure properties hold:*

(i) $X \subseteq \langle X \rangle_{\mathcal{B}}$,   *(extensivity)*;
(ii) $X \subseteq Y \Rightarrow \langle X \rangle_{\mathcal{B}} \; \subseteq \langle Y \rangle_{\mathcal{B}}$,   *(monotonicity)*;
(iii) $\langle X \rangle_{\mathcal{B}} \; = \langle \langle X \rangle_{\mathcal{B}} \rangle_{\mathcal{B}}$,   *(idempotency)*.

Proof: These properties follow immediately from the properties of the operator $\langle \ \rangle$; we leave the details to the reader. ∎

An important problem is the following: given an algebra $\mathcal{B}$ of type τ, and a subset X of B, determine all elements of the carrier set of the subalgebra of $\mathcal{B}$ which is generated by X. To do this, we set

$$E(X): \ = \ X \cup \{ f_i^B(a_1, \ldots, a_{n_i}) \mid i \in I, \ a_1, \ldots, a_{n_i} \in X \}.$$

Then we inductively define $E^0(X) := X$, and $E^{k+1}(X) := E(E^k(X))$, for all $k \in \mathbb{N}$. With this notation we can describe our subalgebra.

Theorem 1.3.8 *For any algebra $\mathcal{B}$ of type τ and for any non-empty subset $X \subseteq B$, we have $\langle X \rangle_\mathcal{B} = \bigcup\limits_{k=0}^{\infty} E^k(X)$.*

Proof: We first give a proof by induction on k that $\langle X \rangle_\mathcal{B} \supseteq \bigcup\limits_{k=0}^{\infty} E^k(X)$. For $k = 0$ it is clear that $E^0(X) = X \subseteq \langle X \rangle_\mathcal{B}$. For the inductive step, assume that the proposition $\langle X \rangle_\mathcal{B} \supseteq E^k(X)$ is true. Let $a \in E^{k+1}(X)$; we may assume that $a \notin E^k(X)$. Then there exists an element $i \in I$ and elements $a_1, \ldots, a_{n_i} \in E^k(X)$ with $a = f_i^B(a_1, \ldots, a_{n_i})$. Since $E^k(X) \subseteq \langle X \rangle_\mathcal{B}$ and $\langle X \rangle_\mathcal{B}$ is the carrier set of a subalgebra of $\mathcal{B}$, we have $a \in \langle X \rangle_\mathcal{B}$ too. Thus $E^{k+1}(X) \subseteq \langle X \rangle_\mathcal{B}$, and therefore by induction on k we have $\langle X \rangle_\mathcal{B} \supseteq \bigcup\limits_{k=0}^{\infty} E^k(X)$.

Now we show that $\langle X \rangle_\mathcal{B} \subseteq \bigcup\limits_{k=0}^{\infty} E^k(X)$, by showing that $\bigcup\limits_{k=0}^{\infty} E^k(X)$ is the carrier set of a subalgebra of $\mathcal{B}$. Let $i \in I$ and $a_1, \ldots, a_{n_i} \in \bigcup\limits_{k=0}^{\infty} E^k(X)$. Then for every $l \in \{1, \ldots, n_i\}$ there is a minimal $k(l) \in \mathbb{N}$ with $a_l \in E^{k(l)}(X)$. Take m to be the maximum of these $k(l)$, for $1 \le l \le n_i$. Then $a_l \in E^m(X)$ for all $l = 1, \ldots, n_i$ and thus

$$f_i^B(a_1, \ldots, a_{n_i}) \in E^{m+1}(X) \subseteq \bigcup\limits_{k=0}^{\infty} E^k(X).$$

This shows that $\bigcup\limits_{k=0}^{\infty} E^k(X)$ is closed under f_i^B for $i \in I$. By Lemma 1.3.3 $\bigcup\limits_{k=0}^{\infty} E^k(X)$ is the carrier set of a subalgebra of $\mathcal{B}$ which contains X. Since by definition $\langle X \rangle_{\mathcal{B}}$ is the least (with respect to subset inclusion) subalgebra of $\mathcal{B}$ which contains X, we get the desired inclusion. Finally from both inclusions we have the equality $\langle X \rangle_{\mathcal{B}} = \bigcup\limits_{k=0}^{\infty} E^k(X)$.   ∎

The proof of Theorem 1.3.8 gives an internal description of the generation process for algebras. In contrast to such a "bottom-up" construction, we have a "Principal of Structural Induction," which can be regarded as a "top-down" description.

Theorem 1.3.9 *(Principle of Structural Induction) Let $\mathcal{A} = (A; (f_i)_{i \in I})$ be an algebra of type τ, generated by a subset X of A. To prove that a property $\mathcal{P}$ holds for all elements of $\mathcal{A}$, it suffices to show the validity of the following two conditions:*

(1) Base of the induction: $\mathcal{P}$ holds for all elements of X.
(2) Induction step: If $\mathcal{P}$ holds for any $a_1, \ldots, a_{n_i}$ in A (the induction hypothesis) then $\mathcal{P}$ holds for $f_i^A(a_1, \ldots, a_{n_i})$, for all $i \in I$.

Proof: Let us denote by $\mathcal{A}_P$ the set of all elements in A for which $\mathcal{P}$ holds. By (1) we know that X is contained in $\mathcal{A}_P$, and hence by monotonicity we have $\langle X \rangle_{\mathcal{A}}$ contained in $\langle \mathcal{A}_P \rangle_{\mathcal{A}}$. Since X generates the algebra $\mathcal{A}$, this means that the algebra generated by $\mathcal{A}_P$ contains $\mathcal{A}$, and so must equal $\mathcal{A}$. But by (2), $\mathcal{A}_P$ is closed under all fundamental operations of $\mathcal{A}$, and hence it is a subalgebra of $\mathcal{A}$. This tells us that $\mathcal{A} = \mathcal{A}_P$; so property $\mathcal{P}$ holds in all of $\mathcal{A}$.   ∎

Now we can consider the structural properties of the set of all subalgebras of an algebra $\mathcal{B}$ of type τ. We denote this set by $Sub(\mathcal{B})$. We know by 1.3.5 that the intersection of two subalgebras of $\mathcal{B}$, if it is non-empty, is again a subalgebra of $\mathcal{B}$, and we would like to use this to define a binary operation on $Sub(\mathcal{B})$. That is, we want to define a binary operation $\wedge$ on $Sub(\mathcal{B})$ by

$$\wedge : (\mathcal{A}_1, \mathcal{A}_2) \mapsto \mathcal{A}_1 \wedge \mathcal{A}_2 : = \mathcal{A}_1 \cap \mathcal{A}_2.$$

However, this operation is not defined on all of $Sub(\mathcal{B})$, because it does not deal with the case where $A_1 \cap A_2$ is the empty set. But we can fix this, either by allowing the empty set to be considered as an algebra (see the remark at the end of Section 1.1), or by adjoining some new element to the set $Sub(\mathcal{B})$ and defining $\mathcal{A}_1 \wedge \mathcal{A}_2$ to be this new element, whenever $A_1 \cap A_2 = \phi$.

A second binary operation $\vee$ can be obtained if we map $(\mathcal{A}_1, \mathcal{A}_2)$ to the subalgebra of $\mathcal{B}$ which is generated by the union $A_1 \cup A_2$:

$$\vee : (\mathcal{A}_1, \mathcal{A}_2) \mapsto \mathcal{A}_1 \vee \mathcal{A}_2 : \ = \langle A_1 \cup A_2 \rangle_{\mathcal{B}}.$$

Our goal in defining these two operations was of course to make $Sub(\mathcal{B})$ into a lattice, and this turns out to be the case.

Theorem 1.3.10 *For every algebra $\mathcal{B}$, the algebra $(Sub(\mathcal{B}); \wedge, \vee)$ is a lattice, called the subalgebra lattice of $\mathcal{B}$.*

Proof: By 1.3.5 and by definition of the subalgebra generated by a set, the symbols $\wedge$ and $\vee$ define binary operations on $Sub(\mathcal{B})$. It is easy to check that all of the axioms required for a lattice are satisfied by these operations. ∎

The subalgebra lattice corresponding to an algebra $\mathcal{B}$ can be a useful tool in studying $\mathcal{B}$ itself. We will also see, in the next chapter, how we can generalize the process we used here to construct a lattice, based on the fact that the intersection of any subalgebras gives us a subalgebra.

We have been considering the problem of finding, for a given algebra $\mathcal{B}$ and a given subset X of B, the subalgebra of $\mathcal{B}$ generated by X. We can also look at the subalgebra generation question from a different point of view, as follows. Given an algebra $\mathcal{B}$, can we find a (proper) subset X of B which generates $\mathcal{B}$? In particular, we are often interested in whether we can find a finite such generating set X. If we can, then $\mathcal{B}$ is said to be *finitely generated*, with X as a finite generating system; otherwise $\mathcal{B}$ is said to be *infinitely generated*.

Of course any finite algebra has a finite generating system (the carrier set itself), so this question is only interesting for infinite algebras. Before addressing the question of when a finite generating system is possible, we consider two examples.

Example 1.3.11 Consider the algebra $(\mathbb{N}; \cdot)$ of type (2), where $\cdot$ is the usual multiplication on the set $\mathbb{N}$ of all natural numbers. Is this finitely generated? Since an arbitrary generating system would have to contain the infinite set of all prime numbers, the answer in this case is no. But the infinite set of all prime numbers is a generating system of this algebra, since every natural number has a representation as a product of powers of prime numbers.

Example 1.3.12 Recall from Example 1.2.14 in Section 1.2 that for any base set A, we can define an algebra of type $(2, 1, 1, 1, 0)$ on the set $O(A)$ of all finitary operations on A. This is the algebra $(O(A); *, \xi, \tau, \Delta, e_1^2)$. It is well known in Logic that for the special case when A is the two-element set $\{0, 1\}$, this algebra is finitely generated. The two-element set $X = \{\wedge, \neg\}$ (conjunction and negation) acts as a generating set for the algebra, since all Boolean functions can be generated from these two using our five operations. In the general case, it is an important problem to decide whether a given set X is a generating system of a subalgebra $\mathcal{C}$ of the algebra $(O(A); *, \xi, \tau, \Delta, e_1^2)$.

To answer the question of whether a given set $X \subseteq A$ is a generating system of an algebra $\mathcal{A}$ of type τ, we define the concept of a maximal subalgebra of an algebra.

Definition 1.3.13 An algebra $\mathcal{A}$ of type τ is called a *maximal subalgebra* of an algebra $\mathcal{B}$ of type τ if there is no subalgebra $\mathcal{C}$ with $\mathcal{A} \subset \mathcal{C} \subset \mathcal{B}$.

Corollary 1.3.14 $\mathcal{A}$ *is a maximal subalgebra of* $\mathcal{B}$ *iff for all* $q \in B \setminus A$, *we have* $\langle A \cup \{q\} \rangle_{\mathcal{B}} = \mathcal{B}$.

Proof: "$\Rightarrow$": Let $\mathcal{A}$ be a maximal subalgebra of $\mathcal{B}$. Since $q \notin \mathcal{A}$ we have $\mathcal{A} \subset \langle A \cup \{q\} \rangle_{\mathcal{B}} \subseteq \mathcal{B}$. From the maximality of $\mathcal{A}$ we obtain our proposition.

"$\Leftarrow$": Assume $\mathcal{A}$ is not maximal in $\mathcal{B}$. Then there is an algebra $\mathcal{C}$ with $\mathcal{A} \subset \mathcal{C} \subset \mathcal{B}$. But then for an arbitrary element $q \in C \setminus A$ we have $\langle A \cup \{q\} \rangle_{\mathcal{B}} \subseteq \mathcal{C} \subset \mathcal{B}$, and the set $A \cup \{q\}$ does not generate the algebra $\mathcal{B}$. ∎

As an important example of a maximal subalgebra, we will return to our example of the algebra of truth-value functions in classical two-valued logic.

Example 1.3.15 Let A be the two-element set $\{0, 1\}$, and let

$$C_2: \ = \ \{f \in O^n(A) \mid f(0, \ldots, 0) \ = \ 0, n \in \mathbb{N}\}.$$

We will use the criterion from Corollary 1.3.14 to verify that $\mathcal{C}_2 = (C_2; *, \xi, \tau, \Delta, e_1^2)$ is a maximal subalgebra of the algebra $(O(\{0, 1\}); *, \xi, \tau, \Delta, e_1^2)$ from Example 1.2.14. For any $f \in O(\{0, 1\}) \setminus C_2$, we must have $f(0, \ldots, 0) = 1$, and applying the operations $*, \xi, \tau, \Delta, e_1^2$ (by superposition) we can produce a unary operation f with $f(0) = 1$. For this f there are only two possibilities: f must be either the negation operation $\neg$ or the constant function with value 1. Since the conjunction $\wedge$ obviously belongs to C_2, in the first case we have

$$\langle C_2 \cup \{f\} \rangle_{O(A)} \supseteq \langle \{\wedge, \neg\} \rangle = O(A),$$

and thus $\langle C_2 \cup \{f\} \rangle_{O(A)} = O(A)$.

For the second case, we use the fact that since $0 + 0 = 0$, the addition modulo 2 (denoted by $+$) also belongs to C_2. From the constant 1 and the addition modulo 2 we get $\neg x = x + 1$ for all $x \in \{0, 1\}$. Now we have both $\neg$ and $\wedge$ in C_2, and we can conclude as we did in the first case.

Every subalgebra $\mathcal{A}$ of a finitely generated algebra $\mathcal{B} = (B; (f_i^A)_{i \in I})$ can be extended to a subalgebra $\mathcal{C}$ which is maximal in $\mathcal{B}$:

$$\mathcal{A} \subseteq \cdots \subseteq \mathcal{C} \subseteq \mathcal{B}$$

(see for instance Gluschkow, Zeitlin and Justschenko, [49]). Using this fact we have the following general completeness criterion for finitely generated algebras.

Theorem 1.3.16 *(General Completeness Criterion for finitely generated algebras) Let $\mathcal{B}$ be a finitely generated algebra. A set $X \subseteq B$ is a generating system of $\mathcal{B}$ iff there is no maximal subalgebra of $\mathcal{B}$ whose universe contains the set X.*

Proof: "$\Rightarrow$": Assume that $\langle X \rangle_{\mathcal{B}} = \mathcal{B}$. If $X \subseteq M$ for some maximal subalgebra $\mathcal{M}$ of $\mathcal{B}$, then by definition of $\langle X \rangle_{\mathcal{B}}$ we have the inclusion $\langle X \rangle_{\mathcal{B}} \subseteq \mathcal{M}$, which contradicts $\langle X \rangle_{\mathcal{B}} = \mathcal{B}$.

"$\Leftarrow$": If X is not a generating system of the algebra $\mathcal{B}$, then $\langle X \rangle_{\mathcal{B}}$ can be extended to a maximal subalgebra $\mathcal{M}$ of $\mathcal{B}$. But then $X \subseteq M$ for this maximal subalgebra. ∎

1.4 Congruence Relations and Quotients

Every function $\varphi : A \to B$ from a set A into a set B defines a partition of the set A into classes of elements having the same image. The equivalence relation $ker\varphi$ corresponding to this partition of A is called the *kernel* of the function φ. That is, for elements $a, b \in A$ we define

$$(a, b) \in ker\varphi : \quad \Leftrightarrow \quad \varphi(a) = \varphi(b).$$

This definition of the kernel makes it clear that the kernel is an equivalence relation on A. It is also easy to show that $ker\varphi$ is equal to the composition $\varphi^{-1} \circ \varphi$.

For instance, let $\mathbb{Z}$ be the set of all integers, let m be a natural number, and let $\mathbb{Z}/(m)$ be the set of all equivalence classes modulo m. Then the kernel of the function $\varphi : \mathbb{Z} \to \mathbb{Z}/(m)$ which is defined by $a \mapsto [a]_m$, that is, which maps every integer to its class modulo m, is obviously the well-known equivalence or congruence modulo m :

$$a \equiv b \ (m) : \quad \Leftrightarrow \quad \exists g \in \mathbb{Z} \ (a - b = gm).$$

(This is also often written as $a \equiv b \ mod \ m$). With respect to addition and multiplication in the ring $(\mathbb{Z}; +, \cdot, -, 0)$ of all integers, this congruence modulo m has the following additional property:

$$a_1 \equiv b_1 \ (m) \ \wedge \ a_2 \equiv b_2(m) \Rightarrow a_1 + a_2 \equiv b_1 + b_2(m) \ \wedge \ a_1 \cdot a_2 \equiv b_1 \cdot b_2(m).$$

This tells us that in this example the kernel of the function φ is compatible with the operations of the ring. This observation motivates the following more general definition.

Definition 1.4.1 Let A be a set, let $\theta \subseteq A \times A$ be an equivalence relation on A, and let f be an n-ary operation from $O^n(A)$. Then f is said to be *compatible* with θ, or to *preserve* θ, if for all $a_1, \ldots, a_n, b_1, \ldots, b_n \in A$,

$$(a_1, b_1) \in \theta, \quad \ldots, \quad (a_n, b_n) \in \theta$$

implies $\quad (f(a_1, \ldots, a_n), f(b_1, \ldots, b_n)) \in \theta.$

Remark 1.4.2 This definition can also be applied to arbitrary relations, not just to binary equivalence relations, and we can speak of a function preserving, or being compatible with, any relation. For instance, the monotone increasing real functions $f : \mathbb{R} \to \mathbb{R}$ are exactly the unary operations on $\mathbb{R}$ which are compatible with the relation $\leq$ on $\mathbb{R}$, since $x_1 \leq x_2 \Rightarrow f(x_1) \leq f(x_2)$ for all x_1, x_2 for which f is defined. The preservation of relations by functions will be studied in more detail in Chapter 2.

Definition 1.4.3 Let $\mathcal{A} = (A; (f_i^A)_{i \in I})$ be an algebra of type τ. An equivalence relation θ on A is called a *congruence relation* on $\mathcal{A}$ if all the fundamental operations f_i^A are compatible with θ. We denote by $Con\mathcal{A}$ the set of all congruence relations of the algebra $\mathcal{A}$. For every algebra $\mathcal{A} = (A; (f_i^A)_{i \in I})$ the trivial equivalence relations

$$\Delta_A : = \{(a,a) \mid a \in A\} \text{ and } \nabla_A = A \times A$$

are congruence relations. An algebra which has no congruence relations except Δ_A and ∇_A is called *simple*.

Example 1.4.4 1. Let $(\mathbb{Z}; +, \cdot, \ -, 0)$ be the ring of all integers and let $m \in \mathbb{Z}$ be an integer with $m \geq 0$. Then the congruence modulo m defined on $\mathbb{Z}$ is a congruence relation on $(\mathbb{Z}; +, \cdot, \ -, 0)$. Note that for $m = 0$ or $m = 1$ we get the trivial relations on $\mathbb{Z}$.

2. On every set A, the constant operations $f_c : A^n \to A$, defined by $f_c(x_1, \ldots, x_n) = c$ for all $x_1, \ldots, x_n \in A$, and the identical operation $id_A : A \to A$ are compatible with all equivalence relations defined on A. The projections $e_i^n : A^n \to A$, defined by $e_i^n(a_1, \ldots, a_n) = a_i$, for all $a_1, \ldots, a_n \in A$ and $1 \leq i \leq n$, also have the same property.

3. Let $\mathcal{A} = (\{a, b, c, d\}; f^A)$ be an algebra of type $\tau = (1)$. Let the unary operation f^A be given by the table

	a	b	c	d
$f(x)$	b	a	d	c

.

To determine all congruence relations on $\mathcal{A}$, we consider all possible partitions on A (and their corresponding equivalence relations), and look for those partitions having the property that the image of every class is also a class. This is easily seen to be the case for the two trivial partitions $\{a\} \cup \{b\} \cup \{c\} \cup \{d\}$ and $\{a, b, c, d\}$, and for the five non-trivial ones

$$\{a\} \cup \{b\} \cup \{c, d\}, \quad \{a, b\} \cup \{c\} \cup \{d\}, \quad \{a, b\} \cup \{c, d\},$$

$$\{a, c\} \cup \{b, d\}, \quad \text{and} \quad \{a, d\} \cup \{b, c\}.$$

Congruence relations have been defined as equivalence relations which are compatible with all the fundamental operations of an algebra. They may also be characterized in other ways, as we shall see at the end of Section 5.3. Here we show that congruences may also be characterized as equivalence relations which are compatible with certain unary functions on an algebra. For $\mathcal{A} = (A; (f_i^A)_{i \in I})$, a *translation* of $\mathcal{A}$ is a mapping of the form

$$x \to f_i^A(a_1, \ldots, a_{i-1}, x, a_{i+1}, \ldots, a_{n_i}),$$

where f_i^A is n_i-ary and a_1, ..., a_{i-1}, a_{i+1}, ..., a_{n_i} are fixed elements of A.

Theorem 1.4.5 *An equivalence relation θ on an algebra $\mathcal{A}$ is a congruence relation on $\mathcal{A}$ iff θ is compatible with all translations of $\mathcal{A}$.*

Proof: It is easy to check that any congruence is compatible with all translations of $\mathcal{A}$. Now suppose that θ is an equivalence relation defined on A which is compatible with all translations. Let f_i^A be n_i-ary, and let (a_j, b_j) be in θ for $j = 1, \ldots, n_i$. Then repeated use of the compatibility with translations gives

$$(f_i^A(a_1, a_2, a_3, \ldots, a_{n_i}), f_i^A(b_1, a_2, \ldots, a_{n_i})) \in \theta,$$
$$(f_i^A(b_1, a_2, a_3, \ldots, a_{n_i}), f_i^A(b_1, b_2, a_3, \ldots, a_{n_i})) \in \theta,$$

$$\ldots,$$

$$(f_i^A(b_1, b_2, \ldots, b_{n_i-1}, a_{n_i}), f_i^A(b_1, b_2, \ldots, b_{n_i})) \in \theta.$$

It follows from this that θ is a congruence on $\mathcal{A}$. ∎

We want now to define lattice operations on the set $Con\mathcal{A}$ of all congruence relations of an algebra $\mathcal{A}$, much as we did on the set $Sub(\mathcal{A})$ for subalgebras. As was the case there, our basic tool is the fact that the intersection of congruences is again a congruence.

Theorem 1.4.6 *The intersection $\theta_1 \cap \theta_2$ of two congruence relations on an algebra $\mathcal{A} = (A; (f_i^A)_{i \in I})$ is again a congruence relation on $\mathcal{A}$.*

Proof: Obviously, the intersection of two equivalence relations on A is again an equivalence relation defined on A. To see that the intersection is again a congruence, let $(a_1, b_1), \ldots, (a_{n_i}, b_{n_i})$ be pairs belonging to the intersection $\theta_1 \cap \theta_2$ and let f_i^A, for $i \in I$, be an arbitrary fundamental operation of $\mathcal{A}$. Then

$$(f_i^A(a_1, \ldots, a_{n_i}), f_i^A(b_1, \ldots, b_{n_i})) \in \theta_l, \quad \text{for} \quad l = 1, 2$$

and thus

$$(f_i^A(a_1, \ldots, a_{n_i}), f_i^A(b_1, \ldots, b_{n_i})) \in \theta_1 \cap \theta_2.$$

Therefore $\theta_1 \cap \theta_2$ is a congruence relation on $\mathcal{A}$. ∎

Remark 1.4.7 Theorem 1.4.6 is also satisfied for arbitrary families of congruence relations on $\mathcal{A}$. But in general, the union of two congruence relations of an algebra $\mathcal{A}$ is not a congruence relation, since this does not hold even for equivalence relations, as the following example shows. Take

$$A = \{1,2,3\}, \quad \theta_1 = \{(1,1),(2,2),(3,3),(1,2),(2,1)\},$$

$$\theta_2 = \{(1,1),(2,2),(3,3),(2,3),(3,2)\}.$$

The relations θ_1 and θ_2 are equivalence relations, but

$$\theta_1 \cup \theta_2 = \{(1,1),(2,2),(3,3),(1,2),(2,1),(2,3),(3,2)\}$$

is not an equivalence relation in A, since it is not transitive:

$$(1,2) \in \theta_1 \cup \theta_2, \ (2,3) \in \theta_1 \cup \theta_2, \ \text{but} \ (1,3) \notin \theta_1 \cup \theta_2.$$

But although the union of two congruence relations θ_1 and θ_2 need not be a congruence relation, as in the subalgebra case we can use intersections of congruences to define a smallest congruence generated by the union. This motivates the following definition.

Definition 1.4.8 Let $\mathcal{A}$ be an algebra, and let θ be a binary relation on A. We define the congruence relation $\langle \theta \rangle_{Con\mathcal{A}}$ on $\mathcal{A}$ generated by θ to be the intersection of all congruence relations θ' on $\mathcal{A}$ which contain θ:

$$\langle \theta \rangle_{Con\mathcal{A}} : = \cap \{\theta' \mid \theta' \in Con\mathcal{A} \text{ and } \theta \subseteq \theta'\}.$$

It is easy to see that $\langle \theta \rangle_{Con\mathcal{A}}$ has the three important properties of a closure operator:

$$\theta \subseteq \langle \theta \rangle_{Con\mathcal{A}} \qquad \text{(extensivity)},$$
$$\theta_1 \subseteq \theta_2 \Rightarrow \langle \theta_1 \rangle_{Con\mathcal{A}} \subseteq \langle \theta_2 \rangle_{Con\mathcal{A}} \quad \text{(monotonicity)},$$
$$\langle \langle \theta \rangle_{Con\mathcal{A}} \rangle_{Con\mathcal{A}} = \langle \theta \rangle_{Con\mathcal{A}} \qquad \text{(idempotency)}.$$

Remark 1.4.9 As we did in Section 1.3 for subalgebras, we can ask how to derive, from a binary relation θ defined on A, the congruence relation of the algebra $\mathcal{A}$ generated by θ. First we have to enlarge the relation θ to its reflexive and symmetric closure $\Delta_A \cup \theta \cup \theta^{-1}$. Then we form the compatible closure $\langle \Delta_A \cup \theta \cup \theta^{-1} \rangle$. Finally the set produced in this way has to be completed by all pairs needed for transitivity. This is done by using the *transitive hull operator*, which maps any relation α to

$$\alpha_T : \ = \ \cap \{ \beta \subseteq A^2 \mid \beta \supseteq \alpha \text{ and } \beta \text{ is transitive} \}.$$

Altogether, then, we need to use

$$\langle \theta \rangle_{Con\mathcal{A}} \ = \ \langle \langle \Delta_A \cup \theta' \cup (\theta')^{-1} \rangle \rangle_T.$$

This construction allows us to produce, for two congruences on $\mathcal{A}$, the least congruence on $\mathcal{A}$ which contains them both. This gives us the two binary operations on $Con(\mathcal{A})$ we wanted,

$$\wedge : \quad Con\mathcal{A} \times Con\mathcal{A} \to Con\mathcal{A} \quad \text{defined by } (\theta_1, \theta_2) \mapsto \theta_1 \cap \theta_2,$$

$$\vee : \quad Con\mathcal{A} \times Con\mathcal{A} \to Con\mathcal{A} \quad \text{defined by } (\theta_1, \theta_2) \mapsto \langle \theta_1 \cup \theta_2 \rangle_{Con\mathcal{A}}.$$

Now we must verify that these operations do make our set into a lattice.

Theorem 1.4.10 *For every algebra $\mathcal{A}$ the structure $(Con\mathcal{A}; \wedge, \vee)$ is a lattice, called the congruence lattice $Con(\mathcal{A})$ of $\mathcal{A}$.*

Proof: We leave it as an exercise for the reader to verify that the binary operations $\wedge$ and $\vee$ on $Con\mathcal{A}$ satisfy all the axioms of a lattice.   ∎

Since congruence relations are equivalence relations, each congruence on an algebra $\mathcal{A}$ induces a partition on the set A. What makes congruence relations significant is that the partition set induced by a congruence can also be made into the universe of an algebra. For each $i \in I$, we define an n_i-ary operation $f_i^{A/\theta}$ on the quotient set A/θ, by

$$f_i^{A/\theta} : \ (A/\theta)^{n_i} \to A/\theta,$$

defined by

$$([a_1]_\theta, \ldots, [a_{n_i}]_\theta) \mapsto f_i^{A/\theta}([a_1]_\theta, \ldots, [a_{n_i}]_\theta) \ := \ [f_i^A(a_1, \ldots, a_{n_i})]_\theta.$$

As always when we define operations on equivalence classes, we must verify that our operations are well-defined, that is, that they are independent of the representatives chosen. But as the following proof shows, this is exactly what the compatibility property of a congruence relation means.

Theorem 1.4.11 *For every algebra $\mathcal{A}$ of type τ and every congruence relation $\theta \in Con\mathcal{A}$, the previous definition produces an algebra $\mathcal{A}/\theta$ of type τ, which is called the quotient algebra or factor algebra of $\mathcal{A}$ by θ.*

Proof: Let $i \in I$. We have to show that if $[a_j]_\theta = [b_j]_\theta$ for all $j = 1, \ldots, n_i$, then $[f_i^A(a_1, \ldots, a_{n_i})]_\theta = [f_i^A(b_1, \ldots, b_{n_i})]_\theta$. But this is exactly what our compatibility property of a congruence guarantees. If $[a_j]_\theta = [b_j]_\theta$, then $(a_j, b_j) \in \theta$ for all $j = 1, \ldots, n_i$, and since θ is a congruence on the algebra $\mathcal{A}$, it follows that

$$(f_i^A(a_1, \ldots, a_{n_i}), f_i^A(b_1, \ldots, b_{n_i})) \in \theta$$

and thus

$$[f_i^A(a_1, \ldots, a_{n_i})]_\theta = [f_i^A(b_1, \ldots, b_{n_i})]_\theta. \qquad \blacksquare$$

Example 1.4.12 Let $m \in \mathbb{N}$, and let $\mathbb{Z}/(m)$ be the set of all congruence classes modulo m. We define two binary operations on these classes, by

$$[a]_m + [b]_m = [a + b]_m \text{ and } [a]_m \cdot [b]_m = [a \cdot b]_m.$$

This makes $(\mathbb{Z}/(m); +, \cdot)$ an algebra of type $(2, 2)$. In fact it is a ring, called the *ring of all congruence classes modulo m*. If m is a prime number, this ring is also a field.

1.5 Exercises

1.5.1. Determine all subalgebras of the algebra

$$\mathcal{A} = (\{0, 1, 2, 3\}; \max(x, y), \min(x, y), x + y(4), 3 - x(4)),$$

where the four binary operations are defined as follows: $max(x, y)$ and $min(x, y)$ are the maximum and the minimum, respectively, with respect to the order relation defined by $0 \leq 1 \leq 2 \leq 3$; $x + y(4)$ is addition modulo 4; and $3 - x(4)$ denotes subtraction modulo 4.

1.5.2. Prove the properties of Theorem 1.3.7.

1.5.3. Is the algebra $(\mathbb{N}; +)$ of type $\tau = (2)$ finitely generated? If so, give a finite generating system.

1.5.4. A generating system $X \subseteq A$ of an algebra $\mathcal{A}$ is called a *basis* of $\mathcal{A}$, if for all $a \in X$ the inequality $\langle X \setminus \{a\} \rangle_{\mathcal{A}} \neq \mathcal{A}$ is satisfied; that is, if no proper subset of X generates $\mathcal{A}$. Prove that every finitely generated algebra has a finite basis.

1.5.5. Let $(\mathbb{N}; \odot)$ be an algebra of type (1), with a unary operation $\odot$ defined by
$$0^{\odot} := 0, \qquad n^{\odot} = n - 1 \text{ for } n > 0.$$
Prove that this algebra has no basis.

1.5.6. Complete the proofs of Theorems 1.3.10 and 1.4.10.

1.5.7. Let $\mathcal{A} = (\{a, b, c, d\}; f)$ be an algebra of type $\tau = (1)$, with

x	a	b	c	d
$f(x)$	c	b	a	$d.$

Find all congruence relations θ on $\mathcal{A}$, and for each one determine the quotient algebra $\mathcal{A}/\theta$.

1.5.8. Let $(A; t)$ be an algebra of type $\tau = (3)$, where t is defined by
$$t(x, y, z) = \begin{cases} z, & \text{if } x = y \\ x, & \text{if } x \neq y. \end{cases}$$
Prove that $(A; t)$ is simple. (Such a function t is called a *ternary discriminator* function.)

1.5.9. Determine all subalgebras and all congruence relations of the algebra $\mathcal{A} = (\{(0), (x), (\square)\}; \times)$ of type $\tau = (2)$, with $\times$ defined by

$\times$	(0)	(x)	$(\square)$
(0)	(0)	(0)	$(\square)$
(x)	(0)	(x)	(x)
$(\square)$	$(\square)$	(x)	$(\square).$

1.5.10. Draw the corresponding directed graph for the finite deterministic automaton $(\{0, 1\}, \{0, 1\}, \{0, 1\}; \delta, \lambda)$, given by

δ	0	1
0	0	1
1	1	0

and

λ	0	1
0	0	0
1	1	1

1.5.11. Let $G = (V, E)$ be a graph, with set V of vertices and set $E \subseteq V \times V$ of edges. Fix a symbol $0 \notin V$, and let $V' = V \cup \{0\}$. We can define a type (2) algebra called a graph-algebra on the set V', with a binary multiplication given by $xy = x$ if $(x, y) \in E$ and $xy = 0$ otherwise. Construct the graph algebra for the graph $G = (\{1, 2\}, \{(1, 2), (2, 1), (2, 2)\})$.

1.5.12. a) Prove that any distributive lattice is modular.
b) Prove that the lattice $\mathcal{L}_1$ shown below is not modular.
c) Prove that the lattice $\mathcal{L}_2$ shown below is modular but not distributive.

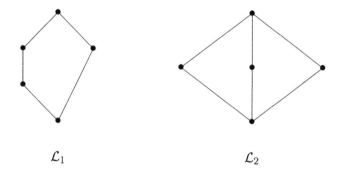

$\mathcal{L}_1$ $\qquad\qquad\qquad\qquad$ $\mathcal{L}_2$

1.5.13. Congruences on an algebra are connected to the concept of a normal subgroup of a group. Let $\mathcal{G}$ be a group, considered as an algebra of type $(2, 1, 0)$, with the binary multiplication written as juxtaposition. For any normal subgroup $\mathcal{N}$ of $\mathcal{G}$, we define a relation θ_N on G by $(a, b) \in \theta_N$ iff $ab^{-1} \in \mathcal{N}$ iff $\exists n \in N(a = nb)$.

a) Prove that the relation θ_N is a congruence relation on $\mathcal{G}$, for which the set N is exactly the equivalence class of the identity element of the group.
b) Prove conversely that if θ is a congruence on a group $\mathcal{G}$, the equivalence class under θ of the identity element of the group forms a normal subgroup of the group.

Chapter 2

Galois Connections and Closures

A Galois-connection is a connection, with certain properties, between two sets of objects, usually of different kinds. Such a connection can provide a useful tool for studying properties of one kind of object, based on the properties of the other (usually more well-known) kind of objects. In the classical Galois theory, for instance, properties of permutation groups are used to study field extensions. Such connections provide a useful tool for studying algebraic structures, and will be a main focus of study in this book. In particular, Galois-connections will be used in later chapters to produce two main examples of complete lattices of closed sets, the lattice of all varieties of a given type in Chapter 6 and the lattice of all clones on a fixed base set in Chapter 10.

Galois-connections are also closely related to closure operators and closure systems. We have already seen two examples of operators with closure properties: the formation of the subalgebra generated by a set, and the formation of the congruence generated by a binary relation. We begin this chapter with a study of closure operators in general. In the second section we define Galois-connections, and relate them to closure operators and systems. The last section in this chapter describes an application of Galois-connections to concept analysis.

2.1   Closure Operators

In the previous chapter we have seen two examples of operators with closure properties. First, when we generate subalgebras of a given algebra $\mathcal{A}$ from subsets X of the carrier set A of $\mathcal{A}$, we have a mapping which takes any $X \subseteq A$ to a unique subset $\langle X \rangle_A$ of A. This gives a unary operation, or *an operator*, $\langle \ \rangle_A : \mathcal{P}(A) \to \mathcal{P}(A)$ on the power set of A, which we showed has the three closure properties of Theorem 1.3.7. Later we saw that the operator which maps any binary relation θ on $\mathcal{A}$ to the congruence $\langle \theta \rangle_{Con\mathcal{A}}$ generated by θ also has these same properties. In fact operators with these properties occur in many areas of Mathematics, and are frequently used as a tool to study other structures.

Definition 2.1.1 Let A be a set. A mapping $C : \mathcal{P}(A) \to \mathcal{P}(A)$ is called a *closure operator* on A, if for all subsets $X, Y \subseteq A$ the following properties are satisfied:

(i) $X \subseteq C(X)$ (extensivity),

(ii) $X \subseteq Y \Rightarrow C(X) \subseteq C(Y)$ (monotonicity),

(iii) $C(X) = C(C(X))$ (idempotency).

Subsets of A of the form $C(X)$ are called *closed* (with respect to the operator C) and $C(X)$ is said to be the closed set generated by X.

Example 2.1.2 1. For every algebra $\mathcal{A}$, we have an operator $\langle \ \rangle_A$ which maps every subset $X \subseteq A$ of the carrier set of $\mathcal{A}$ to the carrier set $\langle X \rangle_A$ of the least subalgebra $\langle X \rangle_A$ of $\mathcal{A}$ which contains X. As we remarked above, Theorem 1.3.7 shows that this is a closure operator.

2. Let $\mathcal{F}$ be a field, and let $(V; +, \circ)$ be an F-vector space. Here we regard $\circ : F \times V \to V$ as the multiplication of the elements or vectors of V on the left by elements, or scalars, from F. The linear hull of a subset $X \subseteq V$ is the F-subspace $\langle X \rangle_V$ of the F-vector space $\mathcal{V}$ generated by X. Again this is an example of a closure operator, with the closure of a set X equal to the set of all linear combinations which can be produced from the vectors of X.

3. Let A be a set. To every binary relation $\theta \subseteq A^2$, we assign the relation $\langle \theta \rangle_T$ generated by its transitive hull, as defined in Remark 1.4.9:

$$\langle \theta \rangle_T : \ = \ \cap \{\theta' \subseteq A^2 \mid \theta' \supseteq \theta \text{ and } \theta' \text{ is transitive}\}.$$

This gives an operator $\langle \ \rangle_T : \mathcal{P}(A^2) \to \mathcal{P}(A^2)$, which can be shown to be a closure operator on A^2.

To characterize sets which are closed with respect to a closure operator, we will use the concept of a closure system.

Definition 2.1.3 Let A be a set. A subset $\mathcal{H}$ of $\mathcal{P}(A)$ is called a *closure system* if it satisfies the following two conditions:

(i) $A \in \mathcal{H}$, and
(ii) $\cap \mathcal{B} \in \mathcal{H}$ for every non-empty subset $\mathcal{B} \subseteq \mathcal{H}$.

The elements of a closure system $\mathcal{H}$ are called *closures*.

Example 2.1.4 1. By Corollary 1.3.5, for every algebra $\mathcal{A}$ the set $Sub(\mathcal{A})$ of all subalgebras of $\mathcal{A}$ is a closure system on A.

2. By Theorem 1.4.6 and Remark 1.4.7, for every algebra $\mathcal{A}$ the set $Con\mathcal{A}$ of all congruences on $\mathcal{A}$ is a closure system on A^2.

3. For every set A, the set EqA of all equivalence relations defined on A is a closure system on A^2.

4. For every set A, the set $\mathcal{P}(A)$ is a closure system on A.

Our goal now is to show that the family of sets which are closed with respect to a given closure operator satisfies the conditions of Definition 2.1.3, and conversely that any closure system induces a closure operator. We start with the following definition.

Definition 2.1.5 Given a closure system $\mathcal{H}$ on a set A, we define an operator

$$C_{\mathcal{H}} : \mathcal{P}(A) \to \mathcal{P}(A)$$

on A, by

$$X \mapsto C_{\mathcal{H}}(X) : \ = \ \cap \{H \in \mathcal{H} \mid H \supseteq X\}, \quad \text{for all } X \subseteq A.$$

Conversely, for any closure operator C on A, we set

$$\mathcal{H}_C: \; = \{X \subseteq A \mid C(X) = X\}.$$

Theorem 2.1.6 *Let $\mathcal{H}$ be a closure system and C be a closure operator on the set A. Then $C_{\mathcal{H}}$ and $\mathcal{H}_C$, as defined above, satisfy the following properties:*

(i) $C_{\mathcal{H}}$ is a closure operator on A and $\mathcal{H}_C$ is a closure system on A.

(ii) The closed sets with respect to $C_{\mathcal{H}}$ are exactly the closures of $\mathcal{H}$.

(iii) The closures of $\mathcal{H}_C$ are exactly the closed sets of C.

(iv) $\mathcal{H}_{C_{\mathcal{H}}} = \mathcal{H}$ and $C_{\mathcal{H}_C} = C$.

Proof: (i) We show first that $C_{\mathcal{H}}$ is a closure operator. Since X is contained in every "component" of the intersection $C_{\mathcal{H}}(X)$, we have $X \subseteq C_{\mathcal{H}}(X)$, and $C_{\mathcal{H}}$ is extensive. $C_{\mathcal{H}}$ is also monotone, because for $X \subseteq Y$, we have

$$C_{\mathcal{H}}(Y) \;=\; \cap \{H \in \mathcal{H} \mid H \supseteq Y\} \supseteq C_{\mathcal{H}}(X) \;=\; \cap \{H \in \mathcal{H} \mid H \supseteq X\}.$$

This follows because all sets H which contain Y also contain X, and therefore the second intersection contains more components than the first one, making it "smaller." For the idempotency of $C_{\mathcal{H}}$, we observe that $C_{\mathcal{H}}(C_{\mathcal{H}}(X))$ is the intersection of all closures from $\mathcal{H}$ which contain $C_{\mathcal{H}}(X)$. But $C_{\mathcal{H}}(X)$ itself is such a closure: it is the intersection of all elements of $\mathcal{H}$ which contain X, and since $\mathcal{H}$ is a closure system this intersection is also in $\mathcal{H}$. This means that $C_{\mathcal{H}}(C_{\mathcal{H}}(X)) = C_{\mathcal{H}}(X)$. Altogether, $C_{\mathcal{H}}$ is a closure operator.

Conversely, let C be a closure operator on A. Obviously, $A = C(A)$; we always have $C(A) \subseteq A$, and $A \subseteq C(A)$ holds because of the extensivity of the closure operator C. But this means that A itself is in our family $\mathcal{H}_C$, as required for the first condition of a closure system. For the second condition, let $\mathcal{B} \subseteq \mathcal{H}_C$, and consider the intersection $\cap \mathcal{B}$. For every $H \in \mathcal{B}$ we have $C(H) = H$ and $\cap \mathcal{B} \subseteq H$. The monotonicity of the closure operator C gives:

$$C(\cap \mathcal{B}) \subseteq C(H) = H \quad \text{and} \quad C(\cap \mathcal{B}) \subseteq \bigcap_{H \in \mathcal{B}} H = \cap \mathcal{B}.$$

Combining this with $\cap \mathcal{B} \subseteq C(\cap \mathcal{B})$ gives $C(\cap \mathcal{B}) \ = \ \cap \mathcal{B}$, and thus $\cap \mathcal{B} \in \mathcal{H}_C$. This shows that the system $\mathcal{H}_C$ is a closure system.

(ii) For any set $X \subseteq A$, we have

$$C_{\mathcal{H}}(X) \ = \ \cap \{H \in \mathcal{H} \mid H \supseteq X\} \ = \ X \Leftrightarrow X \in \mathcal{H}.$$

Thus the sets closed under $C_{\mathcal{H}}$ are exactly the closures of $\mathcal{H}$.

(iii) It follows immediately from the definition of $\mathcal{H}_C$ that the closures of $\mathcal{H}_C$ are exactly the sets which are closed with respect to C.

(iv) Since
$$X \in \mathcal{H}_{C_{\mathcal{H}}} \Leftrightarrow C_{\mathcal{H}}(X) \ = \ X \Leftrightarrow X \in \mathcal{H},$$
we have $\mathcal{H}_{C_{\mathcal{H}}} \ = \ \mathcal{H}$.

To show that $C_{\mathcal{H}_C} \ = \ C$, we start with

$$C_{\mathcal{H}_C}(X) \ = \ \cap \{H \in \mathcal{H}_C \mid H \supseteq X\} \ = \ \cap \{H \mid C(H) = H \text{ and } H \supseteq X\}.$$

For every set H in the latter set, $X \subseteq H$ gives

$$C(X) \subseteq C(H) \ = \ H.$$

Since $C(X)$ is contained in every component of the intersection, it is contained in the intersection itself, giving $C(X) \subseteq C_{\mathcal{H}_C}(X)$.

Since $X \subseteq C(X) \ = \ C(C(X))$, it follows that $C(X)$ is a component of the intersection. Therefore we also have $C_{\mathcal{H}_C}(X) \subseteq C(X)$. The two inclusions then give $C_{\mathcal{H}_C}(X) \ = \ C(X)$. ∎

The following definitions describe some additional properties closure systems may have.

Definition 2.1.7 A non-empty system G of sets is called *upward directed*, if for every pair $X, Y \in G$ there exists a set $Z \in G$ with $X \cup Y \subseteq Z$.

A system $\mathcal{M}$ of sets is called *inductive*, if for every upward directed subsystem G, the union $\cup G$ is also in $\mathcal{M}$.

A closure operator C defined on a set A is called *inductive*, if for all $X \subseteq A$,

$$C(X) \;=\; \cup \{C(E) \mid E \subseteq X \text{ and } E \text{ is finite}\}.$$

If the system of sets being considered is a closure system, the concepts of this definition agree, and we have the following result, which we shall not prove here. (A proof is given by Th. Ihringer in [58]).

Theorem 2.1.8 *A closure system $\mathcal{H}$ is inductive if and only if the corresponding closure operator $C_{\mathcal{H}}$ is inductive.*   ■

We shall need the following important example of an inductive closure system and operator.

Corollary 2.1.9 *For every algebra $\mathcal{A} \;=\; (A; \; (f_i^A)_{i \in I})$, the system $Sub(\mathcal{A})$ is an inductive closure system and $\langle \; \rangle_A$ is an inductive closure operator.*

Proof: Let $G \subseteq Sub(\mathcal{A})$ be upward directed; we have to prove that $\cup G \in Sub(\mathcal{A})$. Let f_i^A be an n_i-ary operation and assume that $b_1, \ldots, b_{n_i} \in \cup G$. Then there exist sets $G_1, \ldots, G_{n_i} \in G$ with $b_1 \in G_1, \ldots, b_{n_i} \in G_{n_i}$. Since G is upward directed, there is a set $G_0 \in G$ with $b_1, \ldots, b_{n_i} \in G_0$. But $G_0 \in Sub(\mathcal{A})$ means that

$$f(b_1, \ldots, b_{n_i}) \in G_0 \subseteq \cup G.$$

This makes $\cup G \in Sub(\mathcal{A})$; so $Sub(\mathcal{A})$ is inductive. The claim for $\langle \; \rangle_A$ follows from Theorem 2.1.6 and the fact that $\langle \; \rangle_A = C_{Sub(\mathcal{A})}$.   ■

Just as we did for $Sub(\mathcal{A})$, we can define a meet and join operation on any closure system. If $\mathcal{H} \subseteq \mathcal{P}(A)$ is a closure system, then for arbitrary sets $\mathcal{B} \subseteq \mathcal{H}$, we use

$$\bigwedge \mathcal{B} : \;=\; \cap \mathcal{B} \text{ and } \bigvee \mathcal{B} : \;=\; \cap \{H \in \mathcal{H} \mid H \supseteq \cup \mathcal{B}\}.$$

In particular, for two-element sets B, this gives us binary meet and join operations on $\mathcal{H}$. Then every closure system is a lattice, under these operations, and in fact a complete lattice (see Definition 1.2.11).

Example 2.1.10 Let $\mathcal{A}$ be any algebra. Since $Sub(\mathcal{A})$ and $Con\mathcal{A}$ are closure systems, the lattices $(Sub(\mathcal{A}); \; \subseteq)$ and $(Con\mathcal{A}; \; \subseteq)$ are complete lattices.

This shows that all subalgebra lattices are complete lattices. We can then ask whether any complete lattice occurs as the subalgebra lattice of some algebra, and if not whether there are some classes of complete lattices which do occur in this way. To answer these questions, we need the following concept.

Definition 2.1.11 Let $(L; \leq)$ be a complete lattice. An element a of L is called *compact*, if for every set $B \subseteq L$ with $a \leq \bigvee B$ there exists a finite subset $B_0 \subseteq B$ with $a \leq \bigvee B_0$. A lattice is called *algebraic* if it is a complete lattice in which every element is the supremum of compact elements.

Theorem 2.1.12 *For every inductive closure system $\mathcal{H}$, the partially ordered set $(\mathcal{H}; \subseteq)$ is an algebraic lattice.*

Proof: In an inductive closure system, every closure is the supremum of finitely generated elements. Therefore inductive closure systems lead to algebraic lattices. ■

We mention here without proof the following important result; a proof may be found for instance in G. Grätzer, [51] or P. Cohn, [10].

Corollary 2.1.13 *A lattice $\mathcal{L}$ is isomorphic to the subalgebra lattice of some algebra iff $\mathcal{L}$ is algebraic.* ■

2.2 Galois Connections

A Galois-connection between two sets of objects is a pair of mappings, with certain properties, between the power sets of the two sets. These mappings allow us to move back and forth between the two kinds of objects, often using information about one kind to learn more about the other. Such connections will be a main focus of study in this book. Before we introduce the general definitions, we look in detail at an example of such a connection. This example builds on the idea of an operation preserving a relation, which was the basis of our definition of a congruence relation. Our connection will be between operations and relations on a given set.

We start with a fixed base set A, and as one of our sets of objects the set $O(A)$ of all operations on A. In Definition 1.4.1, we considered an interconnection between operations from $O(A)$ and binary relations $\theta \subseteq A^2$, namely that an operation could be compatible with, or preserve, a binary relation.

This concept can be generalized to include h-ary relations on A for any $h \geq 1$, as follows. We denote by $R^h(A)$ the set of all h-ary relations defined on A, and by $R(A) = \bigcup_{h=1}^{\infty} R^h(A)$ the set of all finitary relations defined on A.

We say $f \in O^n(A)$ *preserves* the $h-$ary relation $\rho \in R(A)$, if whenever

$$(a_1^1, \ldots, a_h^1) \in \rho, \quad \ldots, \quad (a_1^n, \ldots, a_h^n) \in \rho,$$

it follows that also

$$(f(a_1^1, \ldots, a_1^n), \quad \ldots, \quad f(a_h^1, \ldots, a_h^n)) \in \rho.$$

This connection between operations and relations determines a mapping which associates to any relation a set of operations. For any relation $\rho \in R(A)$, we can consider the set of all operations from $O(A)$ which preserve ρ. This set will be denoted by $Pol_A \rho$, i.e.,

$$Pol_A \rho = \{f \mid f \in O(A) \text{ and } f \text{ preserves } \rho\}.$$

(We remark that in fact $Pol_A \rho$ is a clone on A, in the sense defined in Example 1.2.14: it is a set of operations on A which contains the projections and which is closed under composition of operations.)

Example 2.2.1 Let A be the set $\{0, 1\}$.

1. Let ρ be the unary relation $\{0\}$. Notice that unary relations on a set are simply subsets of this set. Then $Pol_A\{0\}$ is the set of all Boolean functions which preserve $\{0\}$, so

$$Pol_A\{0\} = \{f \in O(A) \mid f(0, \ldots, 0) = 0\}.$$

2. Let

$$\alpha = \{(a, b, c, d) \in A^4 \mid a + b = c + d\},$$

where $+$ is the addition modulo 2. An n-ary operation f on A is called *linear*, if there are elements $a_1, \ldots, a_n, c \in \{0, 1\}$ such that

$$f(x_1, \ldots, x_n) = a_1 x_1 + \cdots + a_n x_n + c$$

for all $x_1, \ldots, x_n \in \{0, 1\}$, where again $+$ is the addition modulo 2. It can be shown that a Boolean function f is linear iff it preserves α. Thus $Pol_A \alpha$ is the set of all linear Boolean functions.

These examples show how, for any given relation ρ on A, we can look for the set of all operations which preserve ρ. In the other direction, for a given operation $f \in O(A)$ one can look for the set of all relations from $R(A)$ which are preserved by f. Such relations are called *invariants* of f, and the set of all such invariants is denoted by $Inv_A f$, that is,

$$Inv_A f = \{\rho \in R(A) \mid f \text{ preserves } \rho\}.$$

We can also extend the maps Pol_A and Inv_A to sets of relations or functions, respectively. If $F \subseteq O(A)$ is a set of operations on A, we define $Inv_A F$ to be the set of all relations which are invariant for all $f \in F$, and similarly for a set $Q \subseteq R(A)$ of relations on A, we denote by $Pol_A Q$ the set of all operations which preserve every relation $\rho \in Q$.

These maps Pol_A and Inv_A, between sets of relations and sets of operations on A, show the basic idea of a Galois-connection, between two sets of objects. The following theorem of Pöschel and Kalužnin shows that in this example, we have the two important properties that will be used shortly to define a Galois-connection. These properties are that our maps between the two sets of objects map larger sets to smaller images, and that mapping a set twice returns us to a set which contains the starting set.

Theorem 2.2.2 *([96]) The following interconnections hold between sets of the form $Pol_A Q$ and $Inv_A F$, for $Q \subseteq R(A)$ and $F \subseteq O(A)$:*

(i) $F_1 \;\subseteq\; F_2 \subseteq O(A) \;\;\;\; \Rightarrow \;\;\; Inv_A F_1 \supseteq Inv_A F_2,$

       $Q_1 \;\subseteq\; Q_2 \subseteq R(A) \;\;\;\; \Rightarrow \;\;\; Pol_A Q_1 \supseteq Pol_A Q_2;$

(ii) $F \;\subseteq\; Pol_A Inv_A F,$

       $Q \;\subseteq\; Inv_A Pol_A Q.$

Proof: These propositions follow immediately from the definitions of Pol_A and Inv_A. ∎

The concept of an invariant, or a set of all objects which are invariant under certain changes, also plays an important role in other branches of Mathematics and Science. From Analytic Geometry we know for example that by *Felix Klein's "Erlanger Programm"* ([64]), different geometries may be regarded as the theories of invariants of different transformation groups. Instead of sets of the form $Pol_A R$, in this setting we have sets of transformations, more exactly the carrier sets of transformation groups. The corresponding invariants can be parallelism, the affine ratio, the cross ratio, the distance, or the angle; again, larger sets of mappings determine smaller sets of invariants. The following table gives a survey of the invariants of three transformation groups:

invariants under affine mappings	invariants under similarity mappings	invariants under motions
parallelism affine ratio	parallelism affine ratio cross ratio angle	parallelism affine ratio angle distance

These examples of mappings between two sets, with the two basic properties from Theorem 2.2.2, form a model for our definition of a Galois-connection.

Definition 2.2.3 A *Galois-connection* between the sets A and B is a pair (σ, τ) of mappings between the power sets $\mathcal{P}(A)$ and $\mathcal{P}(B)$,

$$\sigma : \mathcal{P}(A) \rightarrow \mathcal{P}(B) \text{ and } \tau : \mathcal{P}(B) \rightarrow \mathcal{P}(A),$$

such that for all X, $X' \subseteq A$ and all Y, $Y' \subseteq B$ the following conditions are satisfied:

(i) $X \subseteq X' \Rightarrow \sigma(X) \supseteq \sigma(X')$,   and $Y \subseteq Y' \Rightarrow \tau(Y) \supseteq \tau(Y')$;

(ii) $X \subseteq \tau\sigma(X)$,   and $Y \subseteq \sigma\tau(Y)$.

Galois-connections are also related to closure operators, as the following proposition shows.

Theorem 2.2.4 *Let the pair (σ, τ) with*

$$\sigma : \mathcal{P}(A) \rightarrow \mathcal{P}(B) \text{ and } \tau : \mathcal{P}(B) \rightarrow \mathcal{P}(A)$$

be a Galois-connection between the sets A and B. Then:

(i) $\sigma\tau\sigma = \sigma$ and $\tau\sigma\tau = \tau$;

(ii) $\tau\sigma$ and $\sigma\tau$ are closure operators on A and B respectively;

(iii) The sets closed under $\tau\sigma$ are precisely the sets of the form $\tau(Y)$, for some $Y \subseteq B$; the sets closed under $\sigma\tau$ are precisely the sets of the form $\sigma(X)$, for some $X \subseteq A$.

Proof: (i) Let $X \subseteq A$. By the second Galois-connection property, we have $X \subseteq \tau\sigma(X)$. By the first property, applying σ to this gives $\sigma(X) \supseteq \sigma\tau\sigma(X)$. But we also have $\sigma(X) \subseteq \sigma\tau(\sigma(X))$, by the second Galois-connection property applied to the set $\sigma(X)$. This gives us $\sigma\tau\sigma(X) = \sigma(X)$. The second claim is proved similarly.

(ii) The extensivity of $\tau\sigma$ and $\sigma\tau$ follows from the second Galois-connection property. From the first property we see that

$$X \subseteq X' \Rightarrow \sigma(X) \supseteq \sigma(X') \Rightarrow \tau\sigma(X) \subseteq \tau\sigma(X'),$$

since $\sigma(X)$ and $\sigma(X')$ are subsets of B; and in the analogous way we get from $Y \subseteq Y'$ the inclusion $\sigma\tau(Y) \subseteq \sigma\tau(Y')$. Applying σ to the equation $\tau\sigma\tau = \tau$ from part (i) gives us the idempotency of $\sigma\tau$, and similarly for $\tau\sigma$.

(iii) This is straightforward to verify. ∎

A relation between the sets A and B is simply a subset of $A \times B$. Any relation R between A and B induces a Galois-connection, as follows. We can define the mappings

$$\sigma : \mathcal{P}(A) \to \mathcal{P}(B), \quad \tau : \mathcal{P}(B) \to \mathcal{P}(A),$$

by

$$\sigma(X) : = \{y \in B \mid \forall x \in X \quad ((x,y) \in R)\},$$
$$\tau(Y) : = \{x \in A \mid \forall y \in Y \quad ((x,y) \in R)\}.$$

Then it is easy to verify that the pair (σ, τ) forms a Galois-connection between A and B, called the Galois-connection induced by R. In our example with *Pol* and *Inv*, consider the preservation relation R between $O(A)$ and $R(A)$, defined by

$$R = \{(f, \theta) \in O(A) \times R(A) \mid f \text{ preserves } \theta\}.$$

The reader can verify that the Galois-connection induced by this R is in fact the one we described.

2.3   Concept Analysis

Concept Analysis is a very useful application of Galois-connections. It is based on the assumptions that all human knowledge involves conceptual thinking, and that human reasoning involves manipulation of concepts. Formal concept analysis, developed since 1979 by R. Wille and his collaborators (see [120]), goes back to Aristotle's basic view of a concept as a unit of thought constituted by its extension and its intension.

The extension of a concept is the collection of all objects belonging to the concept, and the intension is the set of all attributes common to those objects. Subconcepts satisfy larger sets of attributes, while subsets of sets of attributes determine superconcepts. This means precisely that we have a Galois-connection between sets of objects and sets of attributes.

Since human thinking and communication always occurs in a context, the first step is the formalization of contexts.

Definition 2.3.1 A *(formal) context* is a triple (G, M, I), in which G and M are sets and $I \subseteq G \times M$ is a relation between G and M. The set G is called the set of *objects*, and M is the set of *attributes* (Gegenstände und Merkmale, respectively, in German). The relation I is defined by

$$I := \{(g, m) \in G \times M \mid \text{object } g \text{ has attribute } m\}.$$

This relation induces a Galois-connection (σ, τ) between the sets G and M, as described in the previous section.

For a given formal context (G, M, I), our philosophical understanding of a concept can be formalized by the following definition.

Definition 2.3.2 A pair (X, Y) is said to be a *formal concept* of the formal context (G, M, I) if $X \subset G$, $Y \subseteq M$, $\sigma(X) = Y$ and $\tau(Y) = X$. The sets X and Y are called the *extent* and the *intent* of the formal concept (X, Y), respectively.

It is clear from this definition that such formal concepts consist of pairs of closed sets, under the closure operators $\tau\sigma$ and $\sigma\tau$.

Let us denote by $\mathcal{B}(G, M, I)$ the set of all concepts of the context (G, M, I). We can define a partial order on this set, by

$$(X_1, Y_1) \leq (X_2, Y_2) :\Leftrightarrow X_1 \subseteq X_2 \ (\Leftrightarrow Y_1 \supseteq Y_2).$$

When $(X_1, Y_1) \leq (X_2, Y_2)$, the pair (X_1, Y_1) is called a *subconcept* of the pair (X_2, Y_2), and conversely (X_2, Y_2) is called a *superconcept* of (X_1, Y_1). It is easy to verify that $(\mathcal{B}(G, M, I), \leq)$ is a complete lattice.

Example 2.3.3 We consider the following concept, using as our set G of objects the set of planets in our solar system and a set of seven attributes concerning size, distance from the sun and whether the planet has moons. The relation I is given by the following table, where an x indicates that the object in that row has the attribute of the given column. For convenience we will abbreviate each planet name by one or two letters, and each attribute as shown in the table below.

To illustrate how the concepts of this context may be identified, let us choose an object, say the planet Jupiter. We find the set of all properties or attributes this object has: large, far from the sun and has moons. Now we look for the set of all objects which has exactly these properties: Jupiter and Saturn. This gives us the concept $(\{J, S\}, \{l, f, y\})$. We can also start with a set of objects instead of with a single object, and dually we can work from a set of attributes. For instance, we obtain the concept $(\{J, S, U, N, P\}, \{f, y\})$, which is a superconcept of the first one.

	small k	*medium* m	*large* l	*close* c	*far* f	*moons* y	*no moons* nm
Me	x			x			x
V	x			x			x
E	x			x		x	
Ma	x			x		x	
J			x		x	x	
S			x		x	x	
U		x			x	x	
N		x			x	x	
P	x				x	x	

This method leads to the following list of concepts:

$(\{Me, V\}, \{k, c, nm\})$, $(\{E, Ma\}, \{k, c, y\})$, $(\{J, S\}, \{l, c, y\})$,
$(\{U, N\}, \{m, c, y\})$, $(\{P\}, \{k, c, y\})$, $(\{Me, C, E, Ma\}, \{k, c\})$,
$(\{J, S, U, N, P\}, \{c, y\})$, $(\{E, Ma, J, S, U, N, P\}, \{y\})$.

Extending this list by taking intersections, we reach the Hasse diagram shown below for the lattice $\mathcal{B}(G, M, I); \leq)$.

The main goal is to find, for a given context, the hierarchy of concepts. The super- and sub-concept relations also give implications between concepts, allowing this method to be used to search for new implications. For more information and examples about the theory of concept analysis we refer the reader to the book [47] by B. Ganter and R. Wille.

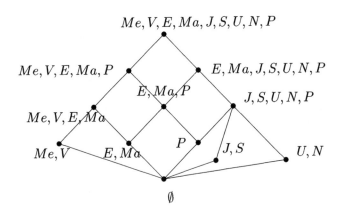

2.4   Exercises

2.4.1. Prove that for any relation $R \subseteq A \times B$, the maps σ and τ defined by

$$\sigma(X) : = \{y \in B \mid \forall x \in X \ ((x, y) \in R)\}$$
$$\tau(Y) : = \{x \in A \mid \forall y \in Y \ ((x, y) \in R)\}$$

define a Galois-connection between A and B.

2.4.2. Let $R \subseteq A \times B$ be a relation between the sets A and B and let (μ, ι) be the Galois-connection between A and B induced by R. Prove that for any families $\{T_i \subseteq A \mid i \in I\}$ and $\{S_i \subseteq B \mid i \in I\}$, the following equalities hold:

a) $\mu(\bigcup_{i \in I} T_i) = \bigcap_{i \in I} \mu(T_i)$.

b) $\iota(\bigcup_{i \in I} S_i) = \bigcap_{i \in I} \iota(S_i)$.

(These properties will be used in Chapter 13.)

2.4.3. Prove Theorem 2.2.2.

2.4.4. A *kernel system* on A is defined as a subset $\mathcal{K} \subseteq \mathcal{P}(A)$ with the property that for all $\mathcal{B} \subseteq \mathcal{K}$, the set $\bigcup \mathcal{B}$ is in $\mathcal{K}$.
A *kernel operator* is a mapping $D : \mathcal{P}(A) \to \mathcal{P}(A)$ with the properties

(i) $\forall M \subseteq A \; (D(M) \subseteq M)$ (intensivity)

(ii) $\forall M, N \subseteq A \; (M \subseteq N \Rightarrow D(M) \subseteq D(N))$ (monotonicity)

(iii) $\forall M \subseteq A \; (D(D(M)) = D(M))$ (idempotency).

Formulate and prove a theorem for kernels analogous to 2.1.6.

2.4.5. We saw in Chapter 1 that given an algebra $\mathcal{A}$, the operation which takes any non-empty subset X of A to the subalgebra of $\mathcal{A}$ generated by X is a closure operator. There is another operator involving subalgebras. For any class K of algebras of a fixed type τ, let $\mathbf{S}(K)$ be the class of all algebras of type τ which are subalgebras of some algebra in K. This defines an operator $\mathbf{S}$ on the class $Alg(\tau)$ of all algebras of type τ. Prove that this operator is a closure operator on $Alg(\tau)$. (This operator will be studied in Chapter 6.)

2.4.6. Choose a set G of objects consisting of four-sided polygons (square, rectangle, parallelogram, etc.) and a set M of attributes of such objects. Prepare a table for the corresponding relation I and draw the concept lattice.

Chapter 3

Homomorphisms and Isomorphisms

Consider the function which assigns to every real number $a \in \mathbb{R}$ its absolute value $|a|$ in the set $\mathbb{R}^+$ of non-negative real numbers. This function $h : \mathbb{R} \to \mathbb{R}^+$ has the property that it is compatible with the multiplicative structure of $\mathbb{R}$, because $|a \cdot b| = |a| \cdot |b|$. We get the same result whether we multiply the numbers first and then use our mapping, or permute these actions, since calculation with the images proceeds in the same way as calculations with the originals. A corresponding observation can be made about assigning to a square matrix A its determinant $|A|$, or to a permutation s on the set $\{1, \ldots, n\}$ its sign. In these cases the compatibility of the mapping with the operation can be described by the equations

$$|A \cdot B| = |A| \cdot |B| \quad \text{and} \quad sgn(s_1 \circ s_2) = sgns_1 \cdot sgns_2.$$

In each of these examples, we have a mapping between the carrier sets of two algebras of the same type, which is compatible with the operations of the algebras. If the mapping between the carrier sets of the two algebras is also a bijection, then the difference between the two algebras amounts only to a relabelling of the elements.

This can be seen for instance in the following example. We consider the algebras

$$\mathcal{A}_3 = (\{(1), (123), (132)\}; \circ) \text{ and } \mathcal{Z}_3 = (\{[0]_3, [1]_3, [2]_3 \}; +),$$

the group of all even permutations (the alternating group) of order three and the cyclic group of order three, respectively. Both algebras have type

$\tau = (2)$, and it is easy to verify that both are groups. From their Cayley tables

$\circ$	(1)	(123)	(132)
(1)	(1)	(123)	(132)
(123)	(123)	(132)	(1)
(132)	(132)	(1)	(123)

$+$	$[0]_3$	$[1]_3$	$[2]_3$
$[0]_3$	$[0]_3$	$[1]_3$	$[2]_3$
$[1]_3$	$[1]_3$	$[2]_3$	$[0]_3$
$[2]_3$	$[2]_3$	$[0]_3$	$[1]_3$

it appears that the bijection

$$h : \{(1),\ (123),\ (132)\} \to \{[0]_3,\ [1]_3,\ [2]_3\ \},$$

defined by the mapping $(1) \mapsto [0]_3$, $(123) \mapsto [1]_3$ and $(132) \mapsto [2]_3$, is compatible with the structure. For instance, we have

$$h((123) \circ (132)) \ = \ h((1)) \ = \ [0]_3 \ = \ [1]_3 + [2]_3 \ = \ h((123)) + h((132)),$$

and similarly for all other pairs of elements.

Functions or bijections between algebras which have this compatibility property are called *homomorphisms* or *isomorphisms*.

The following example shows that this compatibility property occurs in very "practical" cases. An industrial automaton is designed to construct more complex parts of a machine from certain simpler parts. There are working phases in which two parts are combined into a new part, other phases in which three parts are combined into a new one, and so on. These correspond to binary operations, ternary operations, etc. If the parts are combined in a different order, represented by a mapping h, then the automaton will continue to work in the correct way when h has our compatibility property. Thus the concepts of homomorphism and isomorphism may be said to model mathematically the "artificial intelligence" we expect from such an automaton.

In this chapter, we define homomorphisms and isomorphisms, and present several important theorems outlining their properties and connecting them with quotient algebras and congruences.

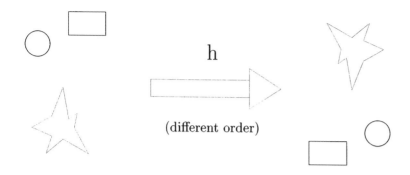

(different order)

3.1 The Homomorphism Theorem

We begin with the definitions of homomorphism and isomorphism.

Definition 3.1.1 Let $\mathcal{A} = (A; (f_i^A)_{i \in I})$ and $\mathcal{B} = (B; (f_i^B)_{i \in I})$ be algebras of the same type τ. Then a function $h : A \to B$ is called a *homomorphism* $h : \mathcal{A} \to \mathcal{B}$ of $\mathcal{A}$ into $\mathcal{B}$ if for all $i \in I$ we have

$$h(f_i^A(a_1, \ldots, a_{n_i})) = f_i^B(h(a_1), \ldots, h(a_{n_i})),$$

for all $a_1, \ldots, a_{n_i} \in A$. In the special case that $n_i = 0$, this equation means that $h(f_i^A(\emptyset)) = f_i^B(\emptyset)$. That is, the element designated by the nullary operation f_i^A in A must be mapped to the corresponding element f_i^B in B.

If the function h is bijective, that is both one-to-one (injective) and "onto" (surjective), then the homomorphism $h : \mathcal{A} \to \mathcal{B}$ is called an *isomorphism* from $\mathcal{A}$ onto $\mathcal{B}$. An injective homomorphism from $\mathcal{A}$ into $\mathcal{B}$ is also called an *embedding* of $\mathcal{A}$ into $\mathcal{B}$.

A homomorphism $h : \mathcal{A} \to \mathcal{A}$ of an algebra $\mathcal{A}$ into itself is called an *endomorphism* of $\mathcal{A}$, and an isomorphism $h : \mathcal{A} \to \mathcal{A}$ from A onto A is called an *automorphism* of $\mathcal{A}$.

Example 3.1.2 1. It is easy to prove that for every algebra $\mathcal{A}$, the identical mapping $id_A : \mathcal{A} \to \mathcal{A}$, defined by $id_A(x) = x$ for all $x \in A$, is an automorphism of $\mathcal{A}$.

2. Let $\mathcal{A}$, $\mathcal{B}$ and $\mathcal{C}$ be algebras of the same type, and let $h_1 : \mathcal{A} \to \mathcal{B}$ and $h_2 : \mathcal{B} \to \mathcal{C}$ be homomorphisms. The composition function $h_2 \circ h_1 : \mathcal{A} \to \mathcal{C}$ is defined by $(h_2 \circ h_1)(x) = h_2(h_1(x))$ for all $x \in A$. The reader should verify that this composition is also a homomorphism, and that when both h_1 and h_2 are surjective, injective or bijective, then the composition has the same property. (See Exercise 3.3.4.)

3. If θ is a congruence relation on $\mathcal{A}$ and if $\mathcal{A}/\theta$ is the corresponding quotient algebra (see 1.4.11) then $h : \mathcal{A} \to \mathcal{A}/\theta$ defined by $a \mapsto [a]_\theta$ is a surjective homomorphism. The definition of operations on $\mathcal{A}/\theta$ from 1.4.11 gives us

$$h(f_i^A(a_1, \ldots, a_{n_i})) = [f_i^A(a_1, \ldots, a_{n_i})]_\theta =$$
$$f_i^{A/\theta}([a_1]_\theta, \ldots, [a_{n_i}]_\theta) = f_i^{A/\theta}(h(a_1), \ldots, h(a_{n_i}))$$

for all $i \in I$. This homomorphism is called the *natural homomorphism* induced by θ on $\mathcal{A}$, and is usually denoted by *nat* θ.

It is useful to consider the behaviour of subalgebras under homomorphic mappings.

Theorem 3.1.3 *Let $h : \mathcal{A} \to \mathcal{B}$ be a homomorphism of the algebra $\mathcal{A}$ of type τ into the algebra $\mathcal{B}$ of type τ. Then we have:*

(i) *The image $\mathcal{B}_1 = h(\mathcal{A}_1)$ of a subalgebra $\mathcal{A}_1$ of $\mathcal{A}$ under the homomorphism h is a subalgebra of $\mathcal{B}$.*

(ii) *The preimage $h^{-1}(\mathcal{B}') = \mathcal{A}'$ of a subalgebra $\mathcal{B}'$ of $h(\mathcal{A}) \subseteq \mathcal{B}$ is a subalgebra of $\mathcal{A}$.*

(iii) *For any subset $X \subseteq A$, we have $\langle h(X) \rangle_\mathcal{B} = h(\langle X \rangle_\mathcal{A})$.*

Proof: (i) By definition,

$$h(\mathcal{A}_1) = \{b \in B \mid \exists\, a \in A_1\ (h(a) = b)\} \subseteq B.$$

Let f_i^B be an n_i-ary operation on $\mathcal{B}$, for $i \in I$, and let $(b_1, \ldots, b_{n_i}) \in h(A_1)^{n_i}$. Then for each $1 \leq j \leq n_i$, we have $b_j = h(a_j)$ for some a_j in A. Then

$$f_i^B(b_1, \ldots, b_{n_i}) = f_i^B(h(a_1), \ldots, h(a_{n_i})) = h(f_i^A(a_1, \ldots, a_{n_i})),$$

and the latter is in $h(A_1)$ since $f_i^A(a_1, \ldots, a_{n_i}) \in A_1$. Application of the subalgebra criterion thus proves that the image $\mathcal{B}_1$ is a subalgebra of $\mathcal{B}$.

(ii) Let $(a_1, \ldots, a_{n_i}) \in (h^{-1}(B'))^{n_i}$ and let f_i^A be n_i-ary. We have

$$h^{-1}(B') = \{a \in A \mid \exists\, b \in B' \ (h(a) = b)\};$$

so for each a_j in $h^{-1}(B')$, there is an element $b_j \in B'$ such that $a_j = h^{-1}(b_j)$. This means that

$$f_i^A(a_1, \ldots, a_{n_i}) = f_i^A(h^{-1}(b_1), \ldots, h^{-1}(b_{n_i})).$$

Then

$$
\begin{aligned}
h(f_i^A(h^{-1}(b_1), \ldots, h^{-1}(b_{n_i}))) \\
= f_i^B(h(h^{-1}(b_1)), \ldots, h(h^{-1}(b_{n_i}))) \\
= f_i^B(b_1, \ldots, b_{n_i}),
\end{aligned}
$$

and the latter is in B', because all of $b_1, \ldots, b_{n_i}$ are in B' and $\mathcal{B}'$ is a subalgebra of $\mathcal{B}$. From this we get

$$f_i^A(h^{-1}(b_1), \ldots, h^{-1}(b_{n_i})) = f_i^A(a_1, \ldots, a_{n_i}) \in h^{-1}(B').$$

(iii) Let E be the operator used in Theorem 1.3.8, so

$$E(X): \ = X \cup \{f_i^A(a_1, \ldots, a_{n_i}) \mid i \in I,\ a_1, \ldots, a_{n_i} \in X\}.$$

We show first that $E(h(X)) = h(E(X))$ for all $X \subseteq A$. The set $E(h(X))$ consists of all elements $h(y)$ with $y \in X$, plus elements of the form

$$f_i^B(h(y_1), \ldots, h(y_{n_i})), \ \text{for} \ i \in I,\ y_1, \ldots, y_{n_i} \in X.$$

The set $h(E(X))$ also consists of the elements $h(y)$ with $y \in X$ plus the elements $h(f_i^A(y_1, \ldots, y_{n_i}))$, which agree with $f_i^B(h(y_1), \ldots, h(y_{n_i}))$.

By induction on k we can prove that $E^k(h(X)) = h(E^k(X))$ for all $k \in \mathbb{N}$. Then we have:

$$
\begin{aligned}
\langle h(X) \rangle_{\mathcal{B}} &= \bigcup_{k=0}^{\infty} E^k(h(X)) = \bigcup_{k=0}^{\infty} h(E^k(X)) \\
&= h(\bigcup_{k=0}^{\infty} E^k(X)) = h(\langle X \rangle_{\mathcal{A}}). \qquad \blacksquare
\end{aligned}
$$

Before we continue with general properties of homomorphisms, we will examine more closely the automorphisms on an algebra. The set of all automorphisms of the algebra $\mathcal{A}$ is denoted by $Aut\mathcal{A}$, and the set of all its endomorphisms is denoted by $End\mathcal{A}$. Since the composition $\circ$ of two automorphisms (endomorphisms) of $\mathcal{A}$ is again an automorphism (endomorphism) of $\mathcal{A}$ and since $\circ$ is associative, $(Aut\mathcal{A};\ \circ, id_A)$ and $(End\mathcal{A};\ \circ, id_A)$ are monoids, with the former a submonoid of the latter. It is also true that if $h : \mathcal{A} \to \mathcal{A}$ is an automorphism of $\mathcal{A}$, the mapping $h^{-1} : \mathcal{A} \to \mathcal{A}$ is again an automorphism of $\mathcal{A}$. This can be seen as follows:

$$
\begin{aligned}
h^{-1}(f_i^A(b_1,\ldots,b_{n_i})) &= h^{-1}(f_i^A(h(a_1),\ldots,h(a_{n_i}))) = \\
h^{-1}(h(f_i^A(a_1,\ldots,a_{n_i}))) &= f_i^A(a_1,\ldots,a_{n_i}) \\
&= f_i^A(h^{-1}(b_1),\ldots,h^{-1}(b_{n_i})).
\end{aligned}
$$

This gives the following result.

Lemma 3.1.4 *The set of all automorphisms of an algebra* $\mathcal{A}$ *forms a group,* $Aut\mathcal{A} = (Aut\mathcal{A};\ \circ,\ ^{-1},\ id_A)$, *called the automorphism group of* $\mathcal{A}$. $\blacksquare$

Let $\mathcal{A} = (A;\ (f_i^A)_{i \in I})$ be an algebra of type τ and let $h : \mathcal{A} \to \mathcal{A}$ be an automorphism of $\mathcal{A}$. An element $a \in A$ is called a *fixed point* of h, if $h(a) = a$. Every $a \in A$ is of course a fixed point of the identical automorphism id_A on $\mathcal{A}$.

Lemma 3.1.5 *The set of all fixed points of an automorphism h of $\mathcal{A}$ is a subalgebra of* $\mathcal{A}$.

Proof: Let h be an automorphism on $\mathcal{A}$. We consider the set F_h of all fixed points of h:

$$F_h :\ =\ \{a \mid a \in A \text{ and } h(a)\ =\ a\}.$$

Let f_i, for $i \in I$, be an n_i-ary operation on $\mathcal{A}$, and assume that $a_1,\ldots,a_{n_i} \in F_h$. Then

$$f_i^A(a_1,\ldots,a_{n_i})\ =\ f_i^A(h(a_1),\ldots,h(a_{n_i}))\ =\ h(f_i^A(a_1,\ldots,a_{n_i})),$$

since h is an automorphism, and thus $f_i^A(a_1,\ldots,a_{n_i}) \in F_h$. By Criterion 1.3.3, F_h is the carrier set of a subalgebra of $\mathcal{A}$. $\blacksquare$

We have just considered, for a given automorphism h on an algebra $\mathcal{A}$, the set of all elements of A which are fixed points of h. In the opposite direction, we can choose a subset B of A and look for the set of all automorphisms on A whose set of fixed points contains B. When B is exactly the set of fixed points of this set of automorphisms, Lemma 3.1.5 tells us that B is a subalgebra of A. To explore this connection further, we will need the following definitions.

Definition 3.1.6 Let A and A' be algebras of the same type τ and let B be a subalgebra of both A and A'. An isomorphism $h : A \to A'$ with $h(b) = b$ for all $b \in B$ is called a *relative isomorphism* between A and A' with respect to B. When h is an automorphism on A, it is then called a *relative automorphism* of A with respect to the subalgebra B.

We can consider such a relative isomorphism, $h : A \to A'$ with respect to a common subalgebra B, in two ways. From "above," we say that the restriction of h to B is the identity isomorphism on B, while from "below," we call h an extension of the identity isomorphism on B.

More generally, let B and B' be subalgebras of the algebras A and A', respectively. Let $g : B \to B'$ and $h : A \to A'$ be isomorphisms. Then we say that the isomorphism h is an extension of g, if for all $b \in B$ the equation $g(b) = h(b)$ is satisfied. The following result is easy to verify.

Lemma 3.1.7 *The set $Aut_{relB}A$ of all relative automorphisms of A with respect to a subalgebra $B \subseteq A$ forms a subgroup of the automorphism group of A.* ∎

We have seen that any automorphism h on A determines a subalgebra of A, consisting of the fixed points of h. If we take h to be a relative automorphism with respect to some subalgebra B, this fixed point subalgebra contains B. Now let G be a subgroup of the group of all relative automorphisms of A with respect to B. We form the set

$$B' := \{b \in A \mid s(b) = b \text{ for all } s \in G\}$$

of the elements of A which are fixed points of all the automorphisms in G. Lemma 3.1.5 tells us that B' is a subalgebra of A, which again has B as a subalgebra. This gives a way of associating to every subgroup $G \subseteq Aut_{relB}A$ an algebra B' between B and A, so $B \subseteq B' \subseteq A$. We can ask whether the converse holds: does every algebra $B \subseteq B' \subseteq A$ determine a subgroup of $Aut_{relB}A$ which consists exactly of the automorphisms of A fixing the elements from B'?

The careful reader will have noticed that these interconnections amount to a Galois-connection between the sets A and $Aut_{relB}A$. We have a basic relation

$$R := \{(a,s) \mid a \in A \text{ and } s \in Aut_{relB}A \text{ and } s(a) = a\},$$

which induces the maps

$$\sigma(X) := \{s \in Aut_{relB}A \mid \forall a \in X(s(a) = a)\}$$
$$\text{and} \quad \tau(Y) := \{a \in A \mid \forall s \in Y(s(a) = a)\},$$

for all $X \subseteq A$ and $Y \subseteq Aut_{relB}A$.

We now return to the general theory of homomorphisms, with some theorems that relate homomorphisms to congruences. We saw in Example 3.1.2 part 3 that every congruence relation θ on an algebra $\mathcal{A}$ determines a homomorphism, namely the natural homomorphism $nat\ \theta : \mathcal{A} \to \mathcal{A}/\theta$ onto the quotient algebra. Now we will show that conversely every homomorphism of an algebra $\mathcal{A}$ also determines a congruence relation on $\mathcal{A}$.

Let $h : \mathcal{A} \to \mathcal{B}$ be a homomorphism. Since the function $h : A \to B$ is in general not injective, we may have different elements with the same image. We investigate the equivalence relation corresponding to the partition of the set A into classes consisting of elements having the same image, that is, the kernel of h.

Definition 3.1.8 Let $\mathcal{A}$ and $\mathcal{B}$ be algebras of the same type τ and let $h : \mathcal{A} \to \mathcal{B}$ be a homomorphism. The following binary relation is called the *kernel of the homomorphism h* :

$$ker\ h : \ = \ \{(a,b) \in A^2 \mid h(a) \ = \ h(b)\}.$$

We may alternately express this as $ker\ h = h^{-1} \circ h$, where h^{-1} is the inverse relation of h.

Lemma 3.1.9 *The kernel of any homomorphism $h : \mathcal{A} \to \mathcal{B}$ is a congruence relation on $\mathcal{A}$.*

Proof: By definition, $ker\ h$ is an equivalence relation on A. To verify the compatibility property needed for a congruence, let f_i^A, for some $i \in I$, be

an n_i-ary fundamental operation of $\mathcal{A}$ and let (a_1, b_1), ..., (a_{n_i}, b_{n_i}) be in
$ker\ h$. This means that $h(a_1) = h(b_1)$, ..., $h(a_{n_i}) = h(b_{n_i})$. Now applying
the operation f_i^B gives the equation:

$$f_i^B(h(a_1), \ldots, h(a_{n_i})) = f_i^B(h(b_1), \ldots, h(b_{n_i})).$$

Since $h : \mathcal{A} \to \mathcal{B}$ is a homomorphism of $\mathcal{A}$ into $\mathcal{B}$, this gives us

$$h(f_i^A(a_1, \ldots, a_{n_i})) = h(f_i^A(b_1, \ldots, b_{n_i})),$$

and thus

$$(f_i^A(a_1, \ldots, a_{n_i}), f_i^A(b_1, \ldots, b_{n_i})) \in ker\ h. \qquad \blacksquare$$

Suppose we have a homomorphism $h : \mathcal{A} \to \mathcal{B}$. We have seen that $ker\ h$
is a congruence on $\mathcal{A}$, so we can form the quotient algebra $\mathcal{A}/ker\ h$, along
with the natural homomorphism $nat(ker\ h) : \mathcal{A} \to \mathcal{A}/ker\ h$ which maps the
algebra $\mathcal{A}$ onto this quotient algebra. Now we have two homomorphic images
of $\mathcal{A}$: the original $h(\mathcal{A})$ and the new quotient $\mathcal{A}/ker\ h$. What connection is
there between these two homomorphic images? The answer to this question
is a special case of the following *General Homomorphism Theorem.*

Theorem 3.1.10 *(General Homomorphism Theorem) Let $h : \mathcal{A} \to \mathcal{B}$ and
$g : \mathcal{A} \to \mathcal{C}$ be homomorphisms, and let g be surjective. Then there exists a
homomorphism $f : \mathcal{C} \to \mathcal{B}$ which satisfies $f \circ g = h$ iff $ker\ g \subseteq ker\ h$.*

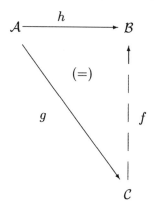

Here $f \circ g = h$ means "commutativity" of the diagram. When this f exists, it has the following properties:

(i) the homomorphism f is uniquely defined by $f = h \circ g^{-1}$;

(ii) f is injective iff $\ker g = \ker h$;

(iii) f is surjective iff h is surjective.

Proof: We first assume that there exists a homomorphism $f : C \to B$ which satisfies $f \circ g = h$. To see that $\ker g \subseteq \ker h$, let $(a, b) \in \ker g$. Then $g(a) = g(b)$, and so $f(g(a)) = f(g(b))$, that is, $(f \circ g)(a) = (f \circ g)(b)$. Since $f \circ g = h$, it follows that $h(a) = h(b)$ and $(a, b) \in \ker h$, as required.

Conversely, let $\ker g \subseteq \ker h$. We define $f : \ = h \circ g^{-1}$, and show that f has the required properties. The domain of f is C because of the surjectivity of g. To see that f is uniquely determined, let c be any element of C, and suppose that both a_1 and a_2 are in $g^{-1}(c)$. Then we have $g(a_1) = g(a_2) = c$, and so $(a_1, a_2) \in \ker g$. But under our assumption this puts $(a_1, a_2) \in \ker h$. Therefore $h(a_1) = h(a_2)$, and $f(c) = h \circ g^{-1}(c)$ gives the same result whether a_1 or a_2 is used for $g^{-1}(c)$. Thus f is a well-defined function of C into B, and clearly f satisfies $f \circ g = h$.

Finally, to see that f is a homomorphism, assume that f_i^C is n_i-ary, for $i \in I$. Let $c_1, \ldots, c_{n_i} \in C$, with $g^{-1}(c_i) = a_i$ for $1 \le i \le n$. Then we have

$$
\begin{aligned}
f(f_i^C(c_1, \ldots, c_{n_i})) &= f(f_i^C(g(a_1), \ldots, g(a_{n_i}))) \\
&= f(g(f_i^A(a_1, \ldots, a_{n_i}))) = (f \circ g)(f_i^A(a_1, \ldots, a_{n_i})) \\
&= f_i^B((f \circ g)(a_1), \ldots, (f \circ g)(a_{n_i})) \\
&= f_i^B(f(g(a_1)), \ldots, f(g(a_{n_i}))) \\
&= f_i^B(f(c_1, \ldots, c_{n_i})),
\end{aligned}
$$

since $h = f \circ g$ is a homomorphism.

(i) We show that any homomorphism $f' : C \to B$ which satisfies $f' \circ g = h$ agrees with $f = h \circ g^{-1}$. If $f \circ g = h$ and $f' \circ g = h$, then we have $f \circ g = f' \circ g$, and hence for all $a \in A$, $f(g(a)) = f'(g(a))$. Thus, because of the surjectivity of g the equation $f(c) = f'(c)$ is satisfied for all $c \in C$, and we have $f = f'$.

(ii) We are assuming that $ker\ g \subseteq ker\ h$, so that f exists. Now we assume that f is injective, and take $(a_1, a_2) \in ker\ h$. So $h(a_1) = h(a_2)$, and using $f \circ g = h$ gives $(f \circ g)(a_1) = (f \circ g)(a_2)$ and $f(g(a_1)) = f(g(a_2))$. Now the injectivity of f gives $g(a_1) = g(a_2)$ and thus $(a_1, a_2) \in ker\ g$. This shows that $ker\ h \subseteq ker\ g$, and hence the two kernels are equal.

Conversely, suppose that $ker\ g = ker\ h$, and let us show that f is injective. Let $f(c_1) = f(c_2)$ for $c_1, c_2 \in C$. Since g is surjective, we can represent c_j in the form $g(a_j)$, for some $a_j \in A$, for $j = 1, 2$. Now we have $f(g(a_1)) = f(g(a_2))$, and hence $h(a_1) = h(a_2)$. This put (a_1, a_2) in $ker\ h$, and by our assumption also in $ker\ g$. But this means $c_1 = g(a_1) = g(a_2) = c_2$, as required.

(iii) When f is surjective, the fact that g is surjective by assumption makes the composition $h = f \circ g$ also surjective. If conversely h is surjective, then of course $f \circ g = h$ is also surjective. Thus for every $b \in B$ there exists an element $a \in A$ with $(f \circ g)(a) = f(g(a)) = b$. Then for every $b \in B$ there exists an element from C, namely $g(a)$, which is mapped by f to b, and f is surjective. ∎

As mentioned above, as a specific case of the General Homomorphism Theorem we can consider the surjective natural homomorphism associated to any congruence. We have the following result, which tells us that any homomorphic image of an algebra is isomorphic to a quotient algebra of that algebra by the kernel of the homomorphism.

Theorem 3.1.11 *(Homomorphic Image Theorem) Let $h : A \to B$ be a surjective homomorphism. Then there exists a unique isomorphism f from $A/ker\ h$ onto B with $f \circ nat(ker\ h) = h$.*

Proof: We have $ker(nat(ker\ h)) = ker\ h$, since

$$(a_1, a_2) \in ker(nat(ker\ h)) \Leftrightarrow (nat(ker\ h))(a_1) = [a_1]_{ker\ h}$$

$$= [a_2]_{ker\ h} = (nat(ker\ h))(a_2) \Leftrightarrow (a_1, a_2) \in ker\ h.$$

Thus Theorem 3.1.10 gives the existence of a uniquely defined homomorphism f with $f \circ nat(ker\ h) = h$. Moreover, parts (ii) and (iii) of Theorem 3.1.10 show that f is an isomorphism. ∎

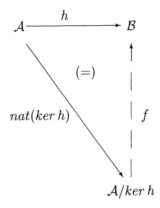

The basic idea of the Homomorphic Image Theorem is that whenever we have a surjective homomorphism h from $\mathcal{A}$ onto $\mathcal{B}$, the image $\mathcal{B}$ is actually isomorphic to the quotient algebra $\mathcal{A}/ker\ h$ of $\mathcal{A}$. This means that any homomorphic image of an algebra, which is usually "outside" of the algebra, can in fact be characterized up to isomorphism "inside" the algebra, as a quotient algebra determined by the kernel of the homomorphism. So to know all homomorphic images of a given algebra, it is enough to find all congruence relations and quotients of this algebra.

3.2   The Isomorphism Theorems

In this section we apply our General Homomorphism Theorem to two other situations, to produce two theorems which are usually called the First and Second Isomorphism Theorems. For the first one, we consider subalgebras of an algebra.

Theorem 3.2.1 *(First Isomorphism Theorem) Let $\mathcal{A}$ and $\mathcal{B}$ be algebras of the same type, and let $h : \mathcal{A} \to \mathcal{B}$ be a homomorphism. Let $\mathcal{A}_1$ be a subalgebra of $\mathcal{A}$ and $h(\mathcal{A}_1) \subseteq \mathcal{B}$ its image. We also assume that $\mathcal{A}_1^*$ is the preimage of $h(\mathcal{A}_1)$ and that $h_1 = h|A_1$ is the restriction of h to A_1, and we take $h_1^* = h|A_1^*$ to be the restriction of h to A_1^*. Then*

$$\varphi : \mathcal{A}_1/kerh_1 \to \mathcal{A}_1^*/kerh_1^*, \quad defined\ by \quad [a]_{ker\ h_1} \mapsto [a]_{ker\ h_1^*},$$

is an isomorphism from $\mathcal{A}_1/ker\ h_1$ onto $\mathcal{A}_1^*/ker\ h_1^*$.

Proof: By definition of the functions h_1 and h_1^*, we start with the commutative diagram shown below. Since h_1 and h_1^* are surjective, we can apply the Homomorphism Theorem twice on them and the corresponding natural homomorphisms, to get the existence of two isomorphisms f_1 and f_1^*, as shown in the second diagram.

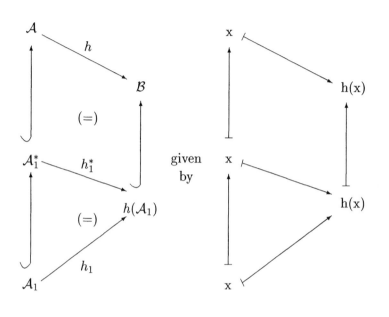

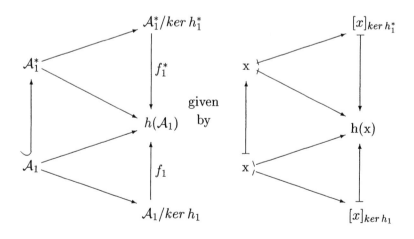

Therefore $\varphi : \ = f_1^{*-1} \circ f_1$ is an isomorphism.   ∎

For the second Isomorphism Theorem, we consider two congruences θ_1 and θ_2 on an algebra $\mathcal{A}$, with $\theta_1 \subseteq \theta_2$. We can define a new relation on $\mathcal{A}/\theta_1$, by

$$\theta_2/\theta_1 : \ = \{([a]_{\theta_1}, [b]_{\theta_1}) \mid (a, b) \in \theta_2\}.$$

Theorem 3.2.2 *(Second Isomorphism Theorem) Let θ_1 and θ_2 be congruences on an algebra $\mathcal{A}$, with $\theta_1 \subseteq \theta_2$. Then the relation θ_2/θ_1 is a congruence relation on $\mathcal{A}/\theta_1$, and the function*

$$\varphi : (\mathcal{A}/\theta_1)/(\theta_2/\theta_1) \to \mathcal{A}/\theta_2, \quad \text{defined by} \quad [[a]_{\theta_1}]_{\theta_2/\theta_1} \mapsto [a]_{\theta_2},$$

is an isomorphism.

Proof: It is clear from the definition that θ_2/θ_1 is a relation and in fact an equivalence relation on A/θ_1. Let f_i^{A/θ_1}, for $i \in I$, be a fundamental operation of the algebra $\mathcal{A}/\theta_1$ and let

$$([a_1]_{\theta_1}, [b_1]_{\theta_1}) \in \theta_2/\theta_1, \ \ldots, \ ([a_{n_i}]_{\theta_1}, [b_{n_i}]_{\theta_1}) \in \theta_2/\theta_1.$$

Then it follows that

$$f_i^{A/\theta_1}([a_1]_{\theta_1}, \ldots, [a_{n_i}]_{\theta_1}) \ = \ [f_i^A(a_1, \ldots, a_{n_i})]_{\theta_1} \text{ and}$$
$$f_i^{A/\theta_1}([b_1]_{\theta_1}, \ldots, [b_{n_i}]_{\theta_1}) \ = \ [f_i^A(b_1, \ldots, b_{n_i})]_{\theta_1}.$$

From the definition of θ_2/θ_1 we also know that

$$(a_1, b_1) \in \theta_2, \ \ldots, \ (a_{n_i}, b_{n_i}) \in \theta_2,$$

and thus

$$(f_i^A(a_1, \ldots, a_{n_i}), \ f_i^A(b_1, \ldots, b_{n_i})) \in \theta_2.$$

This makes

$$([f_i^A(a_1, \ldots, a_{n_i})]_{\theta_1}, \ [f_i^A(b_1, \ldots, b_{n_i})]_{\theta_1}) \in \theta_2/\theta_1,$$

and hence

$$(f_i^{A/\theta_1}([a_1]_{\theta_1}, \ldots, [a_{n_i}]_{\theta_1}), \ f_i^{A/\theta_1}([b_1]_{\theta_1}, \ldots, [b_{n_i}]_{\theta_1})) \in \theta_2/\theta_1.$$

Using the General Homomorphism Theorem on the two surjective homomorphisms

$$nat\theta_2 : \ \mathcal{A} \to \mathcal{A}/\theta_2 \text{ and } nat\theta_1 : \ \mathcal{A} \to \mathcal{A}/\theta_1,$$

we deduce the existence of a surjective homomorphism

$$f : \mathcal{A}/\theta_1 \to \mathcal{A}/\theta_2, \quad \text{which is defined by } [a]_{\theta_1} \mapsto [a]_{\theta_2}.$$

Since θ_2/θ_1 is a congruence relation on $\mathcal{A}/\theta_1$, we also have the corresponding surjective natural homomorphism

$$nat(\theta_2/\theta_1) : \mathcal{A}/\theta_1 \to (\mathcal{A}/\theta_1)/(\theta_2/\theta_1).$$

Again by the General Homomorphism Theorem, there then exists a surjective homomorphism

$$\varphi : (\mathcal{A}/\theta_1)/(\theta_2/\theta_1) \to \mathcal{A}/\theta_2.$$

Furthermore we have:

$$([a_1]_{\theta_1}, [a_2]_{\theta_1}) \in ker \ f \quad \Leftrightarrow \quad f([a_1]_{\theta_1}) = f([a_2]_{\theta_1}) \quad \Leftrightarrow$$
$$[a_1]_{\theta_2} = [a_2]_{\theta_2} \quad \Leftrightarrow \quad (a_1, a_2) \in \theta_2 \quad \Leftrightarrow$$
$$([a_1]_{\theta_1}, [a_2]_{\theta_1}) \in \theta_2/\theta_1 \quad \Leftrightarrow \quad (([a_1]_{\theta_1}, [a_2]_{\theta_1}) \in ker(nat(\theta_2/\theta_1))),$$

because $ker(nat(\theta_2/\theta_1)) = \theta_2/\theta_1$. Therefore part (i) of the General Homomorphism Theorem tells us that our surjective homomorphism is also injective, and so φ is an isomorphism. ∎

3.3 Exercises

3.3.1. Let A be a set, let Θ be an equivalence relation on A and let $f : A \to A$ be a function. Prove that f is compatible with Θ iff there is a mapping $g : A \to A$ with

$$ker g \subseteq \Theta \subseteq ker(f \circ g).$$

Hint: Choose g in such a way that $ker g = \Theta$.

3.3.2. Let $\mathcal{A}$ and $\mathcal{B}$ be algebras of type τ, and let $h : A \to B$ be a function. Prove that h is a homomorphism iff $\{(a, h(a)) \mid a \in A\}$ is a subalgebra of $\mathcal{A} \times \mathcal{B}$.

3.3.3. Let $\mathcal{G} = (\{0, 1, 2, 3\}; \ +, 0)$, where $+$ is the operation of addition modulo 4, and let $\mathcal{A} = (\{e, a\}; \ \cdot, e)$ with $a = e \cdot a = a \cdot e, e \cdot e = a \cdot a = e$, both algebras of type $(2, 0)$. Let $h : G \to A$ be the mapping defined by $0 \mapsto e$, $1 \mapsto e$, $2 \mapsto a$, and $3 \mapsto a$. Is h a homomorphism?

3.3.4. Prove that the composition of two (surjective, injective or bijective) homomorphisms is again a (surjective, injective or bijective) homomorphism.

3.3.5. Prove that the inverse of an isomorphism is also an isomorphism.

3.3.6. We can define an operator $\mathbf{H}$ on the class $Alg(\tau)$ of all algebras of type τ, as follows. For any $K \subseteq Alg(\tau)$, let $\mathbf{H}(K)$ be the class of all algebras of type τ which are homomorphic images of some algebra in K. Prove that this operator $\mathbf{H}$ is a closure operator on $Alg(\tau)$. (This operator will be used again in Chapter 6.)

Chapter 4

Direct and Subdirect Products

In the previous chapters, we have seen three ways to construct new algebras from given algebras: by formation of subalgebras, quotient algebras, and homomorphic images. In this chapter we examine another important construction, the formation of product algebras. One useful feature of this new construction involves the cardinalities of the algebras obtained. The formation of subalgebras or of homomorphic images of a given algebra leads to algebras with cardinality no larger than the cardinality of the given algebra. The formation of products, however, can lead to algebras with bigger cardinalities than those we started with. There are several ways to define a product of given algebras; we shall examine two products, called the direct product and the subdirect product.

4.1 Direct Products

Definition 4.1.1 Let $(\mathcal{A}_j)_{j \in J}$ be a family of algebras of type τ. The *direct product* $\prod\limits_{j \in J} \mathcal{A}_j$ of the $\mathcal{A}_j$ is defined as an algebra with the carrier set

$$P : \quad = \quad \prod_{j \in J} A_j : \quad = \quad \{(x_j)_{j \in J} \mid \forall j \in J \; (x_j \in A_j)\}$$

and the operations

$$(f_i^P(\underline{a}_1, \dots, \underline{a}_{n_i}))(j) \quad = \quad f_i^{A_j}(\underline{a}_1(j), \dots, \underline{a}_{n_i}(j)),$$

63

for $\underline{a}_1, \ldots, \underline{a}_{n_i}$ in P; that is,

$$f_i^P((a_{1j})_{j \in J}, \ldots, (a_{n_ij})_{j \in J}) \quad = \quad (f_i^{A_j}(a_{1j}, \ldots, a_{n_ij})_{j \in J}).$$

If for all $j \in J$, $A_j = A$, then we usually write A^J instead of $\prod_{j \in J} A_j$. If $J = \emptyset$, then $A^\emptyset$ is defined to be the one-element (trivial) algebra of type τ. If $J = \{1, \ldots, n\}$, then the direct product can be written as $A_1 \times \cdots \times A_n$.

The *projections* of the direct product $\prod_{j \in J} A_j$ are the mappings

$$p_k : \prod_{j \in J} A_j \to A_k \text{ defined by } (a_j)_{j \in J} \mapsto a_k.$$

It is easy to check that the projections of the direct product are in fact surjective homomorphisms.

Remark 4.1.2 Let A be an algebra of type τ, let $(A_j)_{j \in J}$ be a family of algebras of type τ and let $(f_j : A \to A_j)_{j \in J}$ be a family of homomorphisms. Then there exists a unique homomorphism $f : A \to \prod_{j \in J} A_j$ such that $p_j \circ f = f_j$ for all $j \in J$, namely $f = (f_j)_{j \in J}$. This homomorphism f makes the following diagram commute.

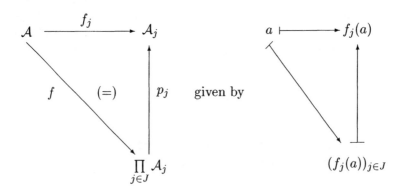

Example 4.1.3 Let us consider the direct product of the two permutation groups S_2 and A_3. Here

$$S_2 = (\{\tau_0, \tau_1\}; \circ, {}^{-1}, (1)) \text{ and } A_3 = (\{\tau_0, \alpha_1, \alpha_2\}; \circ, {}^{-1}, (1)),$$

with

$$\tau_0 : \ = \ (1), \ \tau_1 : \ = \ (12), \ \alpha_1 : \ = \ (123), \ \alpha_2 : \ = \ (132).$$

For the Cartesian product set, we let

$$\gamma_{00} : \ = \ ((1),(1)), \ \gamma_{01} : \ = \ ((1),(123)), \ \gamma_{02} : \ = \ ((1),(132)),$$

$$\gamma_{10} : \ = \ ((12),(1)), \ \gamma_{11} : \ = \ ((12),(123)), \ \gamma_{12} : \ = \ ((12),(132)).$$

Then we have

$$S_2 \times A_3 \ = \ \{\gamma_{00}, \ \gamma_{01}, \ \gamma_{02}, \ \gamma_{10}, \ \gamma_{11}, \ \gamma_{12}\}.$$

The binary operation $\circ$ of the direct product is defined by the following Cayley table:

$\circ$	γ_{00}	γ_{01}	γ_{02}	γ_{10}	γ_{11}	γ_{12}
γ_{00}	γ_{00}	γ_{01}	γ_{02}	γ_{10}	γ_{11}	γ_{12}
γ_{01}	γ_{01}	γ_{02}	γ_{00}	γ_{11}	γ_{12}	γ_{10}
γ_{02}	γ_{02}	γ_{00}	γ_{01}	γ_{12}	γ_{10}	γ_{11}
γ_{10}	γ_{10}	γ_{11}	γ_{12}	γ_{00}	γ_{01}	γ_{02}
γ_{11}	γ_{11}	γ_{12}	γ_{10}	γ_{01}	γ_{02}	γ_{00}
γ_{12}	γ_{12}	γ_{10}	γ_{11}	γ_{02}	γ_{00}	γ_{01}

We now consider a direct product of two factors. In this case we have two projection mappings, p_1 and p_2, each of which has a kernel which is a congruence relation on the product. We will show that these two kernels have some special properties. We recall first the definition of the product (composition) $\theta_1 \circ \theta_2$ of two binary relations θ_1, θ_2 on any set A:

$$\theta_1 \circ \theta_2 \ := \ \{(a,b) \mid \exists c \in A \ ((a,c) \in \theta_2 \wedge (c,b) \in \theta_1)\}.$$

Two binary relations θ_1, θ_2 on A are called *permutable*, if $\theta_1 \circ \theta_2 \ = \ \theta_2 \circ \theta_1$.

Lemma 4.1.4 *Let $\mathcal{A}_1, \mathcal{A}_2$ be two algebras of type τ and let $\mathcal{A}_1 \times \mathcal{A}_2$ be their direct product. Then:*

(i) $\quad ker \ p_1 \wedge ker \ p_2 \ = \ \Delta_{A_1 \times A_2};$

(ii) $\quad ker \ p_1 \circ ker \ p_2 \ = \ ker \ p_2 \circ ker \ p_1;$

(iii) $\quad ker \ p_1 \vee ker \ p_2 \ = \ (A_1 \times A_2)^2.$

Proof: (i) Since $ker\ p_1$ and $ker\ p_2$ are equivalence relations on $A_1 \times A_2$, the relation $ker\ p_1 \wedge ker\ p_2$ is also an equivalence relation on $A_1 \times A_2$, with $\Delta_{A_1 \times A_2} \subseteq ker\ p_1 \wedge ker\ p_2$. Conversely, let $(x, y) \in ker\ p_1 \wedge ker\ p_2$, with $x = (a_1, b_1)$ and $y = (a_2, b_2)$. From $(x, y) \in ker\ p_1$ we have

$$a_1 = p_1((a_1, b_1)) = p_1((a_2, b_2)) = a_2.$$

From $(x, y) \in ker\ p_2$, we have

$$b_1 = p_2((a_1, b_1)) = p_2((a_2, b_2)) = b_2.$$

Thus $x = y$ and $(x, y) \in \Delta_{A_1 \times A_2}$. This shows that

$$ker\ p_1 \wedge ker\ p_2 \subseteq \Delta_{A_1 \times A_2},$$

and altogether we have

$$ker\ p_1 \wedge ker\ p_2 = \Delta_{A_1 \times A_2}.$$

(ii) Assume that $(x, y) \in (A_1 \times A_2)^2$, with $x = (a_1, b_1)$ and $y = (a_2, b_2)$ for some $a_1, a_2 \in A_1$ and $b_1, b_2 \in A_2$. Since

$$((a_1, b_1),\ (a_1, b_2)) \in ker\ p_1 \text{ and } ((a_1, b_2),\ (a_2, b_2)) \in ker\ p_2,$$

we always have

$$((a_1, b_1),\ (a_2, b_2)) \in ker\ p_2 \circ ker\ p_1,$$

giving

$$(A_1 \times A_2)^2 \subseteq ker\ p_2 \circ ker\ p_1.$$

The converse inclusion is true by definition, so we have

$$ker\ p_2 \circ ker\ p_1 = (A_1 \times A_2)^2.$$

Similarly we can show that

$$ker\ p_1 \circ ker\ p_2 = (A_1 \times A_2)^2,$$

making $ker\ p_1 \circ ker\ p_2 = ker\ p_2 \circ ker\ p_1$. This proves (ii).

(iii) Now we use the equality

$$ker\ p_1 \circ ker\ p_2 = ker\ p_2 \circ ker\ p_1 = (A_1 \times A_2)^2$$

from (ii) to show that $ker\ p_1 \vee ker\ p_2 = (A_1 \times A_2)^2$. For this we need the well-known fact that for any two equivalence relations θ_1 and θ_2, the equation

$$\theta_1 \vee \theta_2 = \theta_1 \circ \theta_2$$

is satisfied iff θ_1 and θ_2 are permutable. This is because we always have $ker\ p_1 \vee ker\ p_2$ equal to

$$ker\ p_1 \vee (ker\ p_2 \circ ker\ p_1) \vee (ker\ p_2 \circ ker\ p_1 \circ ker\ p_2) \vee \cdots,$$

(see for instance Th. Ihringer, [58]), so when $ker\ p_1 \circ ker\ p_2 = ker\ p_2 \circ ker\ p_1$ we obtain

$$ker\ p_1 \circ ker\ p_2 = (A_1 \times A_2)^2 \subseteq ker\ p_1 \vee ker\ p_2.$$

Together with $ker\ p_1 \vee ker\ p_2 \subseteq (A_1 \times A_2)^2$, this gives the desired equality. ∎

Thus any direct product of two factors produces two congruences with the three special properties of Lemma 4.1.4. Conversely, the next theorem shows that if we have two congruences on an algebra with these properties, we can use them to write the algebra as a direct product of two factors.

Theorem 4.1.5 *Let $\mathcal{A}$ be an algebra, and let $\theta_1, \theta_2 \in Con\mathcal{A}$ be a pair of congruence relations with the following properties:*

(i) $\theta_1 \wedge \theta_2 = \Delta_A$;

(ii) $\theta_1 \vee \theta_2 = A^2$;

(iii) $\theta_1 \circ \theta_2 = \theta_2 \circ \theta_1$.

Then $\mathcal{A}$ is isomorphic to the direct product $\mathcal{A}/\theta_1 \times \mathcal{A}/\theta_2$, by an isomorphism

$$\varphi : \mathcal{A} \to \mathcal{A}/\theta_1 \times \mathcal{A}/\theta_2$$

given by:

$$\varphi(a) = ([a]_{\theta_1}, [a]_{\theta_2}), \ a \in A.$$

Proof: The given mapping φ is defined using the two natural homomorphisms, and is the unique map determined by them, as in Remark 4.1.2. This makes φ a homomorphism, and we will show that it is also a bijection. First, φ is injective: if $\varphi(a) = \varphi(b)$, then $[a]_{\theta_1} = [b]_{\theta_1}$ and $[a]_{\theta_2} = [b]_{\theta_2}$, so it follows that $(a, b) \in \theta_1 \wedge \theta_2$ and $a = b$ by (i). To see that the map φ is also surjective, let (a, b) be any pair in A^2. Conditions (ii) and (iii) mean that there exists an element $c \in A$ with $(a, c) \in \theta_1$ and $(c, b) \in \theta_2$, and therefore

$$([a]_{\theta_1}, [b]_{\theta_2}) = ([c]_{\theta_1}, [c]_{\theta_2}) = \varphi(c). \qquad \blacksquare$$

This theorem shows that we may use certain congruences on an algebra $\mathcal{A}$ to express $\mathcal{A}$ as a direct product of possibly smaller algebras. An algebra $\mathcal{A}$ is called *directly irreducible* if it cannot be expressed in this way without using $\mathcal{A}$ itself as one of the factors. We use $|A|$ to denote the cardinality of a set A.

Definition 4.1.6 An algebra $\mathcal{A}$ is called *directly irreducible*, if whenever $\mathcal{A} \cong \mathcal{B}_1 \times \mathcal{B}_2$, either $|B_1| = 1$ or $|B_2| = 1$.

Corollary 4.1.7 *An algebra $\mathcal{A}$ is directly irreducible iff $(\Delta_A,\ A \times A)$ is the only pair of congruence relations on $\mathcal{A}$ which satisfies the conditions (i) - (iii) of Theorem 4.1.5.*

Proof: Let $\mathcal{A}$ be directly irreducible and assume that $\theta_1, \theta_2 \in Con\mathcal{A}$ satisfy the three conditions of Theorem 4.1.5. Then $\mathcal{A} \cong \mathcal{A}/\theta_1 \times \mathcal{A}/\theta_2$, and the irreducibility means that one of the factors has cardinality one. Without loss of generality, suppose that $|A/\theta_1| = 1$. Then we must have $\theta_1 = A \times A$, and θ_2 must equal Δ_A by condition (i).

Assume now that conversely $(\Delta_A,\ A \times A)$ is the only pair with the properties (i) - (iii), and let $\mathcal{A} \cong \mathcal{A}_1 \times \mathcal{A}_2$. Then $(\Delta_{A_1 \times A_2}, (A_1 \times A_2)^2)$ is also the only pair of congruence relations on $\mathcal{A}_1 \times \mathcal{A}_2$ to satisfy conditions (i) - (iii). But by Lemma 4.1.4, the kernels of the projection mappings p_1 and p_2 do satisfy the three conditions. Therefore one of $ker\ p_1$ or $ker\ p_2$ must equal $\Delta_{A_1 \times A_2}$, and thus one of A_1 or A_2 must have cardinality one. $\qquad \blacksquare$

4.2   Subdirect Products

There is another way to define a product of algebras, which is different from the direct product.

Definition 4.2.1 Let $(\mathcal{A}_j)_{j \in J}$ be a family of algebras of type τ. A subalgebra $\mathcal{B} \subseteq \prod_{j \in J} \mathcal{A}_j$ of the direct product of the algebras $\mathcal{A}_j$ is called a *subdirect product* of the algebras $\mathcal{A}_j$, if for every projection mapping $p_k : \prod_{j \in J} \mathcal{A}_j \to \mathcal{A}_k$ we have

$$p_k(\mathcal{B}) = \mathcal{A}_k.$$

Example 4.2.2 1. Every direct product is also a subdirect product.

2. For every algebra $\mathcal{A}$ the diagonal $\Delta_A = \{(a, a) \mid a \in A\}$ is easily shown to be the carrier set of a subalgebra Δ_A of $\mathcal{A} \times \mathcal{A}$. Moreover, $p_1(\Delta_A) = p_2(\Delta_A) = \mathcal{A}$, so Δ_A is a subdirect product of $\mathcal{A} \times \mathcal{A}$.

3. Consider the two lattices C_2 and C_3, chains on two and three elements, respectively, shown below:

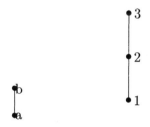

Their direct product is the lattice described by the diagram

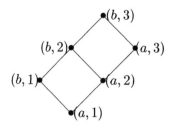

The sublattice $\mathcal{L} \subseteq C_2 \times C_3$ which is described by the diagram

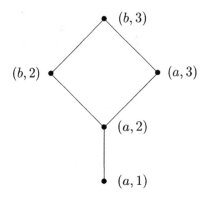

is obviously a subdirect product of C_2 and C_3.

Theorem 4.2.3 *Let B be a subdirect product of the family $(A_j)_{j \in J}$ of algebras of type τ. Then the projection mappings $p_k : \prod_{j \in J} A_j \to A_k$ satisfy the equation $\bigcap_{j \in J} ker(p_j \mid B) = \Delta_B$.*

Proof: Let $(a, b) \in \bigcap_{j \in J} ker(p_j|B)$. This implies that $p_k(a) = p_k(b)$ for all $k \in J$, so that every component of a agrees with the corresponding component of b. This means that $a = b$ and thus $(a, b) \in \Delta_B$. Conversely, it is clear that $\Delta_B \subseteq ker(p_k|B)$ for all p_k.   ∎

As was the case for direct products, it turns out that this property of the kernels of the projection mappings can be used to characterize subdirect products, in the sense that any set of congruences on an algebra with these properties can be used to express the algebra as a subdirect product.

Theorem 4.2.4 *Let A be an algebra. Let $\{\theta_j \mid j \in J\}$ be a family of congruence relations on A, which satisfy the equation $\bigcap_{j \in J} \theta_j = \Delta_A$. Then A is isomorphic to a subdirect product of the algebras A/θ_j, for $j \in J$. In particular, the mapping $\varphi(a) := ([a]_{\theta_j} \mid j \in J)$ defines an embedding $\varphi : A \to \prod_{j \in J} (A/\theta_j)$, whose image $\varphi(A)$ is a subdirect product of the algebras A/θ_j.*

Proof: The map φ is the unique homomorphism determined by the natural homomorphism mappings, as in Remark 4.1.2. Also, φ is injective, since

$\varphi(a) = \varphi(b)$ implies $[a]_{\theta_j} = [b]_{\theta_j}$ and thus $(a, b) \in \theta_j$ for all $j \in J$. Therefore $(a, b) \in \bigcap\limits_{j \in J} \theta_j = \Delta_A$ and so $a = b$. This proves the isomorphism of $\mathcal{A}$ and $\varphi(\mathcal{A})$. Moreover, if $p_k : \prod\limits_{j \in J}(\mathcal{A}/\theta_j) \to \mathcal{A}/\theta_k$ denotes the k-th projection mapping, then by the definition of φ we have $p_k(\varphi(\mathcal{A})) = \mathcal{A}/\theta_k$ for all $k \in J$. Therefore $\varphi(\mathcal{A})$ is a subdirect product of the algebras $\mathcal{A}/\theta_j$. ∎

We remark that the converse of this theorem is also true. If $\mathcal{A}$ is isomorphic to a subdirect product of a family $(\mathcal{A}_j)_{j \in J}$ of algebras, then there exists a family of congruence relations on $\mathcal{A}$ whose intersection is the relation Δ_A. We leave the proof as an exercise for the reader.

In analogy to the definition of irreducible algebras in the direct product case, we want to consider algebras which cannot be expressed as a subdirect product of other smaller algebras, except in trivial ways.

Definition 4.2.5 An algebra $\mathcal{A}$ of type τ is called *subdirectly irreducible*, if every family $\{\theta_j \mid j \in J\}$ of congruences on $\mathcal{A}$, none of which is equal to Δ_A, has an intersection which is different from Δ_A. In this case, the conditions of Theorem 4.2.4 are not satisfied, and no representation of $\mathcal{A}$ as a subdirect product is possible.

Remark 4.2.6 It is easy to see that an algebra $\mathcal{A}$ is subdirectly irreducible if and only if Δ_A has exactly one upper neighbour or cover in the lattice $Con\mathcal{A}$ of all congruence relations on $\mathcal{A}$. Then the congruence lattice has the form shown in the diagram below.

Example 4.2.7 1. Every simple algebra is subdirectly irreducible. Since an arbitrary two-element algebra is simple, such algebras are always subdirectly irreducible.

2. A three-element algebra which has no more than one congruence other than A^2 and Δ_A is subdirectly irreducible.

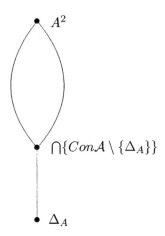

We started this chapter looking at products as a way to produce larger and more complicated algebras out of given algebras. But we have also looked at this process in the other direction: we can try to express any algebra as a product of certain "simpler" algebras, such as irreducible algebras. However, direct products are not the best concept to use here, since not every algebra is isomorphic to a direct product of directly irreducible algebras. Subdirect products, on the other hand, do have the right property, as shown by the following theorem of G. Birkhoff ([8]). We present this important result without proof.

Theorem 4.2.8 *Every algebra is isomorphic to a subdirect product of subdirectly irreducible algebras.*   ∎

4.3   Exercises

4.3.1. Prove that if $\mathcal{A}$ is a finite algebra, then $\mathcal{A}$ is isomorphic to a direct product of directly irreducible algebras.

4.3.2. Prove that every semilattice is isomorphic to a subdirect product of the two-element semilattice $(\{0, 1\}; \wedge)$. This means that (up to isomorphism) the semilattice $(\{0, 1\}; \wedge)$ is the only subdirectly irreducible semilattice.

4.3.3. Let $\mathcal{A}$ and $\mathcal{B}$ be algebras of the same type. Show that if $\mathcal{A}$ has a one-element subalgebra, the direct product $\mathcal{A} \times \mathcal{B}$ has a subalgebra which is isomorphic to $\mathcal{B}$.

4.3.4. Prove that for any algebra $\mathcal{A}$, the diagonal relation Δ_A on the set A is the universe of a subalgebra of the algebra $\mathcal{A} \times \mathcal{A}$.

4.3.5. We can define an operator $\mathbf{P}$ on the class $Alg(\tau)$ of all algebras of type τ, as follows. For any class $K \subseteq Alg(\tau)$, let $\mathbf{P}(K)$ be the class of all algebras of type τ which are products of one or more algebras in K. Prove that this operator $\mathbf{P}$ is extensive and monotone, but not idempotent, and hence is not a closure operator on $Alg(\tau)$. (This operator will be used again in Chapter 6.)

4.3.6. Prove that a finite abelian group is subdirectly irreducible iff it is a cyclic group of prime power order.

4.3.7. Prove the claim made in the remark following Theorem 4.2.4.

Chapter 5

Terms, Trees, and Polynomials

In the previous chapters, we have studied four algebraic constructions on algebras: formation of subalgebras, homomorphic images, quotient algebras and product algebras. Now we begin another approach to the study of algebras, the equational approach. We start by looking at terms and polynomials in this chapter. In the following chapter we will use these concepts to define equations and identities, and connect the algebraic and equational approaches to algebras.

Terms and polynomials on an algebra $\mathcal{A}$ define special kinds of operations on the base set A. We have been studying the properties of the fundamental operations on $\mathcal{A}$, for example that the fundamental operations are compatible with congruences on $\mathcal{A}$ and preserve all subalgebras of $\mathcal{A}$. But for a given algebra $\mathcal{A}$, there are other operations besides the fundamental operations which have these nice properties. Any operation obtained by arbitrary compositions of the fundamental operations will have these properties, and it is these operations which are called term operations. If we also allow the use of constants in our arbitrary compositions, we obtain operations called polynomial operations. Term operations and polynomial operations can also be obtained starting from abstract terms and polynomials, respectively. Terms and polynomials are important as a way to define identities satisfied by an algebra.

5.1   Terms and Trees

In Section 1.2. we defined semigroups as algebras $(G; \cdot)$ of type $\tau = (2)$ where
the associative law

$$x \cdot (y \cdot z) \approx (x \cdot y) \cdot z$$

is satisfied. Satisfaction of this law means that for all elements $x, y, z \in G$,
the equation $x \cdot (y \cdot z) = (x \cdot y) \cdot z$ holds. To write this equation, we need
the symbols x, y and z. But these symbols are not themselves elements of
G; they are only symbols for which elements from G may be substituted.
Such symbols are called variables. To write identities or laws of an algebra
we need a language, which must include such variables as well as symbols to
represent the operations. In the associative law above we did not distinguish
between the operation and the symbol used to denote it, using $\cdot$ for both, but
in our new formal language we shall usually make such a distinction. That
is, we will have formal operation symbols distinct from concrete operations
on a set.

Now we proceed to define this formal language in the general setting. Let
$n \geq 1$ be a natural number. Let $X_n = \{x_1, \ldots, x_n\}$ be an n-element set. The
set X_n is called an *alphabet* and its elements are called *variables*. We also
need a set $\{f_i | i \in I\}$ of operation symbols, indexed by the set I. The sets X_n
and $\{f_i | i \in I\}$ have to be disjoint. To every operation symbol f_i we assign
a natural number $n_i \geq 1$, called the arity of f_i. As in the definition of an
algebra, the sequence $\tau = (n_i)_{i \in I}$ of all the arities is called the *type* of the
language. With this notation for operation symbols and variables, we can
define the terms of our type τ language.

Definition 5.1.1 Let $n \geq 1$. The *n-ary terms* of type τ are defined in the
following inductive way:

(i) Every variable $x_i \in X_n$ is an n-ary term.

(ii) If $t_1, \ldots, t_{n_i}$ are n-ary terms and f_i is an n_i-ary operation symbol, then
$f_i(t_1, \ldots, t_{n_i})$ is an n-ary term.

(iii) The set $W_\tau(X_n) = W_\tau(x_1, \ldots, x_n)$ of all n-ary terms is the smallest
set which contains x_1, ..., x_n and is closed under finite application of
(ii).

Remark 5.1.2 1. It follows immediately from the definition that every n-ary term is also k-ary, for $k > n$.

2. Our definition does not allow nullary terms. This could be changed by adding a fourth condition to the inductive definition, stipulating that every nullary operation symbol of our type is an n-ary term. We could also extend our language to include a third set of symbols, to be used as constants or nullary terms; we shall explore this approach later, in Section 5.3.

Example 5.1.3 Let $\tau = (2)$, with one binary operation symbol f. Let $X_2 = \{x_1, x_2\}$. Then $f(f(x_1, x_2), x_2)$, $f(x_2, x_1)$, x_1, x_2 and $f(f(f(x_1, x_2), x_1), x_2)$ are binary terms. The expression $f(f(x_3, f(x_1, x_2)), x_4)$ is a quaternary or 4-ary term, but $f(f(x_1, x_2), x_3$ is not a term (one bracket is missing).

Example 5.1.4 Let $\tau = (1)$, with one unary operation symbol f. Let $X_1 = \{x_1\}$. Then the unary terms of this type are x_1, $f(x_1)$, $f(f(x_1))$, $f(f(f(x_1)))$, and so on. Note that $W_{(1)}(X_1)$ is infinite. In a specific application such as group theory, we might denote our unary operation by $^{-1}$ instead of f, writing our terms as x_1, x_1^{-1}, $(x_1^{-1})^{-1}$, etc. In the group theory case we might want to consider the terms x_1 and $(x_1^{-1})^{-1}$ as equal. But such an equality depends on a specific application, and does not hold in the most general sense that we are defining here. Thus our terms are often called the "absolutely free" terms, in the sense that we make no restrictions or assumptions about the properties of our operation symbols, beyond their arity as specified in the type.

An important feature of our definition of terms, Definition 5.1.1, is that it is an inductive definition, based on the number of occurrences of operation symbols in a term. This number is sometimes called the complexity of a term, and many of our proofs in this chapter will proceed by induction on the complexity of terms. This means that in order to prove that the set of all terms has a certain property, it will suffice to prove that the variable terms have the property and that if the terms t_1, ..., t_{n_i} have the property, then so does the compound term $f_i(t_1, \ldots, t_{n_i})$.

There are various methods used to measure the complexity of a term, besides the number of operation symbols which occur in it. Another common measure is what is called the depth of the term, defined by the following steps:

(i) $depth(t) = 0$ if $t = x_i$ is a variable,

(ii) $depth(t) = \max\{depth(t_1), \ldots, depth(t_{n_i})\} + 1$ if $t = f(t_1, \ldots, t_{n_i})$.

In Computer Science it is very common to illustrate terms by tree diagrams, called *semantic trees*. The semantic tree of the term t is defined as follows:

(i) If $t = x_i$, then the semantic tree of t consists only of one vertex which is labelled with x_i, and this vertex is called the root of the tree.

(ii) If $t = f_i(t_1, \ldots, t_{n_i})$ then the semantic tree of t has as its root a vertex labelled with f_i, and has n_i edges which are incident with the vertex f_i; each of these edges is incident with the root of the corresponding term $t_1, \ldots, t_{n_i}$ (ordered by $1 \leq 2 \leq \cdots \leq n_i$, starting from the left).

Consider for example the type $\tau = (2,1)$ with a binary operation symbol f_2 and a unary operation symbol f_1, and variable set $X_3 = \{x_1, x_2, x_3\}$. Then the term $t = f_2(f_1(f_2(f_2(x_1, x_2), f_1(x_3))), f_1(f_2(f_1(x_1), f_1(f_1(x_2)))))$ corresponds to the semantic tree shown below.

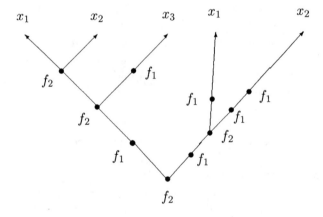

Notice that semantic trees are ordered in such a way that we start with the first variable on the left hand side which occurs in a term t. We can also describe, for any node or vertex in a tree, the path from the root of the tree to that node or vertex. That is, to any node or vertex of a term $t = f_i(t_1, \ldots, t_{n_i})$, we can assign a sequence (or word) over the set $\mathbb{N}^+$ of positive integers, as follows. Suppose the outermost term is n_i-ary. The root of the tree is labelled by the symbol e, which we call the empty sequence. The vertices on the second level up are labelled by $1, \ldots, n_i$, from the left to the right. Continuing in this way, we assign to each branch of the tree a sequence on $\mathbb{N}^+$. For instance, the tree from our example above is labelled as shown below.

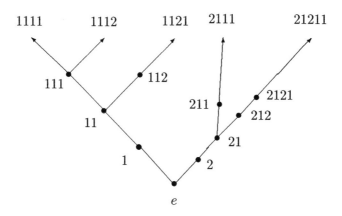

We can think of the words we assign to each branch as determining an "address" for every vertex or node of a tree. This notation is convenient for encoding various operations we can perform on trees. For instance, if t is a tree and u is such an address, then we denote by t/u the subtree of t which starts with the vertex labelled by the sequence u. If s is another tree, then we denote by $t[u/s]$ the tree obtained from t by replacing the subtree t/u by the tree s.

We shall make use of semantic trees in Chapter 7, when we consider term rewriting systems. For now, we return to the main development of the theory of terms. Let τ be a fixed type. Let X be the union of all the sets X_n of variables, so $X = \{x_1, x_2, \ldots\}$.
We denote by $W_\tau(X)$ the set of all terms of type τ over the countably infinite alphabet X:

$$W_\tau(X) \;=\; \bigcup_{n=1}^{\infty} W_\tau(X_n).$$

Now we want to use this set $W_\tau(X)$ as the universe of some algebra, of the same type τ. What operations can we perform on these terms? In fact, for every $i \in I$ we can define an n_i-ary operation $\bar{f}_i$ on $W_\tau(X)$, with

$$\bar{f}_i : W_\tau(X)^{n_i} \to W_\tau(X) \quad \text{defined by} \quad (t_1, \ldots, t_{n_i}) \mapsto f_i(t_1, \ldots, t_{n_i}).$$

Note the distinction here between the concrete operation $\bar{f}_i$ being defined on the set of all terms, and the formal operation symbol f_i, used in the formation of terms. The second step of Definition 5.1.1 shows that the element $f_i(t_1, \ldots, t_{n_i})$ in our definition belongs to $W_\tau(X)$, and so our operation $\bar{f}_i$ is well defined. In this way we make $W_\tau(X)$ into the universe of an algebra of type $\tau = (n_i)_{i \in I}$, since for every operation symbol f_i we have a concrete operation $\bar{f}_i$ on $W_\tau(X)$.

Definition 5.1.5 The algebra $\mathcal{F}_\tau(X) := (W_\tau(X); (\bar{f}_i)_{i \in I})$ is called the *term algebra*, or the *absolutely free algebra*, of type τ over the set X.

The following result is an easy consequence of our inductive definition of terms.

Lemma 5.1.6 *For any type τ, the term algebra $\mathcal{F}_\tau(X)$ is generated by the set X.*

Proof: Definition 5.1.1 (i) shows that $X \subseteq W_\tau(X)$, and 5.1.1 (ii) gives $\langle X \rangle_{\mathcal{F}_\tau(X)} = \mathcal{F}_\tau(X)$. ∎

Instead of $\mathcal{F}_\tau(X)$, we could also consider the algebra $\mathcal{F}_\tau(X_n) := (W_\tau(X_n); (\bar{f}_i)_{i \in I})$ where now $\bar{f}_i$ are the restrictions of the operations defined on $W_\tau(X)$ to the subset $W_\tau(X_n)$. By Definition 5.1.1 these restrictions are also operations defined on $W_\tau(X_n)$. This algebra is called the *absolutely free algebra* or the *term algebra* of type τ over the set X_n of n generators. As in Lemma 5.1.6, the algebra $\mathcal{F}_\tau(X_n)$ is generated by the set X_n.

As we mentioned in Example 5.1.4, the term algebra defined here is the "absolutely free" one, in the sense that we make no assumptions about the operation symbols other than their arities. The phrase "absolutely free" is also used in a more technical sense, as described in the following theorem.

Theorem 5.1.7 *For every algebra $\mathcal{A} \in Alg(\tau)$ and every mapping $f : X \to \mathcal{A}$, there exists a unique homomorphism $\hat{f} : \mathcal{F}_\tau(X) \to \mathcal{A}$ which extends the mapping f and such that $\hat{f} \circ \varphi = f$, where $\varphi : X \to \mathcal{F}_\tau(X)$ is the embedding of X in $\mathcal{F}_\tau(X)$. This homomorphism makes the following diagram commute.*

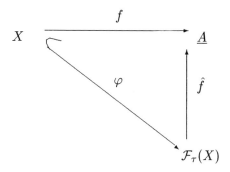

Proof: We define $\hat{f}$ in the following way:

$$\hat{f}(t) := \begin{cases} f(t), & \text{if } t = x \in X \text{ is a variable,} \\ f_i^A(\hat{f}(t_1), \ldots, \hat{f}(t_{n_i})), & \text{if } t = f_i(t_1, \ldots, t_{n_i}). \end{cases}$$

Here f_i^A is the fundamental operation of the algebra $\mathcal{A}$ which corresponds to the operation symbol f_i. This definition is inductive, in that we are assuming in the second condition that each $\hat{f}(t_j)$, for $1 \leq j \leq n_i$, is already defined. Clearly, $\hat{f}(\overline{f}_i(t_1, \ldots, t_{n_i})) = \hat{f}(f_i(t_1, \ldots, t_{n_i})) = f_i^A(\hat{f}(t_1), \ldots, \hat{f}(t_{n_i}))$ shows that $\hat{f}$ is a homomorphism, and it follows directly from the definition that $\hat{f}$ extends f. ∎

We have so far defined term algebras on the finite sets X_n and the countably infinite set X, using variable symbols x_1, x_2, x_3, ... However, it should be clear that we could start with any non-empty set Y of symbols, of any cardinality, and carry out the same process. We define terms on the set Y inductively just as in Definition 5.1.1, with all the variables in Y being terms and then any result of applying the operation symbols f_i to terms giving terms. Then we can form the set $W_\tau(Y)$ of all terms of type τ over Y, and make it into an algebra $\mathcal{F}_\tau(Y)$ generated by Y which has the analogous freeness property of Theorem 5.1.7. Thus we have a free algebra over any set of symbols of any cardinality, although we shall see in Chapter 6 that in some sense the sets of finite or countably infinite cardinality are sufficient for our purposes. Moreover, the next theorem shows that for a fixed cardinality, one may use any choice of variable symbols. For example, the reader may have noticed that in our formal language we have variables x_i for $i \geq 1$, while in our example with the associative law at the beginning of this section we used variables x, y and z. That it is justified to make such a change of variable symbols, where convenient, follows from the following theorem:

Theorem 5.1.8 *Let Y and Z be alphabets with the same cardinality. Then the term algebras $\mathcal{F}_\tau(Y)$ and $\mathcal{F}_\tau(Z)$ are isomorphic.*

Proof: When $|Y| = |Z|$ there exists a bijection $\varphi : Y \to Z$. Since $\mathcal{F}_\tau(Z) \in \text{Alg}\,(\tau)$, by Theorem 5.1.7 we can extend φ to a homomorphism $\hat{\varphi} : \mathcal{F}_\tau(Y) \to \mathcal{F}_\tau(Z)$. Now using Theorem 5.1.7 again on the mapping $\varphi^{-1} : Z \to Y$ gives a homomorphism $(\varphi^{-1})\hat{\ } : \mathcal{F}_\tau(Z) \to \mathcal{F}_\tau(Y)$. We will show by induction on the complexity of the term t that $(\varphi^{-1})\hat{\ } \circ \hat{\varphi} = id_{W_\tau(Y)}$ and $\hat{\varphi} \circ (\varphi^{-1})\hat{\ } = id_{W_\tau(Z)}$. This will prove that $\hat{\varphi}$ is an isomorphism, with $(\varphi^{-1})\hat{\ }$ as its inverse. The claim is clear for the base case that $t = x$ is a variable. Now assume that $t = f_i(t_1, \ldots, t_{n_i})$, and that the claim is true for the terms $t_1, \ldots, t_{n_i}$. Then we have $((\varphi^{-1})\hat{\ } \circ \hat{\varphi})(t) = (\varphi^{-1})\hat{\ }(\hat{\varphi}(t)) = (\varphi^{-1})\hat{\ }(f_i^{\mathcal{F}_\tau(Z)}(\hat{\varphi}(t_1), \ldots, \hat{\varphi}(t_{n_i}))$ $= (\varphi^{-1})\hat{\ }(f_i(\hat{\varphi}(t_1), \ldots, (\hat{\varphi}(t_{n_i}))) = f_i^{\mathcal{F}_\tau(Y)}(((\varphi^{-1})\hat{\ } \circ \varphi)(t_1), \ldots, ((\varphi^{-1})\hat{\ } \circ \varphi)(t_{n_i}))) = f_i(t_1, \ldots, t_{n_i}) = t$. The proof for $\hat{\varphi} \circ (\varphi^{-1})\hat{\ }$ is similar. ∎

5.2   Term Operations

Terms are formal expressions on our formal language of type τ. In order to formulate statements using terms which are true or false in a given algebra $\mathcal{A}$, we have to evaluate the variables in the terms by elements of the concrete set A, and we have to interpret the operation symbols by concrete operations on this set. It is this process which produces term operations from terms. We shall continue to denote by X the countably infinite set $\{x_1, x_2, x_3, \ldots\}$ of variables.

Definition 5.2.1 Let $\mathcal{A}$ be an algebra of type τ and let t be an n-ary term of type τ over X. Then t induces an n-ary operation $t^{\mathcal{A}}$ on $\mathcal{A}$, called the *term operation induced by the term t on the algebra $\mathcal{A}$*, via the following steps:

(i) If $t = x_i \in X_n$, then $t^{\mathcal{A}} = x_j^{\mathcal{A}} = e_j^{n,\mathcal{A}}$; here $e_j^{n,\mathcal{A}}$ is the n-ary projection on A defined by $e_j^{n,\mathcal{A}}(a_1, \ldots, a_n) = a_j$ for all $a_1, \ldots, a_n \in A$.

(ii) If $t = f_i(t_1, \ldots, t_{n_i})$ is an n-ary term of type τ, and $t_1^{\mathcal{A}}, \ldots, t_{n_i}^{\mathcal{A}}$ are the term operations which are induced by $t_1, \ldots, t_{n_i}$, then $t^{\mathcal{A}} = f_i^{\mathcal{A}}(t_1^{\mathcal{A}}, \ldots, t_{n_i}^{\mathcal{A}})$.

In part (ii) of this definition, the right hand side of the equation refers to the composition or superposition of operations, so that

$$t^{\mathcal{A}}(a_1, \ldots, a_n) := f_i^A(t_1^{\mathcal{A}}(a_1, \ldots, a_n), \ldots, t_{n_i}^{\mathcal{A}}(a_1, \ldots, a_n)),$$

for all $a_1, \ldots, a_n \in A$.

Since any concrete operation on a set has an arity attached to it, if we want to induce such an operation from a term, we must also have an arity attached to the term. It is for this reason that our definition of terms begins in Definition 5.2.1 with terms of each fixed arity n.

Roughly speaking, if t is an n-ary term and $\mathcal{A}$ is an algebra of type τ, then to obtain the operation $t^{\mathcal{A}}$ we substitute elements from the set A for the variables occurring in t, and interpret the operation symbols f_i for $i \in I$ as the corresponding fundamental operations f_i^A. More precisely, we map the variables x_j occurring in t to elements $a_1, \ldots, a_n$ of A by a function $f : X \rightarrow A$, and then our term operation $t^{\mathcal{A}}$ is the unique extension $\hat{f} : \mathcal{F}_\tau(X) \rightarrow \mathcal{A}$ from Theorem 5.1.7. We will denote by $W_\tau(X_n)^{\mathcal{A}}$ the set of all n-ary term operations of the algebra $\mathcal{A}$, and by $W_\tau(X)^{\mathcal{A}}$ the set of all (finitary) term operations on $\mathcal{A}$.

There is another way to obtain the set $W_\tau(X)^{\mathcal{A}}$ of all term operations on $\mathcal{A}$, using clone operations. Using A as our base set, we consider the set $O(A)$ of all finitary operations on A. Recall from Example 1.2.14 that the set $O(A)$ is closed under a composition operation

$$O^n(A) \times (O^m(A))^n \rightarrow O^m(A),$$

and contains all the projection operations on A. This makes $O(A)$ a clone on the set A, called the *full clone* on A; any subset of $O(A)$ which contains the projections and is also closed under composition is called a *subclone* of $O(A)$, or a *clone* on A.

Definition 5.2.2 Let $C \subseteq O(A)$ be a set of operations on a set A. Then the *clone generated by C*, denoted by $\langle C \rangle$, is the smallest subset of $O(A)$ which contains C, is closed under composition, and contains all the projections $e_i^{n,A} : A^n \rightarrow A$ for arbitrary $n \geq 1$ and $1 \leq i \leq n$.

Then we have the following connection to our term algebras.

Theorem 5.2.3 Let $\mathcal{A} = (A; (f_i^{\mathcal{A}})_{i \in I})$ be an algebra of type τ, and let $W_\tau(X)$ be the set of all terms of type τ over X. Then $W_\tau(X)^{\mathcal{A}}$ is a clone on $\mathcal{A}$, called the term clone of $\mathcal{A}$. Moreover, the clone $W_\tau(X)^{\mathcal{A}}$ is generated by the set of all fundamental operations of the algebra $\mathcal{A}$. That is, $W_\tau(X)^{\mathcal{A}} = \langle \{f_i^{\mathcal{A}} | i \in I\} \rangle$.

Proof: We prove first that $W_\tau(X)^{\mathcal{A}}$ is indeed a clone. Since $x_i \in X_n$, we have $x_i^{\mathcal{A}} = e_i^{n,\mathcal{A}} \in W_\tau(X)^{\mathcal{A}}$ for all $n \geq 1$. Thus $W_\tau(X)^{\mathcal{A}}$ contains all the projections. Now let $f^{\mathcal{A}}$, $g_1^{\mathcal{A}}$, ..., $g_n^{\mathcal{A}}$ be in $W_\tau(X)^{\mathcal{A}}$, with $f^{\mathcal{A}}$ n-ary and $g_1^{\mathcal{A}}$, ..., $g_n^{\mathcal{A}}$ each m-ary. Then $f(x_1, \ldots, x_n)$ and the $g_i(x_1, \ldots, x_m)$, for $1 \leq i \leq n$, are terms which induce the term operations $f^{\mathcal{A}}$, $g_1^{\mathcal{A}}$, ..., $g_n^{\mathcal{A}}$ respectively. But then $f(g_1(x_1, \ldots, x_n), \ldots, g_n(x_1, \ldots, x_n))$ is also a term, and the induced term operation is $f^{\mathcal{A}}(g_1^{\mathcal{A}}, \ldots, g_n^{\mathcal{A}}) \in W_\tau(X)^{\mathcal{A}}$. Therefore, $W_\tau(X)^{\mathcal{A}}$ is closed under composition of operations, and is a clone.

Clearly, $\{f_i^{\mathcal{A}} | i \in I\} \subseteq W_\tau(X)^{\mathcal{A}}$, and so $\langle \{f_i^{\mathcal{A}} | i \in I\} \rangle \subseteq W_\tau(X)^{\mathcal{A}}$. We will show the converse inclusion by induction on the complexity of a term t. If t is a variable $x_i \in X_n$, then the induced term operation is a projection which belongs to the clone $\langle \{f_i^{\mathcal{A}} | i \in I\} \rangle$. If $t = f_i(t_1, \ldots, t_{n_i})$ and $t^{\mathcal{A}} \in W_\tau(X)^{\mathcal{A}}$ and if we assume that $t_1^{\mathcal{A}}$, ..., $t_{n_i}^{\mathcal{A}} \in \langle \{f_i^{\mathcal{A}} | i \in I\} \rangle$, then $f_i^{\mathcal{A}}(t_1^{\mathcal{A}}, \ldots, t_{n_i}^{\mathcal{A}}) = t^{\mathcal{A}} \in \langle \{f_i^{\mathcal{A}} | i \in I\} \rangle$ since $\langle \{f_i^{\mathcal{A}} | i \in I\} \rangle$ is a clone. Therefore $\langle \{f_i^{\mathcal{A}} | i \in I\} \rangle \supseteq W_\tau(X)^{\mathcal{A}}$, and altogether we have equality. ∎

We point out that Theorem 5.2.3 is no longer true in the case of partial algebras, that is algebras in which the fundamental operations f_i are not totally defined on A.

As we remarked in the introduction to this chapter, an important feature of term operations of an algebra is that they have many of the same useful properties the fundamental operations of the algebra do, with respect to subalgebras, homomorphisms and congruence relations.

Theorem 5.2.4 Let $\mathcal{A}$ be an algebra of type τ and let $t^{\mathcal{A}}$ be the n-ary term operation on $\mathcal{A}$ induced by the n-ary term $t \in W_\tau(X)$.

(i) If $\mathcal{B}$ is a subalgebra of $\mathcal{A}$, then $t^{\mathcal{A}}(b_1, \ldots, b_n) \in B$, for all $b_1, \ldots, b_n \in B$.

(ii) If $\mathcal{B}$ is an algebra of type τ and $\varphi : \mathcal{A} \to \mathcal{B}$ is a homomorphism, then for all a_1, ..., a_n in A,

$$\varphi(t^{\mathcal{A}}(a_1, \ldots, a_n)) = t^{\mathcal{B}}(\varphi(a_1), \ldots, \varphi(a_n)).$$

(iii) If θ is a congruence relation on $\mathcal{A}$, then for all pairs (a_1, b_1), ..., (a_n, b_n) in θ, we have $(t^{\mathcal{A}}(a_1, \ldots, a_n), t^{\mathcal{A}}(b_1, \ldots, b_n)) \in \theta$.

Proof: For (i) and (ii) we give a proof by induction on the complexity of the term $t \in W_\tau(X_n)$. If t is a variable x_i, for $1 \le i \le n$, then

$$x_i^{\mathcal{A}}(b_1, \ldots, b_n) = e_i^{n,\mathcal{A}}(b_1, \ldots, b_n) = b_i \in B \quad \text{and}$$

$$\varphi(e_i^{n,\mathcal{A}}(a_1, \ldots, a_n)) = \varphi(a_i) = e_i^{n,\mathcal{B}}(\varphi(a_1), \ldots, \varphi(a_n)).$$

Inductively, let $t = f_i(t_1, \ldots, t_{n_i})$ and assume that (i) and (ii) are satisfied for the term operations $t_1^{\mathcal{A}}, \ldots, t_{n_i}^{\mathcal{A}}$. Then $t^{\mathcal{A}}(b_1, \ldots, b_n) = f_i^{\mathcal{A}}(t_1^{\mathcal{A}}, \ldots, t_n^{\mathcal{A}})(b_1, \ldots, b_n) = f_i^{\mathcal{A}}(t_1^{\mathcal{A}}(b_1, \ldots, b_n), \ldots, t_n^{\mathcal{A}}(b_1, \ldots, b_n))$, and this is in B since all the $t_i^{\mathcal{A}}(b_1, \ldots, b_n)$ are in B and $\mathcal{B}$ is a subalgebra of $\mathcal{A}$.

For homomorphisms, we have

$$
\begin{aligned}
\varphi(t^{\mathcal{A}}(a_1, \ldots, a_n)) \\
&= \varphi(f_i^{\mathcal{A}}(t_1^{\mathcal{A}}, \ldots, t_{n_i}^{\mathcal{A}})(a_1, \ldots, a_n)) \\
&= \varphi(f_i^{\mathcal{A}}(t_1^{\mathcal{A}}(a_1, \ldots, a_n), \ldots, t_{n_i}^{\mathcal{A}}(a_1, \ldots, a_n))) \\
&= f_i^{\mathcal{B}}(\varphi(t_1^{\mathcal{A}}(a_1, \ldots, a_n)), \ldots, \varphi(t_{n_i}^{\mathcal{A}}(a_1, \ldots, a_n))) \\
&= f_i^{\mathcal{B}}(t_1^{\mathcal{B}}(\varphi(a_1), \ldots, \varphi(a_n)), \ldots, t_{n_i}^{\mathcal{B}}(\varphi(a_1), \ldots, \varphi(a_n))) \\
&= f_i^{\mathcal{B}}(t_1^{\mathcal{B}}, \ldots, t_{n_i}^{\mathcal{B}})(\varphi(a_1), \ldots, \varphi(a_n)) \\
&= t^{\mathcal{B}}(\varphi(a_1), \ldots, \varphi(a_n)).
\end{aligned}
$$

This shows that (i) and (ii) are satisfied.

(iii) Let θ be a congruence on $\mathcal{A}$, and let (a_1, b_1), ..., (a_n, b_n) be in θ. We know from Section 3.1 that θ is the kernel of the corresponding natural homomorphism, nat $\theta : \mathcal{A} \to \mathcal{A}/\theta$. So $(a_1, b_1) \in \theta$, ..., $(a_n, b_n) \in \theta$ means $(a_1, b_1) \in ker\ nat\theta$, ..., $(a_n, b_n) \in ker\ nat\theta$, and then $nat\theta(a_1) = nat\theta(b_1)$, ..., $nat\theta(a_n) = nat\theta(b_n)$. Using (ii) on this homomorphism, we obtain $nat\theta(t^{\mathcal{A}}(a_1, \ldots, a_n)) = t^{\mathcal{A}/\theta}(nat\theta(a_1), \ldots, nat\theta(a_n)) = t^{\mathcal{A}/\theta}(nat\theta(b_1), \ldots, nat\theta(b_n)) = nat\theta(t^{\mathcal{A}}(b_1, \ldots, b_n))$. From this we conclude that $(t^{\mathcal{A}}(a_1, \ldots, a_n), t^{\mathcal{A}}(b_1, \ldots, b_n)) \in \theta$. ∎

5.3 Polynomials and Polynomial Operations

In this section we define polynomials, with corresponding polynomial operations on an algebra. Like terms, polynomials are expressions in a formal language, composed inductively from variables and operation symbols; the difference is that for polynomials we are also allowed to use constants. Thus we introduce a third set of objects, the set of constants, to our language.

There is one subtlety here that the reader should notice. We are defining terms and polynomials in this chapter in a formal or general way, based only on a type, and not on any specific algebra. Thus we should have one set of constants, to be used in the formation of all polynomials of the given type. However, when we consider the induced polynomial operations on a given algebra $\mathcal{A}$, we usually want the constants in our polynomials to represent specific elements of our base set A. There are several ways to deal with this obstacle. We will proceed by fixing one set $\overline{A}$ of constant symbols to be used for all polynomials. Another approach is to associate to every algebra $\mathcal{A}$ of type τ a corresponding set $\overline{A}$ of constants, which has the same cardinality as the universe set A. This approach was used by Denecke and Leeratanavalee in [30] and [31].

Let $\overline{A}$ be our set of constant symbols, pairwise disjoint from both the set X of variables and the set $\{f_i \mid i \in I\}$ of operation symbols. We define *polynomials of type τ over $\overline{A}$* (for short, *polynomials*) via the following inductive steps:

(i) If $x \in X$, then x is a polynomial.

(ii) If $\overline{a} \in \overline{A}$, then $\overline{a}$ is a polynomial.

(iii) If $p_1, \ldots, p_{n_i}$ are polynomials and f_i is an n_i-ary operation symbol, then $f_i(p_1, \ldots, p_{n_i})$ is a polynomial.

(iv) The set $P_\tau(X, \overline{A})$ of all polynomials of type τ over $\overline{A}$ is the smallest set which contains $X \cup \overline{A}$ and is closed under finite application of (iii).

Much of the work we did in Section 5.1 for terms can now be carried out for polynomials. We can define the polynomial algebra $P_\tau(X, \overline{A})$ of type τ over $\overline{A}$, generated by $X \cup \overline{A}$, and prove results similar to 5.1.6 and 5.1.7. We leave the verification as an exercise for the reader.

The next step is to make polynomial operations over an algebra $\mathcal{A}$ out of our formal polynomials. We proceed as for terms, with the addition that we interpret the constant symbols from $\overline{A}$ by elements selected from A as nullary operations. In this case we assume that $|A| \geq |\overline{A}|$, and consider a subset $A_1 \subseteq A$ with $|A_1| = |\overline{A}|$. Then just as for terms we obtain for every polynomial p of type τ over $\overline{A}$ an induced *polynomial operation* $p^{\mathcal{A}}$, induced by the algebra $\mathcal{A}$. Let $P_\tau(X, \overline{A})^{\mathcal{A}}$ be the set of all polynomial operations produced in this way. We have the following analogue of Theorem 5.2.3:

Theorem 5.3.1 $P_\tau(X, \overline{A})^{\mathcal{A}}$ *is a clone, and is generated by the set* $\{f_i^{\mathcal{A}} | i \in I\} \cup \{c_a | a \in A\}$, *where* c_a *is the nullary operation which selects* $a \in A$. *We write* $P_\tau(X, \overline{A})^{\mathcal{A}} = \langle \{f_i^{\mathcal{A}} | i \in I\} \cup \{c_a | a \in A\} \rangle$. ∎

We leave it as an exercise for the reader to prove that every polynomial operation of an algebra $\mathcal{A}$ is compatible with any congruence relation on $\mathcal{A}$. In Theorem 1.4.5 we saw that an equivalence relation on an algebra $\mathcal{A}$ is a congruence iff it is compatible with all translations on the algebra. Since such translations are in fact just unary polynomial operations on the algebra, we have the following useful result.

Lemma 5.3.2 *An equivalence relation* θ *on an algebra* $\mathcal{A}$ *is a congruence relation on* $\mathcal{A}$ *iff* θ *is compatible with all unary polynomial operations on* $\mathcal{A}$. ∎

In Remark 1.4.9 we gave a characterization of the congruence on $\mathcal{A}$ generated by a binary relation on the set A. This characterization can also be rephrased in terms of unary polynomial operations. We shall make use of the following result in Chapters 11 and 12.

Lemma 5.3.3 *Let* $\mathcal{A}$ *be an algebra and let* ϱ *be a binary relation on the set* A, *so that* $\varrho \subseteq A^2$. *Assume that* ϱ *is reflexive and symmetric on* A. *Let* $\langle \varrho \rangle_{Con\mathcal{A}}$ *be the congruence generated by* ϱ. *Then* $(u, v) \in \langle \varrho \rangle_{Con\mathcal{A}}$ *if and only if there are pairs* $(a_1, b_1), \ldots, (a_k, b_k) \in \varrho$ *and unary polynomial operations* $p_1^{\mathcal{A}}, \ldots, p_k^{\mathcal{A}}$ *of* $\mathcal{A}$ *such that*

$$u = p_1^{\mathcal{A}}(a_1),$$
$$p_i^{\mathcal{A}}(b_i) = p_{i+1}^{\mathcal{A}}(a_{i+1}), \quad for \ 1 \le i < k,$$
$$p_k^{\mathcal{A}}(b_k) = v.$$

Proof: We define a set θ, by

$$\theta := \{(u, v) | u, v \in A \text{ and } \exists k \in \mathbb{N} \text{ and } \exists (a_1, b_1), \ldots, (a_k, b_k) \in \varrho \text{ and}$$
$$\exists p_1^{\mathcal{A}}, \ldots, p_k^{\mathcal{A}} \in P^{(A)}(\mathcal{A}) \text{ such that } u = p_1^{\mathcal{A}}(a_1), \ p_i^{\mathcal{A}}(b_i) = p_{i+1}^{\mathcal{A}}(a_{i+1}) \text{ for}$$
$$1 \le i < k, \ p_k^{\mathcal{A}}(b_k) = v\}.$$

Then θ is an equivalence relation with $\varrho \subseteq \theta \subseteq \langle \varrho \rangle_{Con\mathcal{A}}$. If we can show that θ is compatible with all unary polynomial operations of $\mathcal{A}$ then by Lemma 5.3.2 the relation θ is a congruence, and therefore $\theta = \langle \varrho \rangle_{Con\mathcal{A}}$. Let $p^{\mathcal{A}}$ be a unary polynomial operation of $\mathcal{A}$ and let $(u, v) \in \theta$. Then by definition of θ there is a natural number k, there are elements $(a_1, b_1), \ldots, (a_k, b_k) \in \varrho$ and unary polynomial operations $p_1^{\mathcal{A}}, \ldots, p_k^{\mathcal{A}}$ of $\mathcal{A}$ such that $u = p_1^{\mathcal{A}}(a_1)$, $p_i^{\mathcal{A}}(b_i) = p_{i+1}^{\mathcal{A}}(a_{i+1})$ for $1 \le i < k$ and $p_k^{\mathcal{A}}(b_k) = v$. Then we have also $p^{\mathcal{A}}(u) = p^{\mathcal{A}}(p_1^{\mathcal{A}}(a_1)), p^{\mathcal{A}}(p_i^{\mathcal{A}}(b_i)) = p^{\mathcal{A}}(p_{i+1}^{\mathcal{A}}(a_{i+1}))$ for $1 \le i < k$ and $p^{\mathcal{A}}(p_k^{\mathcal{A}}(b_k)) = p^{\mathcal{A}}(b_k)$. Since the composition of two unary polynomial operations of $\mathcal{A}$ is again a unary polynomial operation of $\mathcal{A}$, we have $(p^{\mathcal{A}}(u), p^{\mathcal{A}}(v)) \in \theta$, and θ is a congruence relation on $\mathcal{A}$. $\blacksquare$

5.4 Exercises

5.4.1. Let $\mathcal{L} = (\{a, b, c, d\}; \overset{L}{\wedge}, \overset{L}{\vee})$ be a lattice, with operations $\overset{L}{\wedge}$ and $\overset{L}{\vee}$ given by the following Cayley tables:

$\overset{L}{\wedge}$	a	b	c	d
a	a	a	a	a
b	a	b	a	b
c	a	a	c	c
d	a	b	c	d

$\overset{L}{\vee}$	a	b	c	d
a	a	a	c	d
b	b	b	d	d
c	c	d	c	d
d	d	d	d	d

Let h be the binary operation on the set L given by the Cayley table:

h	a	b	c	d
a	a	b	a	b
b	b	b	b	b
c	a	b	a	b
d	b	b	b	b

Is h a term operation or a polynomial operation on $\mathcal{L}$?

5.4.2. Let $\mathcal{L}$ be the lattice from Exercise 5.4.1. Let f be a function $f : \{x, y, z\} \to \{a, b, c, d\}$, with $x \mapsto b$, $y \mapsto c$, $z \mapsto c$. Let Y be the set $\{x, y, z\}$. By Theorem 5.1.7, there is a unique extension $\hat{f}$ of f to the term algebra

$\mathcal{F}_\tau(Y)$. Calculate $\hat{f}(t)$ for the terms $t = x \overset{L}{\wedge} y$ and $t = (x \overset{L}{\vee} y) \overset{L}{\wedge} z$.

5.4.3. Determine all the term operations and all the polynomial operations of the algebra $(\mathbb{N}; \neg)$, where $\neg x := x + 1$.

5.4.4. Show that if Y and Z are non-empty sets with $|Y| \leq |Z|$, then the algebra $\mathcal{F}_\tau(Y)$ can be embedded in $\mathcal{F}_\tau(Z)$ in a natural way. (One algebra can be embedded in another if the second contains an isomorphic copy of the first.)

5.4.5. Prove Theorem 5.3.1.

5.4.6. Prove that the polynomial algebra of type τ over $\overline{A}$, from Theorem 5.3.1, satisfies properties similar to those of Lemma 5.1.6 and Theorem 5.1.7.

5.4.7. Prove that all polynomial operations on an algebra $\mathcal{A}$ are compatible with all congruence relations on $\mathcal{A}$. That is, prove that for θ a congruence on $\mathcal{A}$ and $p^{\mathcal{A}}$ a polynomial operation, if $(a_1, b_1), \ldots, (a_n, b_n) \in \theta$, then $(p^{\mathcal{A}}(a_1, \ldots, a_n), p^{\mathcal{A}}(b_1, \ldots, b_n)) \in \theta$.

Chapter 6

Identities and Varieties

Our motivation for defining terms and polynomials was to use them to define equations and identities. An equation is a statement of the form $t_1 \approx t_2$, where t_1 and t_2 are terms. We will define what it means for such an equation to be satisfied, or to be an identity, in an algebra $\mathcal{A}$. The relation of satisfaction, of an equation by an algebra, will give us a Galois-connection between sets of equations and classes of algebras, and allow us to consider classes of algebras which are defined by sets of equations. Finally, we show that such equational classes, or model classes, are precisely the same classes of algebras as those we are interested in from the algebraic approach of the first four chapters.

6.1 The Galois Connection (Id, Mod)

We begin by defining formally what is meant by satisfaction of an identity by an algebra. Recall that we have a fixed set $X = \{x_1, x_2, x_3, \ldots\}$ of variable symbols.

Definition 6.1.1 An *equation* of type τ is a pair of terms (s, t) from $W_\tau(X)$; such pairs are more commonly written as $s \approx t$. Such an equation $s \approx t$ is said to be an *identity* in the algebra $\mathcal{A}$ of type τ if $s^{\mathcal{A}} = t^{\mathcal{A}}$, that is, if the term operations induced by s and t on the algebra $\mathcal{A}$ are equal. In this case we also say that the equation $s \approx t$ is *satisfied* or *modelled* by the algebra $\mathcal{A}$, and we write $\mathcal{A} \models s \approx t$.

Remark 6.1.2 Let $\mathcal{A}$ be an algebra and $s \approx t$ an equation of type τ. Let

$X_n = \{x_1, \ldots, x_n\}$ be the set of variables occurring in the equation $s \approx t$. Recall from Chapter 5 that every map $f : X_n \to A$ has a unique extension $\hat{f} : \mathcal{F}_\tau(X_n) \to A$ to the free term algebra on X_n. Then the equation $s^A = t^A$ means that for every mapping $f : X_n \to A$, we have $\hat{f}(s) = \hat{f}(t)$ or $(s, t) \in ker \; \hat{f}$. That is, the pair (s, t) must belong to the intersection of the kernels of all these mappings $\hat{f}$. Thus an identity $s \approx t$ holds in an algebra A (or in a class K of algebras), iff (s, t) is in the intersection of the kernels of $\hat{f}$, for every map $f : X \to A$ (for every algebra A in K). We shall make frequent use in this chapter of this characterization of identities.

We now consider the class $Alg(\tau)$ of all algebras of type τ, and the class $W_\tau(X) \times W_\tau(X)$ of all equations of type τ. Satisfaction of an equation by an algebra gives us a fundamental relation between these two sets. Formally, we have the relation $\models$ of all pairs $(A, s \approx t)$ for which $A \models s \approx t$. As discussed in Section 2.2, this relation induces a Galois-connection between $Alg(\tau)$ and $W_\tau(X) \times W_\tau(X)$. We will use the names Id and Mod for the two associated mappings. That is, for any subset $\Sigma \subseteq W_\tau(X) \times W_\tau(X)$ and any subclass $K \subseteq Alg(\tau)$ we define

$$Mod\Sigma := \{A \in Alg(\tau) \mid \forall s \approx t \in \Sigma, \; (A \models s \approx t)\} \text{ and}$$

$$IdK := \{s \approx t \in W_\tau(X)^2 \mid \forall A \in K, \; (A \models s \approx t)\}.$$

Then the pair (Id, Mod) is the Galois-connection induced by the satisfaction relation $\models$. This gives us the various properties, common to any Galois-connection, which we proved in Theorem 2.2.4. Since these properties will be needed for our further work in this chapter, we list them again here for our new example of a Galois-connection.

Theorem 6.1.3 *Let τ be a fixed type.*

(i) *For all subsets Σ and Σ' of $W_\tau(X) \times W_\tau(X)$, and for all subclasses K and K' of $Alg(\tau)$, we have*
   $\Sigma \subseteq \Sigma' \Rightarrow Mod\Sigma \supseteq Mod\Sigma'$ *and* $K \subseteq K' \Rightarrow IdK \supseteq IdK'$;

(ii) *For all subsets Σ of $W_\tau(X) \times W_\tau(X)$ and all subclasses K of $Alg(\tau)$, we have $\Sigma \subseteq IdMod\Sigma$ and $K \subseteq ModIdK$;*

(iii) *The maps $IdMod$ and $ModId$ are closure operators on $W_\tau(X) \times W_\tau(X)$ and on $Alg(\tau)$, respectively.*

(iv) The sets closed under $ModId$ are exactly the sets of the form $Mod\Sigma$, for some $\Sigma \subseteq W_\tau(X) \times W_\tau(X)$, and the sets closed under $IdMod$ are exactly the sets of the form IdK, for some $K \subseteq Alg(\tau)$.

Proof: (i) and (ii) correspond to the definition of a Galois-connection, while (iii) and (iv) correspond to Theorem 2.2.4, parts (ii) and (iii). ∎

Definition 6.1.4 A class $K \subseteq Alg(\tau)$ is called an *equational class*, or is said to be *equationally definable*, if there is a set Σ of equations such that $K = Mod\Sigma$. A set $\Sigma \subseteq W_\tau(X) \times W_\tau(X)$ is called an *equational theory* if there is a class $K \subseteq Alg(\tau)$ such that $\Sigma = IdK$.

From Theorem 6.1.3, part (iv), we see that the equational classes are exactly the closed sets, or fixed points, with respect to the closure operator $ModId$, and, dually, the equational theories are exactly the closed sets, or fixed points, with respect to the closure operator $IdMod$. As we mentioned in Chapter 2, the collections of such closed sets form complete lattices.

Theorem 6.1.5 *The collection of all equational classes of type τ forms a complete lattice $\mathcal{L}(\tau)$, and the collection of all equational theories of type τ forms a complete lattice $\mathcal{E}(\tau)$. These lattices are dually isomorphic: there exists a bijection $\varphi : \mathcal{L}(\tau) \to \mathcal{E}(\tau)$, satisfying $\varphi(K_1 \vee K_2) = \varphi(K_1) \wedge \varphi(K_2)$ and $\varphi(K_1 \wedge K_2) = \varphi(K_1) \vee \varphi(K_2)$.*

Proof: As remarked above, the fact that these collections are complete lattices follows from the general results on Galois-connections and closure operators in Chapter 2. For arbitrary subclasses K of $\mathcal{L}(\tau)$, the infimum of K is the set-theoretical intersection, $\wedge K = \cap K$, and the supremum is the variety which is generated by the set-theoretical union, so $\vee K = \cap\{K' \in \mathcal{L}(\tau)| K' \supseteq \cup K\}$. Again by Theorem 2.1.6 we get $\vee K = $ Mod Id $(\cup K)$. The meet and join for $\mathcal{E}(\tau)$ are obtained similarly.

Now we define a mapping $\varphi : \mathcal{L}(\tau) \to \mathcal{E}(\tau)$ by $K \mapsto IdK$ for every equational class K of $\mathcal{L}(\tau)$. By Definition 6.1.4, the image IdK is an equational theory of type τ; so the mapping φ is well defined. By Definition 6.1.4 and Theorem 6.1.3 (iv), we know that φ is also surjective. Furthermore, if $IdK_1 = IdK_2$ then we have $K_1 = ModIdK_1 = ModIdK_2 = K_2$, using Theorem 6.1.3 (iii) and the fact that equational classes are exactly the fixed points with respect to the closure operator $ModId$. Therefore, φ is a bijection on $\mathcal{L}(\tau)$.

Next we show that our map φ has the property claimed on meets. We have $\varphi(K_1 \wedge K_2) = Id(K_1 \cap K_2)$, by definition. Since the operator Id reverses inclusions, our set $Id(K_1 \cap K_2)$ contains both IdK_1 and IdK_2. Since it is an equational theory, it also contains their join, $IdK_1 \vee IdK_2$, which equals $\varphi(K_1) \vee \varphi(K_2)$. This gives us the inclusion $\varphi(K_1 \wedge K_2) \supseteq \varphi(K_1) \vee \varphi(K_2)$.

For the opposite inclusion, we start with the fact that for each $i = 1, 2$, we have $IdK_i \subseteq IdK_1 \vee IdK_2$. Applying the operator Mod to this, and using the fact that $K_i = ModIdK_i$ for equational classes, we have $K_i \supseteq Mod(IdK_1 \vee IdK_2)$. Thus we have

$$Mod(IdK_1 \vee IdK_2) \subseteq K_1 \cap K_2.$$

Applying Id to this gives

$$IdMod(IdK_1 \vee IdK_2) \supseteq Id(K_1 \cap K_2).$$

But $IdK_1 \vee IdK_2$ is an equational theory, and therefore it is closed under $IdMod$. This gives us

$$Id(K_1 \cap K_2) \subseteq IdK_1 \vee IdK_2.$$

Thus we have our inclusion $\varphi(K_1 \wedge K_2) \subseteq \varphi(K_1) \vee \varphi(K_2)$, and hence the equality we needed.

Finally, we verify the claim for joins. Since φ reverses inclusions, the inclusion $K_i \subseteq K_1 \vee K_2$, for $i = 1, 2$, implies $\varphi(K_1 \vee K_2) \subseteq \varphi(K_i)$, for $i = 1, 2$. This gives us one direction, namely that $\varphi(K_1 \vee K_2) \subseteq \varphi(K_1) \wedge \varphi(K_2)$. Conversely, we know that $IdK_1 \wedge IdK_2 = IdK_1 \cap IdK_2 \subseteq IdK_i$, for $i = 1, 2$; so applying Mod gives $Mod(IdK_1 \wedge IdK_2) \supseteq ModIdK_i$, for $i = 1, 2$. Since K_1 and K_2 are closed under $ModId$, we get $Mod(IdK_1 \wedge IdK_2) \supseteq K_1 \vee K_2$. Applying Id once more gives $IdMod(IdK_1 \wedge IdK_2) \subseteq Id(K_1 \vee K_2)$. But $IdK_1 \wedge IdK_2$ is an equational theory and closed under $IdMod$; so finally we have $IdK_1 \wedge IdK_2 \subseteq Id(K_1 \vee K_2)$. This amounts to $\varphi(K_1) \wedge \varphi(K_2) \subseteq \varphi(K_1 \vee K_2)$. This completes our proof of the equality $\varphi(K_1) \cap \varphi(K_2) = \varphi(K_1 \vee K_2)$.   ∎

We point out that in fact the claims of Theorem 6.1.5 are true for the lattices of closed sets of any Galois-connection. A careful reading of the previous proof will show that we used only properties of the Galois connection, and not any properties of the particular connection $Id - Mod$.

If K is an arbitrary class of algebras, then its closure under the closure operator $ModId$ is the equational class $ModIdK$. This class is called the *equational class generated by* K, and usually denoted by $E(K) := ModIdK$.

Remark 6.1.6 We have developed our Galois-connection (Id, Mod) in this section using the countably infinite alphabet X of variable symbols for terms. Since any individual identity $s \approx t$ contains at most finitely many variables, it is clear that we need at most a countably infinite alphabet to discuss equational classes and equational theories. Moreover, since absolutely free algebras on different sets of the same cardinality are isomorphic, by Theorem 5.1.8, it is enough to use the set X. However, we could of course carry out the same process for any set Y of variables, with the same theorems holding. This will be significant in Sections 6.4 and 6.5, where we will want to consider free algebras and sets of identities on arbitrary sets Y.

6.2 Fully Invariant Congruence Relations

As we saw in the previous section, for a given type τ the collection of all equational theories of type τ forms a complete lattice, obtained as the lattice of closed sets from our Galois-connection. In this section we characterize such equational theories, using the concept of a fully invariant congruence relation.

Definition 6.2.1 A congruence relation θ on an algebra $\mathcal{A}$ of type τ is said to be *fully invariant* if whenever $(x, y) \in \theta$, we also have $(\varphi(x), \varphi(y)) \in \theta$, for every endomorphism φ of $\mathcal{A}$; that is, if θ is compatible with all endomorphisms φ of $\mathcal{A}$.

Theorem 6.2.2 *Let* $\Sigma \subseteq W_\tau(X) \times W_\tau(X)$ *be a set of equations of type* τ. *Then* Σ *is an equational theory if and only if it is a fully invariant congruence relation on the term algebra* $\mathcal{F}_\tau(X)$.

Proof: If Σ is an equational theory of type τ, then there exists a class $K \subseteq Alg(\tau)$ of algebras of type τ such that $\Sigma = IdK$. The set IdK is a binary relation on the set $W_\tau(X)$. For every term $t \in W_\tau(X)$ we have $t \approx t \in IdK$, since of course $t^{\mathcal{A}} = t^{\mathcal{A}}$ for any algebra $\mathcal{A}$ in the class K. The symmetry and transitivity of the binary relation IdK are similarly easy to verify, so that $\Sigma = IdK$ is at least an equivalence relation on $W_\tau(X)$. For the congruence property, suppose that $s_1 \approx t_1, \ldots, s_{n_i} \approx t_{n_i}$ are identities satisfied in K. This means that $s_1^A = t_1^A, \ldots, s_{n_i}^A = t_{n_i}^A$ for every algebra $\mathcal{A}$ from K. Let f_i be an n_i-ary operation symbol. Then our assumption means that $f_i^A(s_1^A, \ldots, s_{n_i}^A) = f_i(t_1^A, \ldots, t_{n_i}^A)$. By the inductive step of the definition of a term operation induced by a term, this means that $[f_i(s_1, \ldots, s_{n_i})]^A = [f_i(t_1, \ldots, t_{n_i})]^A$, and the definition of satisfaction gives $f_i(s_1, \ldots, s_{n_i}) \approx f_i(t_1, \ldots, t_{n_i}) \in IdA$. Thus $f_i(s_1, \ldots, s_{n_i}) \approx f_i(t_1, \ldots, t_{n_i}) \in IdK$, as required for a congruence.

For the fully invariant property, we have to show that IdK is preserved by an arbitrary endomorphism φ of $\mathcal{F}_\tau(X)$. So we take $(s, t) \in IdK = \Sigma$, and show that $(\varphi(s), \varphi(t))$ is also in Σ. For this we use the property from Remark 6.1.2, that $\Sigma = IdK$ is equal to the intersections of the kernels of the homomorphisms $\hat{f}$, for all maps $f : X \to A$ and all algebras $\mathcal{A}$ in K. But for any such $\mathcal{A}$ and map f, the map $\hat{f} \circ \varphi$ is also a homomorphism from $\mathcal{F}_\tau(X)$ to $\mathcal{A}$, and is the extension of some map g from X to A. Thus our pair (s, t) from Σ must also be in the kernel of this new homomorphism $\hat{f} \circ \varphi$. This means precisely that the pair $(\varphi(s), \varphi(t))$ must be in $ker\ \hat{f}$. Since this is true for all algebras $\mathcal{A}$ and maps $f : X \to A$, we have $(\varphi(s), \varphi(t))$ in Σ. This shows that Σ is a fully invariant congruence.

Conversely, assume that θ is a fully invariant congruence relation on $\mathcal{F}_\tau(X)$. We show that θ is the equational theory IdK, for K the class consisting of the quotient algebra $\mathcal{F}_\tau(X)/\theta$.

First, let $s \approx t \in Id\mathcal{F}_\tau(X)/\theta$. Consider the mapping $f : X \to \mathcal{F}_\tau(X)/\theta$ defined by $f(x) = [x]_\theta$ for all $x \in X$. Since $\mathcal{F}_\tau(X)/\theta$ is an algebra of type τ, this mapping f has a unique homomorphic extension $\hat{f} : \mathcal{F}_\tau(X) \to \mathcal{F}_\tau(X)/\theta$. But the natural homomorphism $nat\theta : \mathcal{F}_\tau(X) \to \mathcal{F}_\tau(X)/\theta$ has the same property, so the uniqueness of $\hat{f}$ gives $\hat{f} = nat\theta$. Now our assumption that $s \approx t$ is an identity in $\mathcal{F}_\tau(X)/\theta$ means, by Remark 6.1.2, that (s, t) is in $ker\hat{f}$. This gives $nat\theta(s) = nat\theta(t)$, so $(s, t) \in \theta$. Thus we have $Id(\mathcal{F}_\tau(X)/\theta) \subseteq \theta$.

For the opposite inclusion, we use Remark 6.1.2 again; it will suffice to show that for any map $f : X \to \mathcal{F}_\tau(X)/\theta$, we have $\theta \subseteq ker\ \hat{f}$, where as usual $\hat{f}$ is the unique extension of f. To show this, we first define a map $g : X \to \mathcal{F}_\tau(X)$ by $x \mapsto s$ in such a way that we assign to x a representative $s \in \hat{f}(x) = [x]_\theta$. Then we get the commutative diagram shown below, with φ the natural embedding. Combining $\hat{g} \circ \varphi = g$ and $nat\theta \circ g = \hat{f} \circ \varphi$ gives us $nat\theta \circ \hat{g} \circ \varphi = \hat{f} \circ \varphi$, and since φ is an embedding we have $nat\theta \circ \hat{g} = \hat{f}$. Now for any $(s, t) \in \theta$, the full invariance of θ means that $(\hat{g}(s), \hat{g}(t)) \in \theta$. Therefore $\hat{f}(s) = [\hat{g}(s)]_\theta = [\hat{g}(t)]_\theta = \hat{f}(t)$, and $(s, t) \in ker\hat{f}$. ∎

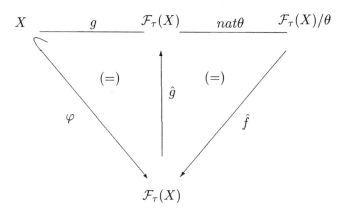

As a consequence of this theorem, we see that the set of all fully invariant congruences on the free algebra $\mathcal{F}_\tau(X)$ forms a complete lattice, called $Con_{fi}\,\mathcal{F}_\tau(X)$, and that this lattice is a sublattice of the congruence lattice $Con\mathcal{F}_\tau(X)$.

6.3 The Algebraic Consequence Relation

In this section we define a consequence relation on sets of equations. There are two ways we can do this, which turn out to be equivalent. The first approach is a logical one, based on deductions of some equations from others according to certain rules of deduction. We write $\Sigma \vdash s \approx t$, read as "$\Sigma$ yields $s \approx t$," if there is a formal deduction of $s \approx t$ starting with identities in Σ, using the following five *rules of consequence* or *derivation* or *deduction rules*:

(1) $\emptyset \vdash s \approx s$,

(2) $\{s \approx t\} \vdash t \approx s$,

(3) $\{t_1 \approx t_2, t_2 \approx t_3\} \vdash t_1 \approx t_3$,

(4) $\{t_j \approx t_j' : 1 \leq j \leq n_i\} \vdash f_i(t_1, \ldots, t_{n_i}) \approx f_i(t_1', \ldots, t_{n_i}')$,   for every operation symbol f_i $(i \in I)$ (the *replacement rule*),

(5) Let $s, t, r \in W_\tau(X)$ and let $\tilde{s}, \tilde{t}$ be the terms obtained from s, t by replacing every occurrence of a given variable $x \in X$ by r. Then $s \approx t \vdash \tilde{s} \approx \tilde{t}$. (This is called the *substitution rule*.)

It should be clear that these rules (1) - (5) reflect the properties of a fully invariant congruence relation: the first three are the properties of an equivalence relation, the fourth describes the congruence property and the fifth the fully invariant property. Thus by Theorem 6.2.2 equational theories Σ are precisely sets of equations which are closed with respect to finite application of the rules (1) - (5). This equational approach will be used in Chapter 7, when we study term-rewriting systems.

The second approach to consequences is a more algebraic one. Here we say that $s \approx t$ follows from a set Σ of identities if $s \approx t$ is satisfied as an identity in every algebra $\mathcal{A}$ of type τ in which all equations from Σ are satisfied as identities. In this case we write $\Sigma \models s \approx t$.
The connection between these approaches is given by the Completeness and Consistency Theorem of equational logic: For $\Sigma \subseteq W_\tau(X) \times W_\tau(X)$ and $s \approx t \in W_\tau(X) \times W_\tau(X)$, we have

$$\Sigma \models s \approx t \quad \Leftrightarrow \quad \Sigma \vdash s \approx t.$$

The "$\Rightarrow$"-direction of this theorem is usually called completeness, since it means that any equation which is "true" is derivable; the "$\Leftarrow$"-direction is called consistency, in the sense that every derivable equation is "true." The equivalence here follows directly from Theorem 6.2.2.

6.4 Relatively Free Algebras

Let Y be any non-empty set of variable symbols, and let $\mathcal{F}_\tau(Y)$ be the absolutely free algebra of type τ generated by Y. Let $K \subseteq Alg(\tau)$ be a class of algebras, and let IdK be the set of all identities on the alphabet Y satisfied in K (that is, satisfied in every algebra $\mathcal{A}$ of K). Since IdK is a (fully invariant) congruence relation on the free algebra $\mathcal{F}_\tau(Y)$, we can form the quotient algebra $\mathcal{F}_\tau(Y)/IdK$, as we did in the proof of Theorem 6.2.2. This quotient of the absolutely free algebra also has some "freeness" properties, and is called the *relatively free algebra*, with respect to the class K, over the set Y. Notice that since the absolutely free algebra $\mathcal{F}_\tau(Y)$ is generated by Y, the quotient algebra is generated by the image $\overline{Y}$ of Y under the natural homomorphism $natIdK$.

Definition 6.4.1 Let Y be a non-empty set of variables, and K be a class of algebras of type τ. The algebra $\mathcal{F}_K(Y) := \mathcal{F}_\tau(Y)/IdK$ is called the *K-free algebra over Y* or the *free algebra relative to K* generated by $\overline{Y} = Y/IdK$.

Remark 6.4.2 1. Since $\mathcal{F}_K(Y)$ is generated by the set $\overline{Y} = Y/IdK = \{[y]_{IdK} | y \in Y\}$, instead of $\mathcal{F}_K(Y)$ we should actually write $\mathcal{F}_K(\overline{Y})$; but for notational convenience this is not usually done.

2. $\mathcal{F}_K(Y)$ exists iff $\mathcal{F}_\tau(Y)$ exists iff $Y \neq \emptyset$. Thus we have a K-free algebra of type τ over any non-empty set Y.

3. If $|Y| = |Z| \neq 0$, then $\mathcal{F}_K(Y) \cong \mathcal{F}_K(Z)$ under an isomorphism mapping $\overline{Y}$ to $\overline{Z}$. The proof of this fact is similar to the proof of Theorem 5.1.8. This means that (up to isomorphism) only the cardinality of the generating set Y is important, and not the particular choice of variable symbols.

4. In the case that Y is the set $X_n = \{x_1, \ldots, x_n\}$, we will write $\mathcal{F}_K(n)$ instead of $\mathcal{F}_K(X_n)$. In this case, our algebra is called the *K-free algebra on n generators*.

The algebra $\mathcal{F}_K(Y)$ also satisfies a relative "freeness" property corresponding to the absolutely free property of Theorem 5.1.7:

Theorem 6.4.3 *Let Y be any non-empty set of variables. For every algebra $A \in K \subseteq Alg(\tau)$ and every mapping $f : \overline{Y} \to A$, there exists a unique homomorphism $\hat{f} : \mathcal{F}_K(Y) \to A$ which extends f.*

Proof: Let $\mathcal{A}$ be in K, with a map $f : \overline{Y} \to A$. Let φ be the inclusion mapping from $\overline{Y}$ to the absolutely free algebra $\mathcal{F}_\tau(Y)$. By the freeness of this latter algebra, Theorem 5.1.7, there is a unique homomorphism $\overline{f} : \mathcal{F}_\tau(Y) \to \mathcal{A}$, which extends f, so that $\overline{f} \circ \varphi = f$. Now consider the homomorphism $nat\ IdK$ from $\mathcal{F}_\tau(Y)$ to $\mathcal{F}_K(Y)$. Its kernel is precisely IdK, and by Remark 6.1.2 this is contained in $ker\overline{f}$. Since we have two homomorphisms defined on $\mathcal{F}_\tau(Y)$ with the kernel of one contained in the kernel of the other, we can use the General Homomorphism Theorem to conclude that there is a homomorphism $\hat{f}$ from $\mathcal{F}_K(Y)$ to $\mathcal{A}$, with $\hat{f} \circ nat\ IdK = \overline{f}$. (See the diagram below.) Moreover $\hat{f} \circ natIdK|Y = \hat{f} \circ natIdK \circ \varphi = \hat{f} \circ \varphi = f$, and $\hat{f}$ extends f. By the homomorphism theorem $\hat{f}$ is uniquely determined by K and f. ∎

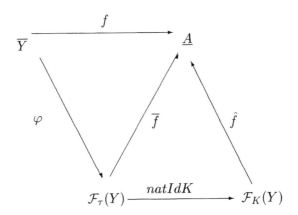

The free algebra with respect to a class K is in fact uniquely determined (up to isomorphism) by the relative freeness property expressed in Theorem 6.4.3.

Theorem 6.4.4 *Let K be a class of algebras of type τ. Let $\mathcal{F}$ be an algebra in K with the properties that $\mathcal{F}$ is generated by a subset $Y \subseteq F$ and that for any algebra $\mathcal{A} \in K$ and for any mapping $f : Y \to A$ there exists a homomorphism $\hat{f} : \mathcal{F} \to \mathcal{A}$ extending f. Then $\mathcal{F}$ is isomorphic to $\mathcal{F}_K(Y)$.*

Proof: Let φ be the inclusion mapping of Y into F. Since $\mathcal{F}$ is an algebra of type τ, by Theorem 5.1.7 the mapping $\varphi : Y \to F$ can be extended to a unique homomorphism $\overline{\varphi} : \mathcal{F}_\tau(Y) \to F$. This homomorphism is surjective, since its image contains $\varphi(Y)$, which generates $\mathcal{F}$. Thus by the Homomorphic Image Theorem, we see that $\mathcal{F}$ is isomorphic to the quotient $\mathcal{F}_\tau(Y)/ker\ \overline{\varphi}$. Since our relatively free algebra is just $\mathcal{F}_\tau(Y)/IdK$, it will suffice to prove that $ker\overline{\varphi} = IdK$.

First, since $\mathcal{F} \in K$ we have $IdK \subseteq ker\ \overline{\varphi}$, by Remark 6.1.2. Conversely, to show that $ker\overline{\varphi} \subseteq IdK$, we use Remark 6.1.2 again, and take any algebra $\mathcal{A}$ in K and any map $f : Y \to A$. Using the inclusion map $\varphi : Y \to F$ and the freeness assumption on $\mathcal{F}$, we see that there is a unique homomorphism $\hat{f} : \mathcal{F} \to \mathcal{A}$, extending f. But then $\hat{f} \circ \overline{\varphi}$ is a uniquely determined homomorphism from $\mathcal{F}_\tau(Y)$ to $\mathcal{A}$, and we have $ker\overline{\varphi} \subseteq ker(\hat{f} \circ \overline{\varphi})$. This shows that $ker\overline{\varphi}$ is contained in IdK, which completes our proof. ∎

Theorem 6.4.4 means that free algebras relative to $K \subseteq Alg(\tau)$ can be characterized and therefore defined by the freeness property from Theorem 6.4.3.

We conclude this section with an example of a relatively free algebra which is very useful in automata theory and other areas of theoretical Computer Science, and which we shall use in Chapter 8.

Example 6.4.5 Let τ be the type (2), so that we have one binary operation symbol which we will denote by f. Let K be the class of all semigroups, that is, of type (2) algebras which satisfy the associative identity, and let Y be any non-empty set of variables. We consider the K-free algebra $\mathcal{F}_K(Y)$ over Y. It is customary in this case to indicate the binary operation f by juxtaposition, writing xy for the term $f(x, y)$. Moreover, since $x(yz) \approx (xy)z$ is an identity of K, we see that any two terms in $W_\tau(Y)$ which have the same variables occurring in the same order are equivalent under the congruence IdK. This means that we can write any term in a *normal form* in which we omit the brackets. For example, the term $f(f(f(x, y), f(y, x)), f(z, y))$ can be written as $xyyxzy$. We refer to terms in this normal form as *words* on the alphabet Y. It is easy to verify that the set of all such words forms a semigroup under the operation of concatenation of words, called the *free semigroup on Y*, and usually denoted by Y^+. By adjoining an empty word e to Y^+ to act as an identity element, we can also form the *free monoid* Y^* on the alphabet Y.

6.5 Varieties

In this section we link together our two approaches to classes of algebras, the equational approach from the preceding sections and the algebraic approach from Chapters 1,3 and 4. We introduce operators **H**, **S** and **P** on classes of algebras, corresponding to the algebraic constructions of homomorphic images, subalgebras and product algebras studied earlier. A class of algebras which is closed under these operators is called a variety. Our main theorem in this section will show that in fact varieties are equivalent to equational classes.

Definition 6.5.1 We define the following operators on the set $Alg(\tau)$ of all algebras of a fixed type τ. For any class $K \subseteq Alg(\tau)$,

S(K) is the class of all subalgebras of algebras from K,
H(K) is the class of all homomorphic images of algebras from K,
P(K) is the class of all direct products of families of algebras from K,

I(K) is the class of all algebras which are isomorphic to algebras from K, **P**$_S(K)$ is the class of all subdirect products of families of algebras from K.

We can also combine these operators to produce new ones; we write **IP** for instance for the composition of **I** and **P**. Recall from Chapter 2 that an operator is a closure operator if it is extensive, monotone and idempotent. We first verify that some of our operators are in fact closure operators.

Lemma 6.5.2 *The operators* **H**, **S** *and* **IP** *are closure operators on the set* $Alg(\tau)$.

Proof: We will give a proof only for **H**; the others are quite similar. It is clear from the definition that for any subclasses K and L of $Alg(\tau)$, the inclusion $K \subseteq L$ implies **H**$(K) \subseteq$ **H**(L), and since any algebra is a homomorphic image of itself under the identity homomorphism, we always have $K \subseteq$ **H**(K). For the idempotency of **H**, we note that by the extensivity and monotonicity we have **H**$(K) \subseteq$ **H**(**H**$(K))$. Conversely, let $\mathcal{A}$ be in **H**(**H**$(K))$. Then there exists an algebra $\mathcal{B} \in$ **H**$(\mathcal{K})$ and a surjective homomorphism $\varphi : \mathcal{B} \to \mathcal{A}$. For $\mathcal{B} \in$ **H**$(\mathcal{K})$ there exists an algebra $\mathcal{C} \in K$ and a surjective homomorphism $\psi : \mathcal{C} \to \mathcal{B}$. Then the composition $\varphi \circ \psi : \mathcal{C} \to \mathcal{A}$ is also a surjective homomorphism, and thus $\mathcal{A} \in$ **H**$(\mathcal{K})$.   ■

Note however that the operator **P** is not a closure operator, since it is not idempotent: $A_1 \times (A_2 \times A_3)$ is not equal to $(A_1 \times A_2) \times A_3$, although they are isomorphic.

Definition 6.5.3 A class $K \subseteq Alg(\tau)$ is called a *variety* if K is closed under the operators **H**, **S** and **P**; that is, if **H**$(K) \subseteq K$, **S**$(K) \subseteq K$ and **P**$(K) \subseteq K$.

We investigate now the properties of these operators, especially how they combine with each other.

Lemma 6.5.4 *Let K be a class of algebras of type τ. Then*

(i) **SH**$(K) \subseteq$ **HS**(K),

(ii) **PS**$(K) \subseteq$ **SP**(K),

(iii) **PH**$(K) \subseteq$ **HP**(K).

Proof: (i) Let $\mathcal{A}$ be an element of **SH**(K). Then $\mathcal{A}$ is a subalgebra of an algebra $\mathcal{B}$, which in turn is a homomorphic image of an algebra $\mathcal{C}$ in K, under a surjective homomorphism $\varphi : \mathcal{C} \to \mathcal{B}$. Since $\mathcal{A} \subseteq \mathcal{B}$, by Theorem 3.1.3 the preimage $\varphi^{-1}(\mathcal{A})$ is a subalgebra of $\mathcal{C}$. This preimage satisfies $\varphi(\varphi^{-1}(\mathcal{A})) = \mathcal{A}$, making $\mathcal{A}$ a homomorphic image of the subalgebra $\varphi^{-1}(\mathcal{A})$ of $\mathcal{C}$. Therefore $\mathcal{A} \in \textbf{HS}(\mathcal{K})$.

(ii) If $\mathcal{A} \in \textbf{PS}(\mathcal{K})$ then $\mathcal{A} = \underset{j \in J}{\Pi} \mathcal{B}_j$ for some algebras $\mathcal{B}_j$ in $\textbf{S}(K)$, and for each $j \in J$ there is an algebra $\mathcal{C}_j$ in K with $\mathcal{B}_j \subseteq \mathcal{C}_j$. Since $\underset{j \in J}{\Pi} \mathcal{B}_j$ is then a subalgebra of $\underset{j \in J}{\Pi} \mathcal{C}_j$, we have $\mathcal{A} \in \textbf{SP}(\mathcal{K})$.

(iii) If $\mathcal{A} \in \textbf{PH}(\mathcal{K})$ then $\mathcal{A} = \underset{j \in J}{\Pi} \mathcal{B}_j$ for some algebras $\mathcal{B}_j \in \textbf{H}(K)$, and for each $j \in J$ there is an algebra $\mathcal{C}_j$ in K and a surjective homomorphism $\varphi_j : \mathcal{C}_j \to \mathcal{B}_j$. For the projection mapping $p_\ell : \underset{j \in J}{\Pi} \mathcal{C}_j \to \mathcal{C}_\ell$, the composition $\varphi_j \circ p_\ell : \underset{j \in J}{\Pi} \mathcal{C}_j \to \mathcal{B}_\ell$ is a surjective homomorphism. By Remark 4.1.2 there exists a homomorphism $\underset{j \in J}{\Pi} \mathcal{C}_j \to \underset{j \in J}{\Pi} \mathcal{B}_j$ which is surjective. This makes $\mathcal{A} = \underset{j \in J}{\Pi} \mathcal{B}_j \in \textbf{HP}(K)$. ∎

Using this Lemma we can obtain a characterization of the smallest variety to contain a class K of algebras.

Theorem 6.5.5 *For any class K of algebras of type τ, the class* $\textbf{HSP}(K)$ *is the least (with respect to set inclusion) variety which contains K.*

Proof: We show first that $\textbf{HSP}(K)$ is indeed a variety, that is, that it is closed under application of **H**, **S** and **P**. We have $\textbf{H}(\textbf{HSP}(K)) = \textbf{HSP}(K)$ by the idempotence of **H**, and $\textbf{S}(\textbf{HSP}(K)) \subseteq \textbf{H}(\textbf{SSP}(K)) = \textbf{HSP}(K)$ by Lemma 6.5.4 (i) and the idempotence of **S**. For **P**, we have $\textbf{P}(\textbf{HSP}(K)) \subseteq$ $(\textbf{HPSP}(K)) \subseteq \textbf{HSPP}(K) \subseteq \textbf{HSIPIP}(K) = \textbf{HSIP}(K) \subseteq \textbf{HSHP}(K) \subseteq$ $\textbf{HHSP}(K) = \textbf{HSP}(K)$, using properties from 6.5.2 and 6.5.4.

Thus $\textbf{HSP}(K)$ is a variety. Now let K' be any variety which contains K. Then $\textbf{HSP}(K) \subseteq \textbf{HSP}(K') \subseteq K'$, since as a variety K' is closed under all three operators. ∎

For any class K of algebras of the same type, the variety $\mathbf{HSP}(K)$ from Theorem 6.5.5 is called the variety generated by K. It is often denoted by $V(K)$. When K consists of a single algebra $\mathcal{A}$, we usually write $V(\mathcal{A})$ for the variety generated by K.

By Theorem 6.5.5, we have $K \subseteq \mathbf{HSP}(K)$ for any class K. When K is a variety, then closure of K under each of $\mathbf{H}$, $\mathbf{S}$ and $\mathbf{P}$ gives us $\mathbf{HSP}(K) \subseteq K$ as well, and hence $\mathbf{HSP}(K) = K$ for K a variety. Conversely, whenever $\mathbf{HSP}(K) = K$, we must have K a variety, as shown in the first part of the proof of Theorem 6.5.5. This gives us the following Corollary.

Corollary 6.5.6 *A class K of algebras of type τ is a variety if and only if* $\mathbf{HSP}(K) = K$. ∎

In Chapter 4 (Theorem 4.2.8) we have already remarked that every algebra is isomorphic to a subdirect product of subdirectly irreducible algebras. Now we can prove a stronger result, that every algebra of a variety K is isomorphic to a subdirect product of subdirectly irreducible algebras from K.

Theorem 6.5.7 *Every algebra of a variety K is isomorphic to a subdirect product of subdirectly irreducible algebras from K.*

Proof: By Theorem 4.2.8 every algebra $\mathcal{A}$ in K is isomorphic to a subdirect product of some subdirectly irreducible algebras $\mathcal{A}_j$. By the remark following Theorem 4.2.4, each $\mathcal{A}_j$ is in fact isomorphic to a quotient algebra of $\mathcal{A}$, and is thus a homomorphic image of $\mathcal{A}$. Since $\mathcal{A} \in K$ and K is a variety, we see that $\mathcal{A}_j \in \mathbf{H}(K) \subseteq K$ too. ∎

Our goal is to prove that equational classes and varieties are the same thing. The next Lemma gives one direction of this equivalence, that any equational class of algebras is a variety.

Lemma 6.5.8 *Let K be an equationally definable class of algebras of type τ. Then K is a variety.*

Proof: Let K be an equationally definable class of algebras of type τ. This means that there exists a set Σ of equations of type τ, over the alphabet X, such that $K = Mod\Sigma$. To see that $Mod\Sigma$ is closed under $\mathbf{H}$, $\mathbf{S}$ and $\mathbf{P}$, let $s \approx t \in \Sigma$ be any identity in Σ. Then we have $s^{\mathcal{A}} = t^{\mathcal{A}}$ for any algebra $\mathcal{A} \in \mathcal{K}$.

If $\mathcal{B}$ is a subalgebra of an $\mathcal{A}$ in K, then by Theorem 5.2.4 $t^{\mathcal{B}} = t^{\mathcal{A}}|B$ and therefore $s^{\mathcal{B}} = t^{\mathcal{B}}$; thus $s \approx t$ holds in $\mathcal{B}$ and $\mathcal{B} \in \mathcal{K}$. Therefore $\mathbf{S}(K) \subseteq K$. In a similar way, using Theorem 5.2.4 (ii), we can show that $\mathbf{H}(K) \subseteq K$. Lastly, suppose $\mathcal{A}_j \in K$ and that $\mathcal{A}_j$ satisfies $s \approx t$ for each $j \in J$. Then for $\underline{a}_1, \ldots, \underline{a}_n \in C = \prod_{j \in J} \mathcal{A}_j$ we have $s^{\mathcal{A}_j}(a_1(j), \ldots, a_n(j)) = t^{\mathcal{A}_j}(a_1(j), \ldots, a_n(j)) \Rightarrow (s^C(\underline{a_1}, \ldots, \underline{a_n}))(j) = (t^C(\underline{a_1}, \ldots, \underline{a_n}))(j)$ for all $j \in J$ and thus $s^C = t^C$. Therefore the product C of algebras in K is in K, which shows that $\mathbf{P}(K) \subseteq K$. ∎

Before we can show the other direction of our equivalence, we need one more fact. We show that for any class K of algebras of type τ, and for any set Y of variables, the free algebra $\mathcal{F}_K(Y)$ with respect to K belongs to $\mathbf{ISP}(K)$.

Theorem 6.5.9 *For every class* $K \subseteq Alg(\tau)$ *and every non-empty set* Y *of variables, the relatively free algebra* $\mathcal{F}_K(Y)$ *is in* $\mathbf{ISP}(K)$.

Proof: We want to use Theorem 4.2.4 to write our algebra $\mathcal{F}_K(Y)$ as a sub-direct product. To do this, we first need to verify that the intersection of all the congruences on $\mathcal{F}_K(Y)$ is the identity relation. First note that $\mathcal{F}_K(Y)$ is the quotient $\mathcal{F}_\tau(Y)/IdK$. Any congruence on this algebra is the kernel of a homomorphism onto some algebra $\mathcal{A}$ in K, and any such homomorphism is the extension $\hat{f}$ of some mapping $f : Y \to A$. Thus it is enough to show that the intersection of the kernels of all such $\hat{f}$ on $\mathcal{F}_\tau(Y)$ is the identity relation on $\mathcal{F}_\tau(Y)/IdK$. But this holds, by Remark 6.1.2 and the definition of IdK.

Now by Theorem 4.2.4 the algebra $\mathcal{F}_\tau(Y)/IdK$ is isomorphic to a subdirect product of algebras $(\mathcal{F}_\tau(Y)/IdK)/((\ker \hat{f})/IdK)$. For each of these algebras, using the Homomorphic Image Theorem and the Second Isomorphism Theorem, and the definition of $\hat{f}$, we obtain $(\mathcal{F}_\tau(Y)/IdK)/((\ker \hat{f})/IdK) \cong \mathcal{F}_\tau(Y)/\ker \hat{f} \cong \hat{f}(\mathcal{F}_\tau(Y))$. This last algebra is a subalgebra of the algebra $\mathcal{A}$, which is in K, putting it in $\mathbf{S}(K)$. Thus the relatively free algebra is isomorphic to a subdirect product of algebras which in turn are isomorphic to subalgebras of K. Altogether we have $\mathcal{F}_\tau(Y)/IdK = \mathcal{F}_K(Y) \in \mathbf{ISP}(\mathbf{IS}(K))$. Using the properties of Lemma 6.5.4 (which hold for $\mathbf{I}$ as well as $\mathbf{H}$) and Lemma 6.5.2, we get $\mathcal{F}_K(Y) \subseteq \mathbf{ISP}(\mathbf{S}(K)) \subseteq \mathbf{ISP}(K)$. ∎

Since varieties are also closed under the operators $\mathbf{I}$, $\mathbf{S}$ and $\mathbf{P}$, it follows from this theorem that every variety K contains all the relatively free algebras $\mathcal{F}_K(Y)$, for each non-empty set Y.

Next we show that the identities satisfied in a class K of algebras of type τ are exactly the identities satisfied in the free algebra $\mathcal{F}_K(X)$, where X as usual is our countably infinite set of variables.

Lemma 6.5.10 *Let K be a variety of algebras of type τ, and let s and t be terms in $W_\tau(X)$. Then*

$$K \models s \approx t \quad \Leftrightarrow \quad \mathcal{F}_K(X) \models s \approx t.$$

Proof: Since K is a variety, we know by the remark just above that the relatively free algebra $\mathcal{F}_K(X)$ is in K. This means that any identity of K must in particular hold in $\mathcal{F}_K(X)$, giving us one direction of the claim.

If conversely $\mathcal{F}_K(X)$ satisfies $s \approx t$, then $s^{\mathcal{F}_K(X)} = t^{\mathcal{F}_K(X)}$. Since $\mathcal{F}_K(X)$ is the quotient of $\mathcal{F}_\tau(X)$ by IdK, this forces $[s]_{IdK} = [t]_{IdK}$; and from this, we have $(s,t) \in \ker natIdK = IdK$ and so K satisfies $s \approx t$.   ∎

With these results, we are ready to prove our main theorem, sometimes referred to as Birkhoff's Theorem.

Theorem 6.5.11 *(Main Theorem of Equational Theory) A class K of algebras of type τ is equationally definable if and only if it is a variety.*

Proof: We have already proved one direction of this theorem, in Lemma 6.5.8. We will now prove the converse, that any variety K is also an equational class; specifically, we will show that K equals the equational class K' := $ModIdK$. Again by Lemma 6.5.8, this class is a variety, and we clearly have $K \subseteq ModIdK = K'$. Applying the operator Id to $K' = ModIdK$ gives us $IdK = IdModIdK' = IdK'$, since $IdMod$ is a closure operator. But then we have $\mathcal{F}_K(Y) = \mathcal{F}_\tau(Y)/IdK = \mathcal{F}_\tau(Y)/IdK' = \mathcal{F}_{K'}(Y)$, for every non-empty set Y of variables. Now let $\mathcal{A}$ be any algebra in K', and choose a set Y of variables such that $|Y| = |A|$. There will be a mapping $f : Y \to A$ which is surjective, and we know by Theorem 6.4.3 that this mapping has a unique extension to a homomorphism from $\mathcal{F}_{K'}(Y)$ to $\mathcal{A}$, also surjective. This makes $\mathcal{A}$ a homomorphic image of $\mathcal{F}_{K'}(Y)$, which in turn is equal to $\mathcal{F}_K(Y)$. But this relatively free algebra is in the variety K, and K is closed under homomorphic images, so we have $\mathcal{A}$ in K. This shows that $K' \subseteq K$, and finishes our proof that $K = K' = ModIdK$.   ∎

Note that in the proof of Theorem 6.5.11, we needed the existence of relatively free algebras over non-empty variable sets of arbitrary cardinality. However, as we commented in Chapter 5, it is in some sense sufficient to have absolutely or relatively free algebras over variable sets of finite or countably infinite cardinality. The following Lemma explains this more precisely.

Lemma 6.5.12 *Let K be a variety of type τ and let Y be any non-empty variable set. The relatively free algebra $\mathcal{F}_K(Y)$ with respect to K over Y is isomorphic to a subdirect product of the algebras $\mathcal{F}_K(E)$, for $E \subseteq Y$ non-empty and finite.*

Proof: For each variable y in Y we define $\overline{y} := [y]_{IdK}$. For every subset $E \subseteq Y$, we set $\overline{E} := \{\overline{y} \mid y \in E\}$. Let $\mathcal{U}(\overline{E})$ be the subalgebra of $\mathcal{F}_K(Y)$ generated by $\overline{E}$. It is easy to see that $\mathcal{U}(\overline{E})$ and $\mathcal{F}_K(\overline{E})$ are isomorphic. Therefore, it is enough to show that $\mathcal{F}_K(Y)$ is isomorphic to a subdirect product of the algebras $\mathcal{U}(\overline{E})$.

For each such E consider a mapping $\varphi_E : \overline{Y} \to U(\overline{E})$ with φ_E defined to be the identity mapping on $\overline{E}$. This mapping φ_E has a unique homomorphic extension $\hat{\varphi}_E$, which is surjective and also the identity mapping on both $\overline{E}$ and the subalgebra $\mathcal{U}(\overline{E})$ it generates. Every term depends only on finitely many variables. Therefore, for every pair s, t in $F_K(Y)$ there exists a finite subset $E \subseteq X$ with $s, t \in U(\overline{E})$. If $s \neq t$ then $(s, t) \notin \ker \hat{\varphi}_E$, since $\hat{\varphi}_E(s) = s$ and $\hat{\varphi}_E(t) = t$. This shows that the intersections of the kernels of all such homomorphisms $\hat{\varphi}_E$ is the identity relation on $\mathcal{F}_K(Y)$. This means that we can use Theorem 4.2.4 to express our algebra $\mathcal{F}_K(Y)$ as isomorphic to a subdirect product of the algebras $\mathcal{F}_K(Y)/\ker(\hat{\varphi}_E)$. Finally, since $\mathcal{F}_K(Y)/\ker(\hat{\varphi}_E) \cong \mathcal{U}(\overline{E})$, our claim is proved. ∎

As a consequence of this theorem, we have the following useful result about generating sets for a variety K. Any variety is generated by its relatively free algebra on a countably infinite set of generators, or by the collection of relatively free algebras on n generators, for each natural number $n \geq 1$.

Theorem 6.5.13 *For every variety K,*

$$K = \mathbf{HSP}(\{\mathcal{F}_K(n) \mid n \in \mathbb{N}, n \geq 1\}) = \mathbf{HSP}(\{\mathcal{F}_K(X)\}).$$

Proof: As we saw following Theorem 6.5.9, the fact that K is a variety means that it contains the relatively free algebras with respect to K on any finite or countably infinite set of variables. Thus the two generating sets are contained in K. For the converse, let $\mathcal{A}$ be any algebra in K. If we choose a set Y whose cardinality is greater than the cardinality of A, we can make a surjective homomorphism from $\mathcal{F}_K(Y)$ onto $\mathcal{A}$. Then using Lemma 6.5.12 we can express $\mathcal{F}_K(Y)$ as a subdirect product of relatively free algebras with respect to K on sets E of finite cardinality. Thus $\mathcal{A}$ is a homomorphic image of a subdirect product of the algebras in our generating set. This shows that K is contained in $\mathbf{HSP}(\{\mathcal{F}_K(n) \mid n \in \mathbb{N}\})$, and hence we have equality.

Finally, we know that for every $n \in \mathbb{N}$ the algebra $\mathcal{F}_K(n)$ is isomorphic to a subalgebra of $\mathcal{F}_K(X)$. Thus $\mathbf{HSP}(\{\mathcal{F}_K(X)\})$ contains all the elements of the first generating set, and so contains all of K as well.   ∎

A useful consequence of this theorem is that two varieties K and K' are equal if all the free algebras $\mathcal{F}_K(n) = \mathcal{F}_{K'}(n)$ are equal, for every natural number n.
We conclude this section with another application of free algebras.

Definition 6.5.14 An algebra $\mathcal{A}$ of type τ is called *locally finite* if every finitely generated subalgebra of $\mathcal{A}$ is finite. A class K of algebras of the same type is called *locally finite* if every member of K is locally finite.

Theorem 6.5.15 *A variety K is locally finite iff the relatively free algebra $F_K(Y)$ is finite for every non-empty finite set Y.*

Proof: Since K is a variety, it contains the K-free algebra on any non-empty set Y. Moreover, this algebra is generated by the set Y. So if K is locally finite, then by definition any finite set Y determines a finite algebra $\mathcal{F}_K(Y)$.

For the converse, let $\mathcal{A}$ be a finitely generated algebra from K, with a finite set $B \subseteq A$ of generators. Now we choose an alphabet Y in such a way that there exists a bijection $\alpha : Y \to B$. This bijection can be extended to a homomorphism $\hat{\alpha} : \mathcal{F}_K(Y) \to \mathcal{A}$. The image $\hat{\alpha}(\mathcal{F}_K(Y))$ is then a subalgebra of $\mathcal{A}$ containing B, and hence must be equal to $\mathcal{A}$. Therefore $\hat{\alpha}$ is surjective, and as $\mathcal{F}_K(Y)$ is finite so is $\mathcal{A}$.   ∎

Theorem 6.5.16 *Let $K \subseteq Alg(\tau)$ be a finite set of finite algebras. Then the variety $V(K)$ generated by K is a locally finite variety.*

Proof: We will verify first that the class $\mathbf{P}(K)$ is locally finite. We define an equivalence relation $\sim$ on $W_\tau(\{x_1, \ldots, x_n\})$ by $p \sim q$ iff the term operations corresponding to p and q are the same for each member of K. The finiteness condition shows that $\sim$ has finitely many equivalence classes. Subalgebras of $\mathbf{P}(K)$ are also finite, since only finite sets can be produced from finite sets using finitary operations. Since every finitely generated member of $V(K) = \mathbf{HSP}(K)$ is a homomorphic image of a finitely generated member of $\mathbf{SP}(K)$, we see that $V(K)$ is locally finite. ∎

6.6 The Lattice of All Varieties

In Section 6.1 we proved that the collection of all equational classes of algebras of a fixed type τ (over a countably infinite alphabet) forms a complete lattice $\mathcal{L}(\tau)$, which is dually isomorphic to the lattice of all equational theories of type τ. One can prove that these lattices are in fact algebraic (see Section 2.1). The greatest element of the lattice $\mathcal{L}(\tau)$ of all varieties of type τ is the variety consisting of all algebras of type τ; this variety is denoted by $Alg(\tau)$. Clearly, $Alg(\tau) = Mod\{x \approx x\}$. The least element in the lattice $\mathcal{L}(\tau)$ is the trivial variety T consisting exactly of all one-element algebras of type τ; we have $T = Mod\{x \approx y\}$. Dually, the greatest element in the lattice $\mathcal{E}(\tau)$ of all equational theories of type τ is the equational theory generated by $\{x \approx y\}$ (using the five derivation rules), and the least element in $\mathcal{E}(\tau)$ is the equational theory generated by $\{x \approx x\}$. Clearly, the former consists of all equations of type τ.

A subclass W of a variety V which is also a variety is called a *subvariety* of V. The variety V is a *minimal*, or *equationally complete*, variety if V is not trivial but the only subvariety of V not equal to V is the trivial variety. We show now that every non-trivial $V \in \mathcal{L}(\tau)$ contains a minimal subvariety.

Theorem 6.6.1 *Let V be a non-trivial variety. Then V contains a minimal subvariety.*

Proof: Since $V = ModIdV$, the set of all identities of V defines V and by Theorem 6.2.2 the set IdV is a fully invariant congruence relation on $\mathcal{F}_\tau(X)$. Since V is non-trivial this fully invariant congruence relation is not all of $F_\tau(X) \times F_\tau(X)$. Since $F_\tau(X) \times F_\tau(X)$ is the fully invariant congruence generated by any pair (x, y) with $x \neq y$, it follows that as a fully invariant congruence $F_\tau(X) \times F_\tau(X)$ is finitely generated. Using Zorn's Lemma we can extend IdV to a maximal fully invariant congruence relation. By Theorem 6.2.2, the fact that the set $Con_{fi}\mathcal{F}_\tau(X)$ forms a sublattice of $Con\mathcal{F}_\tau(X)$ and the properties of the Galois-correspondence (Id, Mod), this gives a minimal variety which is included in V.   ∎

It is known that the variety of all Boolean algebras and the variety of all distributive lattices are minimal. A variety of groups is minimal if and only if it is abelian of some prime exponent (that is, consists of all abelian groups satisfying the identity $x^p \approx e$ where p is a fixed prime number). For semigroup varieties of type (2), using the convention of replacing the binary operation symbol by juxtaposition, we have the following minimal varieties:

$SL = Mod\{x(yz) \approx (xy)z,\ xy \approx yx,\ x^2 \approx x\}$, the variety of semilattices,
$LZ = Mod\{xy \approx x\}$, the variety of left-zero semigroups,
$RZ = Mod\{xy \approx y\}$, the variety of right-zero semigroups,
$Z = Mod\{xy \approx zt\}$, the variety of zero semigroups,   and
$A_p = Mod\{x(yz) \approx (xy)z, xy \approx yx, x^p y \approx y\}$, the variety of abelian groups of prime exponent p.

If V is a given variety of type τ, then the collection of all subvarieties of V forms a complete lattice $\mathcal{L}(V) := \{W | W \in \mathcal{L}(\tau) \text{ and } W \subseteq V\}$. This lattice is called the *subvariety lattice* of V.

6.7   Finite Axiomatizability

An old question in universal algebra is whether or not the identities of a finite algebra can be derived (using the deduction rules of Section 6.3) from a finite set of identities of the algebra. When this can be done, we say that the algebra, or the variety it generates, is *finitely axiomatizable* or *finitely based*, and we refer to the finite set of identities as a *basis* for the identities of the algebra or variety. R. C. Lyndon proved in [72] that every two-element algebra is finitely axiomatizable. We know that finite groups (see S. Oates and M. B. Powell, [84]), finite rings (see R. L. Kruse, [68]) and finite algebras generating a variety in which all congruence lattices are distributive (see K. A Baker, [5]) are all finitely axiomatizable. It was quite surprising when in 1954 R. C. Lyndon constructed in [73] a seven-element algebra with one binary and one nullary operation whose identities are not finitely based. The smallest such example is a non-finitely axiomatizable three-element algebra of type (2) found by V. L. Murskij in [83]; this is the algebra with base set $A = \{0, 1, 2\}$ and binary operation given by

	0	1	2
0	0	0	0
1	0	0	1
2	0	2	2

P. Perkins also constructed a six-element semigroup whose identities are not finitely based in [86]. An example of a finite non-associative ring whose identities are not finitely based was constructed by S. V. Polin in [93].

Although the set of all identities of an algebra $\mathcal{A}$ may not be finitely axiomatizable, we can show that the set of all identities which use only a finite number of variables is finitely based.

Theorem 6.7.1 *Let $\mathcal{A}$ be a finite algebra of type τ, and let X_m be a finite set of variables. Then $W_\tau(X_m)^2 \cap Id\mathcal{A}$ is finitely based.*

Proof: We set $\theta := W_\tau(X_m)^2 \cap Id\mathcal{A}$. Then θ is a congruence of the absolutely free algebra $\mathcal{F}_\tau(X_m)$ which defines the relatively free algebra $\mathcal{F}_{V(\mathcal{A})}(X_m)$. Since $\mathcal{A}$ is finite, and since a pair of terms (p, q) is in θ iff the induced term operations satisfy $p^{\mathcal{A}} = q^{\mathcal{A}}$, there are only finitely many equivalence classes of θ. We choose one representative from each equivalence class, and form the set $A = \{q_1, \ldots, q_n\}$ of representatives. Using this set we define the following finite set Σ of identities:

$$\Sigma = \{x \approx y \mid x, y \in X_m \text{ and } (x, y) \in \theta\} \cup$$

$$\{q_i \approx x \mid x \in X_m, q_i \in Q \text{ and } (x, q_i) \in \theta\}$$

$$\cup \{f_i(q_{i_1}, \ldots, q_{i_{n_i}}) \approx q_{i_{n_i}+1} \mid q_j \in Q \text{ and } (f_i(q_{i_1}, \ldots, q_{i_{n_i}}), q_{i_{n_i}+1}) \in \theta\}.$$

Clearly Σ is finite, and is contained in $Id\mathcal{A}$, so it will suffice to prove that Σ is a basis for $Id\mathcal{A}$. By induction on the number of operation symbols occurring in a term p, it can be shown that

$$\text{if } (p, q_i) \in \theta, \text{ then } p \approx q_i \text{ is derivable from } \Sigma.$$

It follows from this that for arbitrary terms p and q,

if $(p, q) \in \theta$, then $p \approx q$ is derivable from Σ.   ■

We conclude this section by stating without proof some important theorems in the area of finite axiomatizability. These theorems involve some properties of a variety, usually properties of the subdirectly irreducible elements and of the congruence lattices of the algebras in the variety. Properties of the congruence lattices of a variety are often definable by what are called Mal'cev-type conditions, and will be studied in more detail in Chapter 9. We define here only the two properties we need to state our theorems.

Definition 6.7.2 A variety V is called *congruence distributive* if for every algebra $\mathcal{A}$ in V, the congruence lattice $Con\mathcal{A}$ satisfies the distributive law. V is called *congruence meet-semidistributive* if for every algebra $\mathcal{A}$ in V, the congruence lattice $Con\mathcal{A}$ satisfies the following *meet-semidistributive law*:

$$\theta \wedge \psi = \theta \wedge \varphi \implies \theta \wedge \psi = \theta \wedge (\psi \vee \varphi).$$

Note that congruence distributive varieties are also congruence meet-semidistributive.

We also need the concept of the residual bound of a variety or an algebra.

Definition 6.7.3 Let V be a variety. We denote by $\kappa(V)$ the least cardinal number λ such that every subdirectly irreducible algebra in V has cardinality less than λ, if there is such a cardinal number; in this case we say that V is *residually small*. If no such cardinal number exists, we let $\kappa(V) = \infty$, and V is said to be residually large. The cardinal number $\kappa(V)$ is called the *residual bound* of the variety V. The residual bound of an algebra $\mathcal{A}$ is defined to be the residual bound of the variety $V(\mathcal{A})$ generated by $\mathcal{A}$. A variety V is called *residually finite* if all its subdirectly irreducible algebras are finite.

An important result in this area is Baker's Theorem.

Theorem 6.7.4 *(Baker's Theorem)([5]) A congruence distributive variety of finite type which is residually finite is finitely based.*   ■

R. McKenzie proved in [77] that a locally finite variety V having only finitely many subdirectly irreducible elements and having an additional property called *definable principal congruences* is finitely axiomatizable. McKenzie also proved in [81] that there are only countably many values possible for the residual bound of a finite algebra. This residual bound must be either ∞ or one of the following cardinals:

$$0, \ 3, \ 4, \ \ldots, \ \omega, \ \omega_1, \ (2^\omega)^+,$$

where ω is the cardinal number of the set of natural numbers (that is, the first infinite cardinal), $\omega_1 = \omega^+$ is the next largest cardinal number after ω, and $(2^\omega)^+$ is the successor cardinal of the cardinal of the continuum. Recently R. Willard proved the following important theorem about finite axiomatizability, which generalizes Baker's Theorem.

Theorem 6.7.5 ([119]) *If a variety is both congruence meet-semidistributive and residually finite, then it is finitely axiomatizable.* ∎

6.8 Exercises

6.8.1. Verify that the pair (Id, Mod) forms a Galois-connection between the sets $Alg(\tau)$ and $W_\tau(X)^2$.

6.8.2. Prove that the set of all fully invariant congruence relations $Con_{fi}\mathcal{A}$ of an algebra $\mathcal{A}$ forms a sublattice of the lattice $Con\mathcal{A}$ of all congruence relations on $\mathcal{A}$.

6.8.3. Determine all elements of $\mathcal{F}_{RB}(\{x,y\})$, where $RB = Mod\{x(yz) \approx (xy)z, xyz \approx xz, x^2 \approx x\}$ is the type (2) variety of all rectangular bands.

6.8.4. Let L be the variety of all lattices. Determine all elements of $\mathcal{F}_L(\{x\})$ and of $\mathcal{F}_L(\{x,y\})$.

6.8.5. Prove that $Id\mathbf{HSP}(\mathcal{A}) = Id\mathcal{A}$.

6.8.6. Show that $\mathbf{ISP}(K)$ is the smallest class containing K and closed under $\mathbf{I}$, $\mathbf{S}$ and $\mathbf{P}$.

6.8.7. Let V be a variety and let Y and Z be non-empty sets with $|Y| \leq |Z|$. Show that $\mathcal{F}_V(Y)$ can be embedded in $\mathcal{F}_V(Z)$ in a natural way.

6.8.8. Prove that in the variety of all algebras of type $\tau = (3)$ defined by the identities

$$f(x, x, z) \approx f(x, z, x) \approx f(z, x, x) \approx z,$$
$$f(f(x_1, y_1, z_1), f(x_2, y_2, z_2), f(x_3, y_3, z_3)) \approx$$
$$f(f(x_1, x_2, x_3), f(y_1, y_2, y_3), f(z_1, z_2, z_3)),$$

all algebras are free.

6.8.9. Using the normal form for semigroup terms from Example 6.4.5, describe the free semilattice on the set X_n of n generators.

Chapter 7

Term Rewriting Systems

When a relation θ is a congruence relation on an algebra $\mathcal{A}$ of type τ, we can form the quotient algebra $\mathcal{A}/\theta$. The Homomorphic Image Theorem tells us that any homomorphic image of an algebra $\mathcal{A}$ is isomorphic to such a quotient algebra of $\mathcal{A}$. Another important use of the quotient algebra was seen in Section 6.4: taking $\mathcal{A}$ to be the absolutely free algebra $\mathcal{F}_\tau(X)$ of type τ and θ to be the set IdK of identities of a class K of algebras of type τ, we formed the quotient algebra $\mathcal{F}_\tau(X)/IdK$, called the relatively free algebra with respect to K over the set X.

Quotient algebras and relatively free algebras are examples of a more general construction. Given any set A and any equivalence relation on A, we can form the quotient set A/θ of all the equivalence classes with respect to θ. Elements of A/θ are classes or sets of equivalent elements of A, but calculations on such classes are always done by choosing a representative element from each class, and calculating with these representatives. This means that it is important to be able to check whether two elements belong to the same equivalence class. Given any two elements a and b of A, we must check whether the pair (a, b) is in our original equivalence relation.

In this chapter we consider this basic problem from a constructive point of view. In general, a problem is said to be *effectively solvable* if there is an algorithm that provides the answer in a finite number of steps, no matter what the particular inputs are. We want the maximum number of steps the algorithm will take to be predictable in advance. An effective solution to a problem that has a "yes" or "no" answer is called a *decision procedure*, and a problem which has a decision procedure is said to be *decidable*.

Thus for an equivalence relation θ on a set A we may ask whether it is decidable for any two elements $a, b \in A$ if $(a, b) \in \theta$ is true or not. In the particular case that $\mathcal{A} = W_\tau(X)$ and $\theta = IdK$ for a variety K of algebras of type τ, our question becomes: given two terms s and t of type τ, is it decidable whether $s \approx t$ is an identity in K or not? This problem is also called the *word problem* for the variety K or for the fully invariant congruence IdK. Only in cases where the word problem is decidable can computations be carried out using representatives. This shows the fundamental significance of the word problem. In the more general setting, for an arbitrary equivalence relation on a set A, the equivalence problem has been shown to be undecidable (see M. Davis, [15]). In this chapter we will examine methods for deciding the word problem of a variety.

Since every equivalence or congruence is a binary relation, we begin with a description in Section 7.1 of the equivalence relation generated by an arbitrary binary relation. Then we shift from the algebraic model of a set ρ of ordered pairs on a set A to a more machine-oriented model: we think of a pair (a, b) in ρ as a rule $a \to b$ which says that we can transform or reduce a to b. Writing $\to$ for the relation ρ, we consider *reduction systems* $(A; \to)$. Sections 7.1 and 7.2 study the properties of termination and confluence for such systems. Are infinite sequences of reductions based on $\to$ possible? Is it possible to reduce an element a by more than one sequence of reductions, and if so do such reductions lead to equivalent results? Section 7.3 examines the special case that our base set A is the set $W_\tau(X)$ of all terms of a fixed type, in which case a reduction system is called a term rewriting system. The important problem of testing for termination of a reduction system is considered in Section 7.4.

7.1   Confluence

We shall be interested in testing for equivalence of elements with respect to an equivalence relation on a set A. Any equivalence relation on A is a binary relation on A, and any binary relation generates an equivalence relation. To describe the equivalence relation generated by a relation ρ on A, we need the following notation:

$\rho^R := \rho \cup \triangle_A$ is the reflexive closure of ρ,

$\rho^S := \rho \cup \rho^{-1}$ is the symmetric closure of ρ,

$\rho^{(0)} := \triangle_A, \qquad \rho^{(i)} := \rho \circ \rho^{(i-1)},$

where $\circ$ is the relational product defined by

$\rho_1 \circ \rho_2 := \{(x,y) \mid \exists z \in A((x,z) \in \rho_2 \text{ and } (z,y) \in \rho_1)\},$

$\rho^T := \underset{i \geq 1}{\cup} \rho^{(i)}$ is the transitive closure of ρ,

$\rho^{RT} := \underset{i \geq 0}{\cup} \rho^{(i)}$ is the reflexive and transitive closure of ρ,

$\rho^{SRT} := \rho^{RT} \cup (\rho^{RT})^{-1}$ is the symmetric, reflexive, transitive closure of ρ.

We shall be particularly interested in the reflexive, transitive closure ρ^{RT} of a relation ρ. Note that the symmetric, reflexive, transitive closure of ρ is just the reflexive, transitive closure of ρ^S; we leave it as an exercise (see Exercise 7.5.1) to verify that

$$\rho^{SRT} = (\rho \cup \rho^{-1} \cup \triangle_A)^T.$$

Lemma 7.1.1 *For any two binary relations ρ_1 and ρ_2 on a set A, we have*

$$\rho_2 = \triangle_A \cup \rho_1 \circ \rho_2 \quad \Rightarrow \quad \rho_1^{RT} \subseteq \rho_2.$$

Proof: Let $\rho_2 = \triangle_A \cup \rho_1 \circ \rho_2$. Then by definition both $\rho_1^{(0)} = \triangle_A \subseteq \rho_2$ and $\rho_1 \circ \rho_2 \subseteq \rho_2$. Inductively, if $\rho_1^{(n-1)} \subseteq \rho_2$ then $\rho_1^{(n)} = \rho_1 \circ \rho_1^{(n-1)} \subseteq \rho_1 \circ \rho_2 \subseteq \rho_2$. Thus $\rho_1^{(n)} \subseteq \rho_2$ for all natural numbers n, and we have

$$\rho_1^{RT} = \underset{i \geq 0}{\cup} \rho^{(i)} \subseteq \rho_2.$$

■

Lemma 7.1.2 *For any two binary relations ρ_1 and ρ_2 on a set A,*

$$(\rho_1 \cup \rho_2)^{RT} = \rho_1^{RT} \circ (\rho_2 \circ \rho_1^{RT})^{RT}.$$

Proof: We show first the inclusion $(\rho_1 \cup \rho_2)^{RT} \subseteq \rho_1^{RT} \circ (\rho_2 \circ \rho_1^{RT})^{RT}$. By Lemma 7.1.1, it suffices to show that $\rho_1^{RT} \circ (\rho_2 \circ \rho_1^{RT})^{RT} = \triangle_A \cup (\rho_1 \cup \rho_2) \circ (\rho_1^{RT} \circ (\rho_2 \circ \rho_1^{RT})^{RT})$. The following calculation shows that this equation is satisfied:

$$\rho_1^{RT} \circ (\rho_2 \circ \rho_1^{RT})^{RT} = (\rho_2 \circ \rho_1^{RT})^{RT} \cup \rho_1 \circ \rho_1^{RT} \circ (\rho_2 \circ \rho_1^{RT})^{RT}$$
$$= \triangle_A \cup \rho_2 \circ \rho_1^{RT} \circ (\rho_2 \circ \rho_1^{RT})^{RT} \cup \rho_1 \circ \rho_1^{RT} \circ (\rho_2 \circ \rho_1^{RT})^{RT}$$
$$= \triangle_A \cup (\rho_1 \cup \rho_2) \circ (\rho_1^{RT} \circ (\rho_2 \circ \rho_1^{RT})^{RT}).$$

To show the opposite inclusion, $\rho_1^{RT} \circ (\rho_2 \circ \rho_1^{RT})^{RT} \subseteq (\rho_1 \cup \rho_2)^{RT}$, we have to show that $\rho_1^{(m)} \subseteq (\rho_1 \cup \rho_2)^{RT}$ and $\rho_1^{(m)} \circ (\rho_2 \circ \rho_1^{(l_1)}) \circ \cdots \circ (\rho_2 \circ \rho_1^{(l_n)}) \subseteq (\rho_1 \cup \rho_2)^{RT}$ for all $m, l_1, \ldots, l_n \in I\!\!N$ and $n \geq 1$. But this follows from the definition of $(\rho_1 \cup \rho_2)^{RT}$. ∎

From Lemma 7.1.2 we have the inclusion $\rho_1^{RT} \circ \rho_2^{RT} \subseteq (\rho_1 \cup \rho_2)^{RT}$ for any two binary relations ρ_1 and ρ_2. We look now for conditions under which we get equality here. First, we observe that the map ^{RT} taking any relation ρ to its reflexive transitive closure ρ^{RT} is a closure operator. In particular, we have

$$\rho_1 \subseteq \rho_2 \quad \Rightarrow \quad \rho_1^{RT} \subseteq \rho_2^{RT} \quad \text{and} \quad (\rho_1^{RT})^{RT} = \rho_1^{RT}.$$

Also, transitivity of ρ^{RT} means that $\rho_1^{RT} \circ \rho_1^{RT} = \rho_1^{RT}$. This shows us that

$$\rho_1 \circ \rho_2^{RT} \subseteq \rho_2^{RT} \circ \rho_1^{RT} \quad \Rightarrow \quad (\rho_1 \circ \rho_2^{RT})^{(n)} \subseteq \rho_2^{RT} \circ \rho_1^{RT},$$

for all $n \in I\!\!N$ (see Exercise 7.5.2). We have now proved the following result.

Lemma 7.1.3 *For any two binary relations ρ_1 and ρ_2 on set A,*

$$\rho_1 \circ \rho_2^{RT} \subseteq \rho_2^{RT} \circ \rho_1^{RT} \quad \Rightarrow \quad (\rho_1 \circ \rho_2^{RT})^{RT} \subseteq \rho_2^{RT} \circ \rho_1^{RT}. \quad ∎$$

Using this we prove the following equivalences.

Proposition 7.1.4 *For any two binary relations ρ_1 and ρ_2 on a set A the following conditions are equivalent:*

(i) $\rho_1 \circ \rho_2^{RT} \subseteq \rho_2^{RT} \circ \rho_1^{RT}$;

(ii) $\rho_1^{RT} \circ \rho_2^{RT} \subseteq \rho_2^{RT} \circ \rho_1^{RT}$; *and*

(iii) $(\rho_1 \cup \rho_2)^{RT} \subseteq \rho_1^{RT} \circ \rho_2^{RT}$.

Proof: (i) $\Rightarrow$ (ii): Clearly, $\rho_1^{(0)} \circ \rho_2^{RT} = \triangle_A \circ \rho_2^{RT} = \rho_2^{RT} \subseteq \rho_2^{RT} \circ \rho_1^{RT}$. Inductively, if for some $n \geq 0$ we have $\rho_1^{(n-1)} \circ \rho_2^{RT} \subseteq \rho_2^{RT} \circ \rho_1^{RT}$, then also $\rho_1^{(n)} \circ \rho_2^{RT} = \rho_1 \circ (\rho_1^{(n-1)} \circ \rho_2^{RT}) \subseteq (\rho_1 \circ \rho_2^{RT}) \circ \rho_1^{RT} \subseteq \rho_2^{RT} \circ (\rho_1^{RT} \circ \rho_1^{RT}) = \rho_2^{RT} \circ \rho_1^{RT}$. This means that for all $n \geq 0$, we have $\rho_1^{(n)} \circ \rho_2^{RT} \subseteq \rho_2^{RT} \circ \rho_1^{RT}$. From this we have $\bigcup\limits_{i \geq 0} \rho_1^{(i)} \circ \rho_2^{RT} = \rho_1^{RT} \circ \rho_2^{RT} \subseteq \rho_2^{RT} \circ \rho_1^{RT}$.

(ii) $\Rightarrow$ (i): By definition of ρ_1^{RT} we have $\rho_1 \circ \rho_2^{RT} \subseteq \rho_1^{RT} \circ \rho_2^{RT} \subseteq \rho_2^{RT} \circ \rho_1^{RT}$.

(i) $\Rightarrow$ (iii): If (i) is satisfied then $(\rho_1 \circ \rho_2^{RT})^{RT} \subseteq \rho_2^{RT} \circ \rho_1^{RT}$ by Lemma 7.1.3. Now by Lemma 7.1.2, $(\rho_2 \cup \rho_1)^{RT} \subseteq \rho_2^{RT} \circ (\rho_2^{RT} \circ \rho_1^{RT}) = \rho_2^{RT} \circ \rho_1^{RT}$. The fact that $\rho_1 \cup \rho_2 = \rho_2 \cup \rho_1$ then gives the claim.

(iii) $\Rightarrow$ (i): Using $\rho_2 \circ \rho_1^{RT} \subseteq \rho_1^{RT} \circ (\rho_2 \circ \rho_1^{RT})^{RT}$ we get $\rho_2 \circ \rho_1^{RT} \subseteq (\rho_1 \cup \rho_2)^{RT}$ by Lemma 7.2.2. This implies $\rho_2 \circ \rho_1^{RT} \subseteq \rho_1^{RT} \circ \rho_2^{RT}$, by (iii). ∎

Our aim is to obtain a decision procedure for the equivalence relation ρ^{SRT} generated by a binary relation ρ on a base set A. We shall see that this will require some finiteness restrictions. We introduce now a change of notation: instead of the algebraic notation ρ for a binary relation on a set A, we will use the notation $\rightarrow$ more commonly used in Computer Science. The statement $(a, b) \in \rho$ becomes $a \rightarrow b$. For the equivalence relation generated by $\rightarrow$, we first form the inverse relation $\leftarrow := \rightarrow^{-1}$ and the symmetric closure $\longleftrightarrow$ defined as $\rightarrow \cup \leftarrow$; then we use the reflexive transitive closure $\stackrel{RT}{\longleftrightarrow}$ of this.

We think of a statement $a \rightarrow b$, corresponding to a pair (a, b) in our relation, as a *rule* which allows us to transform a into b. Usually in practice these rules have the property that the element b is somehow simpler (by some measure of complexity) than a, and we think of the rule as a *reduction rule*. The next definition sets out some basic properties of such reduction systems $\rightarrow$.

Definition 7.1.5 Let $\rightarrow$ be a binary relation on A. If there is an element y such that $x \rightarrow y$, then x is said to be *reducible* and y is called a *reduct* of x, with respect to $\rightarrow$. If no such y exists, then x is said to be *irreducible* or in *normal form* with respect to $\rightarrow$.

An element $a \in A$ has a *terminating reduction* if there are elements $x_0, \ldots, x_m$ of A such that $a = x_0$, $x_i \rightarrow x_{i+1}$ for $i = 1, \ldots, m - 1$ and

x_m is irreducible. In this case we write

$$a = x_0 \to x_1 \to \cdots \to x_m \downarrow .$$

We say that an element a of A has a *non-terminating reduction* if there is an infinite sequence $(x_n : n \in I\!N)$ in A such that $a = x_0$ and $x_n \to x_{n+1}$ for all $n \in I\!N$. We write such a reduction as

$$a = x_0 \to x_1 \to \cdots x_s \to \cdots .$$

The relation $\to$ is called *noetherian* or *terminating* if there are no infinite sequences of the form $x_1 \to x_2 \to x_3 \to \cdots$.

It is easy to see that if the relation $\to$ contains any pairs (a, a), or any pairs (a, b) and (b, a), for elements a and b of A, non-terminating reductions will be possible, and so $\to$ will not be terminating. As we will see in Section 7.3, we usually start with relations $\to$ which are both irreflexive and anti-symmetric, to avoid this problem, and later form the reflexive, symmetric, transitive closure of the relation.

Now suppose that we have a noetherian relation $\to$, with the equivalence relation $\overset{RT}{\longleftrightarrow}$ it generates. We would like to be able to decide whether two elements a and b from A are equivalent under $\overset{RT}{\longleftrightarrow}$. One possibility might be to reduce a and b as much as possible, to normal forms c and d, respectively, and then to test whether these normal forms are equal. If they are equal, then we have $a \overset{RT}{\longrightarrow} c = d \overset{RT}{\longleftarrow} b$ and we can conclude that $a \overset{RT}{\longleftrightarrow} b$. But if $c \neq d$, no conclusion as to whether $a \overset{RT}{\longleftrightarrow} b$ can be reached, since in general it is impossible to infer $c = d$ from $a \overset{RT}{\longleftrightarrow} b$ and $a \overset{RT}{\longrightarrow} c$, $b \overset{RT}{\longrightarrow} d$.

Example 7.1.6 Let $A = \{a, b, c, d\}$ and let $\to$ be the binary relation on A given by the diagram below.

$$
\begin{array}{ccc}
a & \longleftrightarrow & b \\
\downarrow & \searrow\!\swarrow & \\
c & d &
\end{array}
$$

Here c and d are in normal form, and we have $a \overset{RT}{\longleftrightarrow} b$, $a \overset{RT}{\longrightarrow} c$ and $b \overset{RT}{\longrightarrow} d$, but $c \neq d$.

In general then we want to start with an element a in our base set, and reduce it to some normal form c. However, it is possible, in an arbitrary relation $\rightarrow$, that there may be many different reductions starting from a, which may or may not converge to the same result. In particular, an element a may have no normal form, if the reduction relation $\rightarrow$ is not terminating, or it may have more than one normal form reached by different reductions from a. Some kinds of restrictions used to ensure a unique normal form for every element are described in the next definition.

Definition 7.1.7 A relation $\rightarrow$ on a set A has the *Church-Rosser property* if for all $x, y \in A$, if $x \xleftrightarrow{RT} y$ then there is an element $z \in A$ such that $x \xrightarrow{RT} z \xleftarrow{RT} y$. We write $x \downarrow^{RT} y$ to indicate this relationship between x and y. Thus the Church-Rosser property may be expressed as the fact that $\xleftrightarrow{RT} \subseteq \downarrow^{RT}$. Note also that $\downarrow^{RT} = \xrightarrow{RT} \circ \xleftarrow{RT}$. We will also write $x \uparrow_{RT} y$ to mean that there is an element $u \in A$ such that $x \xleftarrow{RT} u \xrightarrow{RT} y$.
A relation $\rightarrow$ on A is called *confluent* if for all $x, y \in A$,

$$\text{if } x \uparrow_{RT} y \text{ then } x \downarrow^{RT} y.$$

The property of confluence means that if we can go from an element u to two different elements x and y by $\xrightarrow{RT}$, then we can also go from x and y to some common point z. This can be illustrated by the following picture:

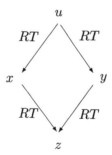

Lemma 7.1.8 *Let $\rightarrow$ be a relation on a set A. Then $\rightarrow$ has the Church-Rosser property iff for all $a, b \in A$ we have $a \xleftrightarrow{RT} b$ iff there is an element z such that $a \xrightarrow{RT} z \xleftarrow{RT} b$; that is, iff $\xleftrightarrow{RT} = \downarrow^{RT}$.*

Proof: As we remarked in Definition 7.1.7, the Church-Rosser property means precisely that $\xleftrightarrow{RT} \subseteq \downarrow^{RT}$. Thus equality of these two relations guarantees that $\to$ is Church-Rosser, and it is enough to show that any Church-Rosser relation satisfies $\downarrow^{RT} \subseteq \xleftrightarrow{RT}$. By Lemma 7.1.2 we have $\xleftrightarrow{RT} = \xrightarrow{RT}$ $\circ (\leftarrow \circ \xrightarrow{RT})^{RT}$. Since $\xleftarrow{RT} \subseteq (\leftarrow \circ \xrightarrow{RT})^{RT}$, it follows that $\xrightarrow{RT} \circ \xleftarrow{RT} \subseteq$ $\xleftrightarrow{RT}$, and $\downarrow^{RT} \subseteq \xleftrightarrow{RT}$.   ■

The central role of the Church-Rosser property is shown by the following theorem.

Theorem 7.1.9 *Let $\to$ be a relation on A which has the Church-Rosser property. Let a, b, c and d be elements of A such that $a \xrightarrow{RT} c$ and $b \xrightarrow{RT} d$, where c and d are in normal form. Then $a \xleftrightarrow{RT} b$ if and only if $c = d$.*

Proof: If $c = d$ then by Lemma 7.1.8 we have $a \xleftrightarrow{RT} b$. Conversely, suppose that $a \xleftrightarrow{RT} b$. Since $a \xrightarrow{RT} c$ and $b \xrightarrow{RT} d$, and $\xleftrightarrow{RT}$ is the equivalence relation generated by $\longrightarrow$, we also have $c \xleftrightarrow{RT} d$. Using the Church-Rosser property we have an element $e \in A$ such that $c \xrightarrow{RT} e \xleftarrow{RT} d$. But c and d are in normal form and thus $c = e = d$.   ■

Theorem 7.1.8 can be used to decide whether any pair is in the equivalence relation $\xleftrightarrow{RT}$. But in order to use this test we have to know that $\to$ has the Church-Rosser property, and in general it is a hard problem to show that a relation has this property. In the next sections some lemmas are derived which allow us to reduce the "global" problem of proving the Church-Rosser property to a "local" problem. But first we prove that having the Church-Rosser property is equivalent to being confluent, a result known as the "Church-Rosser-Theorem."

Theorem 7.1.10 *(Church-Rosser Theorem) A relation $\to$ defined on a set A has the Church-Rosser property if and only if it is confluent.*

Proof: First, suppose that the relation $\to$ has the Church-Rosser property. Consider elements x, y and z in A such that $x \xleftarrow{RT} z \xrightarrow{RT} y$. This implies that $x \xleftrightarrow{RT} y$; so by the Church-Rosser property we have $x \downarrow^{RT} y$. Thus $\to$ is confluent.

Conversely, if $\to$ is a confluent relation, then we have to show that $x \downarrow^{RT} y$ for arbitrary elements $x, y \in A$ with $x \xleftrightarrow{RT} y$. By Lemma 7.1.2 we have $\xleftrightarrow{RT}$ $= \xrightarrow{RT} \circ \cup \{(\leftarrow \circ \xrightarrow{RT})^n \mid n \in I\!N\}$. We will show by induction on n that

$(\leftarrow \circ \xrightarrow{RT})^n \subseteq \xrightarrow{RT} \circ \xleftarrow{RT}$ for all $n \in I\!\!N$. This is clearly satisfied for $n = 0$. If $(\leftarrow \circ \xrightarrow{RT})^{n-1} \subseteq \xrightarrow{RT} \circ \xleftarrow{RT}$, then

$$
\begin{aligned}
(\leftarrow \circ \xrightarrow{RT})^n &= \leftarrow \circ \xrightarrow{RT} \circ (\leftarrow \circ \xrightarrow{RT})^{n-1} \\
&\subseteq \leftarrow \circ \xrightarrow{RT} \circ \xrightarrow{RT} \circ \xleftarrow{RT} \quad \text{by induction hypothesis} \\
&= \leftarrow \circ \xrightarrow{RT} \circ \xleftarrow{RT} \quad \text{since } \xrightarrow{RT} \circ \xrightarrow{RT} = \xrightarrow{RT} \\
&\subseteq \xrightarrow{RT} \circ \xleftarrow{RT} \circ \xleftarrow{RT} \quad \text{by confluence} \\
&= \xrightarrow{RT} \circ \xleftarrow{RT} \quad \text{since } \xleftarrow{RT} \circ \xleftarrow{RT} = \xleftarrow{RT} .
\end{aligned}
$$

Therefore, we get $(\leftarrow \circ \xrightarrow{RT})^n \subseteq \xrightarrow{RT} \circ \xleftarrow{RT}$ for all $n \in I\!\!N$, and so $\bigcup \{(\leftarrow \circ \xrightarrow{RT})^n \mid n \in I\!\!N\} \subseteq \xrightarrow{RT} \circ \xleftarrow{RT}$. Consequently $\xleftrightarrow{RT} \subseteq \xrightarrow{RT} \circ \xleftarrow{RT}$, which means that $\rightarrow$ has the Church-Rosser property. ∎

7.2 Reduction Systems

A pair $(A; \rightarrow)$ consisting of a set A and a binary relation $\rightarrow$ on A is called a *reduction system*. In Definition 7.1.5 we introduced the concept of a terminating reduction. The most useful reduction systems are those which are both confluent and terminating, and in this section we characterize such special systems.

Definition 7.2.1 A reduction system which is both confluent and terminating is called *complete*.

If $(A; \rightarrow)$ is a confluent reduction system, each element of A has at most one normal form. But when $(A; \rightarrow)$ is terminating, each element of A has at least one normal form. Therefore, in a complete reduction system every element has exactly one normal form, which we will call the *canonical normal form* of the element. It is this property that makes complete reduction systems so useful. In order to characterize such systems, we will need the property of local confluence.

Definition 7.2.2 A reduction system $(A; \rightarrow)$ is called *locally confluent* if for any two elements $x, y \in A$, whenever there exists an element u of A such that $u \rightarrow x$ and $u \rightarrow y$, then we have $x \downarrow^{RT} y$.

Local confluence thus means that whenever there is an element u which we can reduce in one step to both x and y, then we can go from each of x and y in one or more steps to a common element z. This situation is illustrated by the following picture.

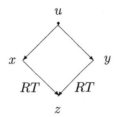

Clearly, every confluent reduction system is locally confluent. The converse is not true in general, and the following picture shows a relation which is locally confluent but not confluent.

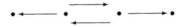

For the proof of our next result we need the following *Principle of Noetherian Induction.*

Theorem 7.2.3 *Let $\to$ be a noetherian relation on A. Let $\mathfrak{P}$ be a property and let $\mathfrak{P}(a)$ be the statement that the property $\mathfrak{P}$ holds for the element a. Assume that $\to$ satisfies the following condition: If $\mathfrak{P}(x)$ for all x with $a \to x$, then $\mathfrak{P}(a)$. Then $\mathfrak{P}$ holds for all a in A.*

Proof: Let us denote by $A_{\mathfrak{P}}$ the set of all elements in A for which $\mathfrak{P}$ holds. By definition $A_{\mathfrak{P}} \subseteq A$. Suppose that $A_{\mathfrak{P}} \subset A$. Then there are elements in A for which $\mathfrak{P}$ is not satisfied. Consider $B = A \setminus A_{\mathfrak{P}}$. Since $\to$ is noetherian, there are elements m in B such that no element x of A with $m \to x$ belongs to B, for otherwise we would have infinite chains in B. Therefore every element x with $m \to x$ belongs to $A_{\mathfrak{P}}$ and satisfies $\mathfrak{P}$. But then by assumption we have $\mathfrak{P}(m)$, a contradiction. ∎

Theorem 7.2.4 *A terminating reduction system is confluent if and only if it is locally confluent.*

Proof: We know that any confluent reduction system is locally confluent. For the opposite direction, suppose that $(A; \to)$ is a locally confluent and terminating reduction system. We consider the following property $\mathfrak{P}$:

$$\mathfrak{P}(a) \quad \text{iff} \quad \forall x, y \in A \; (a \xrightarrow{RT} x \text{ and } a \xrightarrow{RT} y \; \Rightarrow \; x \downarrow_{RT} y).$$

We verify that

$$\forall a \in A(\forall x \in A(a \to x \Rightarrow \mathfrak{P}(x)) \Rightarrow \mathfrak{P}(a))$$

is satisfied. Consider an arbitrary element a of A. We can assume that a is reducible, since otherwise $\mathfrak{P}(a)$ is trivially satisfied. To show that $\mathfrak{P}$ holds for a, we prove that for all $x, y \in A$ if $a \xrightarrow{RT} x$ and $a \xrightarrow{RT} y$ then we have $x \downarrow_{RT} y$. The reducibility of a means that there are elements $b, c \in A$ such that $a \to b \xrightarrow{RT} x$ and $a \to c \xrightarrow{RT} y$. Then by the local confluence there is an element $d \in A$ such that $b \xrightarrow{RT} d$ and $c \xrightarrow{RT} d$.

By our induction hypothesis, $\mathfrak{P}$ holds for b and c and therefore there is an element z such that $d \xrightarrow{RT} z$ and $c \xrightarrow{RT} z$. Now we have $c \xrightarrow{RT} z$ and $c \xrightarrow{RT} y$. By the induction hypothesis applied to y and z we also get an element z' such that $y \xrightarrow{RT} z'$ and $z \xrightarrow{RT} z'$. But then for x and y there is an element, namely $z' \in A$, such that $x \xrightarrow{RT} z'$ and $y \xrightarrow{RT} z'$. This means that $\mathfrak{P}(a)$ is satisfied. ∎

Since a complete reduction system is one which is both confluent and terminating, we can rephrase Theorem 7.2.4 as follows.

Corollary 7.2.5 *A reduction system is complete if and only if it is locally confluent and terminating.* ∎

Reduction systems will be used to help us to solve the word problem. Given an equivalence relation $\sim$ on A, we have to find a complete reduction system which generates the relation $\sim$. Of course, different reduction systems can generate the same equivalence relation. Such reduction systems are called equivalent. Thus we want to find, among all the equivalent reduction systems which generate our equivalence relation, any complete reduction systems.

Definition 7.2.6 Two reduction systems $(A; \to)$ and $(A; \Rightarrow)$ are called *equivalent* if $\xleftrightarrow{RT} = \stackrel{RT}{\Longleftrightarrow}$. A complete reduction system which is equivalent to $(A; \to)$ is called a *completion* of $(A; \to)$.

As a first step in finding a completion of a given reduction system, we must show that every reduction system does indeed have a completion. To see this, consider an arbitrary reduction system $(A; \rightarrow)$. Let $\overset{RT}{\longleftrightarrow}$ be the equivalence relation generated by $\rightarrow$, and let $A/\overset{RT}{\longleftrightarrow}$ be the quotient set with respect to $\overset{RT}{\longleftrightarrow}$. A choice function $\Phi : A/\underset{RT}{\longleftrightarrow} \rightarrow A$ is a function which selects exactly one element from each equivalence class with respect to $\overset{RT}{\longleftrightarrow}$. (By the axiom of choice such a mapping always exists.) The choice function Φ induces a mapping $S : A \rightarrow A$ for which $S(x) \overset{RT}{\longleftrightarrow} x$ and $S(x) = S(x')$ for all $x, x' \in A$ whenever $x \overset{RT}{\longleftrightarrow} x'$. (Essentially, S is defined by assigning the value used by Φ to every element of each equivalence class, so that the classes are viewed not so much as new objects but as subsets of A, all of whose elements take the same value.)

Then we define a new relation $\Rightarrow$ on A, by

$$x \Rightarrow y \quad \text{iff} \quad x \neq y \text{ and } y = S(x),$$

for all $x, y \in A$.

Clearly $\overset{RT}{\Longleftrightarrow} = \overset{RT}{\longleftrightarrow}$, so that the reduction systems $(A; \rightarrow)$ and $(A; \Rightarrow)$ are equivalent. Moreover, if $x \overset{RT}{\longleftrightarrow} y$ then there is an element z with $x \overset{RT}{\Longrightarrow} z \overset{RT}{\Longleftarrow} y$, namely $z = S(x)$ if $x \neq y$ and $x = z = y$ otherwise. Therefore $\Rightarrow$ has the Church-Rosser property and is confluent by Theorem 7.1.10. The relation $\Rightarrow$ is also terminating and therefore complete. This shows that any reduction system does have a completion.

For a terminating reduction system $(A; \rightarrow)$, we know by Corollary 7.2.5 that local confluence is enough to guarantee completeness. To find a completion, therefore, we can try the following construction, based on the strategy of locating all cases in which local confluence is violated. We consider the set of all so-called "critical pairs," that is, pairs $(x, y) \in A \times A$ for which there is an element $z \in A$ such that $z \rightarrow x$ and $z \rightarrow y$, but no element $z' \in A$ satisfying $x \overset{RT}{\longrightarrow} z'$ and $y \overset{RT}{\longrightarrow} z'$. If there are no such pairs, then the relation $\rightarrow$ is locally confluent and hence confluent and complete. Otherwise, if there exists at least one critical pair (x, y) we can try to fix the local confluence violation by adding either (x, y) or (y, x) to the reduction relation $\rightarrow$. That is, we add either $x \rightarrow y$ or $y \rightarrow x$. However, it is possible that with this addition the new enlarged relation is no longer terminating. If both possible additions destroy the termination property, our procedure stops with "failure."

Otherwise, we are successful in adjoining a critical pair to $\rightarrow$, and getting a new larger relation which still terminates. Then we repeat the process on this new relation, finding its critical pairs and enlarging again if necessary. If this procedure terminates, at some relation $\Rightarrow$, then no critical pairs are left, and $(A; \Rightarrow)$ is complete. By construction the system $(A; \Rightarrow)$ is then a complete reduction system which is a completion of $(A; \rightarrow)$.

The problem with this procedure is that in general there can be infinitely many critical pairs. So this procedure does not give an algorithm for construction of the completion of a terminating reduction system, and in general there is no such algorithm.

We conclude this section with an example to illustrate our procedure. As we remarked earlier, we want to use this procedure in the special case that our equivalence relation is the fully invariant congruence consisting of the set of identities of some variety. Our example starts with a term clone as our base set A.

Example 7.2.7 Taking S to be the type (2) variety of all semigroups, we consider as our set A the base set of the free semigroup $\mathcal{F}_S(X_2)$ over the two-element alphabet $X_2 = \{x, y\}$. As we saw in Example 6.4.5, we usually omit both the binary operation symbol and brackets from our terms, writing them as words on the alphabet X_2. Thus our set A consists of all such words composed from the two letters x and y. We define a relation $\rightarrow$ on A as follows. For any two words w and w', we set $w \rightarrow w'$ iff there exist words $w_1, w_2 \in W(X_2)$ such that $w = w_1 yyxyw_2$ and $w' = w_1 w_2$. Notice that the relation $\rightarrow$ is compatible with the semigroup operation.

The reduction system $(A; \rightarrow)$ is terminating since any reduction step decreases the length of the words. But it is not confluent, since we have the reductions

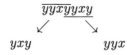

but it is not possible to reduce yxy and yyx to a common element. Applying our procedure to the critical pair (yxy, yyx), we check whether we should add $yxy \to yyx$ or $yyx \to yxy$. Since both words have equal length, we use the lexicographical order induced by $x > y$ to make our decision: since $yxy > yyx$ we add the rule $yxy \to yyx$. Now we consider the relation $\to_1$ generated by the rules $r_0 = yyxy \to_1 e$ and $r_1 = yxy \to_1 yyx$, where e is the empty word (having the property $we \approx ew \approx w$ for all words w). Our new relation $\to_1$ is defined by $w \to_1 w'$ if there are words $w_1, w_2 \in A$ such that either

(1)　$w = w_1 yyxy w_2$ and $w' = w_1 w_2$,　　or

(2)　$w = w_1 yxy w_2$ and $w' = w_1 yyx w_2$.

Since our relation is generated as a compatible relation from these rules, we only have to check all possible "overlappings" of $yyxy$ and yyx. By construction the reduction system $(A; \to_1)$ is terminating, and we check now that it is equivalent to $(A; \to)$. First, since $\to$ is a subset of $\to_1$ we get $\overset{RT}{\longleftrightarrow} \subseteq \overset{RT}{\longleftrightarrow}_1$. Conversely, if (w, w') is a critical pair with respect to $\to_1$ then w and w' are equivalent modulo $\overset{RT}{\longleftrightarrow}$. Thus $\to_1 \subseteq \overset{RT}{\longleftrightarrow}$, and from this it follows that $\overset{RT}{\longleftrightarrow}_1 \subseteq \overset{RT}{\longleftrightarrow}$. Altogether we have equality.

The reduction system $(A; \to_1)$ is still not confluent. The overlappings of the left hand sides of the two rules give new rules. The overlapping of the left hand side of r_0 with itself was already examined. Consider now r_0 and r_1.

$$\begin{array}{ccc} & \overline{yyxy} & \\ \swarrow & & \searrow \\ e & & yyyx \end{array}$$

$$\begin{array}{ccc} & yy x\overline{yxy} & \\ \swarrow & & \searrow \\ yyxyxy & & yx \end{array}$$

$$\begin{array}{ccc} & \overline{yyxy}xy & \\ \swarrow & & \searrow \\ xy & & yyxyyx \end{array}$$

If we consider overlappings of the left hand side of r_1 with itself we obtain only

We add the following new rules:

$$r_2 = yyyx \to_2 e, \qquad r_3 = yyxyxy \to_2 yx,$$
$$r_4 = yyxyyx \to_2 xy, \qquad r_5 = yxyyx \to_2 yyxxy.$$

Let $\to_2$ be the relation generated by the rules r_0, r_1, r_2, r_3, r_4 and r_5. Then $(A; \to_2)$ is terminating and equivalent to $(A; \to)$, but is still not confluent, since for instance the overlapping of the left hand side of r_0 with the left hand side of r_3 gives

But if we now add the rule $r_6 = xy \to_3 yx$, it is easy to show that no other overlappings produce critical pairs. We denote by $\to_3$ the relation generated by r_0, r_1, r_2, r_3, r_4, r_5 and r_6. Then $(A; \to_3)$ is a completion of $(A; \to)$. In fact we can omit the rules r_0 to r_5, since $(A; \to_3)$ is equivalent to the complete reduction system $(A; \Rightarrow)$ where $\Rightarrow$ is generated by $yyyx \to e$ and $xy \to yx$. Thus $(A; \Rightarrow)$ is also a completion of $(A; \to)$.

7.3 Term Rewriting

In the previous section we studied reduction systems $(A; \to)$ in general. Now we turn to the specific case we are interested in, when A is actually the universe set of the free algebra, or term clone, of a fixed type τ. That is, we take A to be the set $W_\tau(X)$ of all terms of type τ over an alphabet X. We will need to impose some conditions on the relation $\to$ to guarantee termination. When the relation $\to$ meets these conditions, the resulting reduction system $(W_\tau(X); \to)$ is called a *term reduction system*. We usually write pairs (t, t') in the relation $\to$ as rules $t \to t'$, and refer to them as *reduction rules*. A set of such reduction rules is called a *term rewriting system* for $W_\tau(X)$.

As an autonomous research field, term rewriting dates back to the introduction of the so-called Knuth-Bendix completion procedure ([66]) in 1970. The aim of the Knuth-Bendix algorithm is to transform a given input, a set of equations, into a convergent rewriting system: a system for rewriting terms which is always terminating and which produces a unique possible result for any term, called its normal form. In this system, two terms are equal in the input theory if and only if they have the same normal form.

The motivation for this algorithm lies in the deductive method for equational theories in Universal Algebra. That is, we want to form our deduction rules for terms from some identities of an equational theory. There is an important difference, however, in that identities $t \approx t'$ are symmetric, while our deduction rules $t \to t'$ are one-way only. We usually start with an irreflexive and anti-symmetric set of rules as our set $\to$, and later form the reflexive, symmetric and transitive closure of this, as in the previous sections.

As we saw in Section 6.3, there are five rules of deduction for equational theories, which allow us to deduce new equations from given ones. Of these rules, the first three correspond to the three basic properties of an equivalence relation. As before, these are taken care of by taking the closure $\xleftrightarrow{RT}$ of $\to$. We will focus now on the other two rules, the substitution rule and the so-called "replacement" rule which allows us to apply fundamental operation symbols or (from Theorem 5.2.4) arbitrary terms to equations to produce new equations. We will call a relation *invariant* if it is closed under the replacement rule, rule (4), and *fully invariant* if it is closed under both the replacement rule and the substitution rule. The word problem for fully invariant congruences can thus be transformed into a reduction problem for fully invariant relations, and the first condition we impose on our relation $\to$ is that it should be fully invariant.

To express the algebraic properties of fully invariant relations in our new language of reduction rules, and to describe another necessary condition for $\to$, we recall some notation from Chapters 5 and 6 which we can use to describe the effects of the substitution and replacement rules. A *substitution* (of type τ) is any map $s : X \mapsto W_\tau(X)$, and any such substitution has a unique endomorphic extension $\hat{s} : W_\tau(X) \mapsto W_\tau(X)$. In Section 5.1 we assigned to any vertex or node of a term t, regarded as a tree, a sequence of positive integers called its address. We use the notation t/u for the subtree of t which starts with the vertex labelled with address u, and $t[u/s]$ for the tree obtained from t by replacing the subtree t/u by the tree s.

Then a relation $\to$ on the free algebra $\mathcal{F}_\tau(X)$ is fully invariant iff for all terms t, t' and t'', for all substitutions $s : X \to W_\tau(X)$, and all addresses u in t'',

(1) $t \to t'$ implies $t''[u/t] \to t''[u/t']$, and

(2) $t \to t'$ implies $\hat{s}[t] \to \hat{s}[t']$.

It is clear that these two conditions express the deduction rules (4) and (5), the replacement rule and the substitution rule, respectively. The following proposition gives a necessary condition for a relation on $W_\tau(X)$ to be terminating. We use the notation $var(t)$ for the set of all variables which occur in a term t.

Proposition 7.3.1 *If $\to$ is a terminating reduction on $W_\tau(X)$, then for all terms $t, t' \in W_\tau(X)$, if $t \to t'$ then $var(t') \subseteq var(t)$.*

Proof: Let t be a terminating reduction on $W_\tau(X)$. Suppose that there are terms $t, t' \in W_\tau(X)$ with $t \to t'$ but $var(t') \not\subseteq var(t)$. Then there is a variable z in t' which does not occur in t. Consider the substitution $s : X \to W_\tau(X)$ defined by

$$s(x) = \begin{cases} x & \text{if} & x \neq z \\ t & \text{if} & x = z. \end{cases}$$

Since z does not occur in t it is clear that $\hat{s}[t] = t$ for the extension of s. We set $t_0 := \hat{s}[t]$ and $t_1 := \hat{s}[t']$. Then from $t \to t'$ and condition (2) for term reductions we get $t_0 \to t_1$.

Since $z \in var(t')$ there is an address u in t' with $t'/u = z$. Therefore for the subtree t_1/u of t_1 starting with the vertex addressed by u we have

$$t_1/u = \hat{s}[t']/u = \hat{s}[t'/u] = \hat{s}[z] = t,$$

and then $t_1 = t_1[u/t]$. Now we set $t_2 := t_1[u/t_1]$. We have $t = t_0 \to t_1$, and applying condition (1) to this gives $t_1 = t_1[u/t] \to t_1[u/t_1] = t_2$. The term t_1 is a subterm of t_2 since $t_2/u = t_1$. Continuing in this way, we produce a non-terminating reduction

$$t_0 \to t_1 \to t_2 \to \cdots.$$

This contradicts the assumption that $\rightarrow$ is terminating.   ■

This necessary condition for a terminating reduction is built into our definition of a term reduction system, along with the requirement of full invariance.

Definition 7.3.2 Let τ be a fixed type, and let $\rightarrow$ be a set of pairs on $W_\tau(X)$. The pair $(W_\tau(X); \rightarrow)$ is called a *term reduction system* if $\rightarrow$ is fully invariant and for all terms $t, t' \in W_\tau(X)$, if $t \rightarrow t'$ then $\text{var}(t') \subseteq \text{var}(t)$. A pair $(t, t') \in W_\tau(X)^2$, or a rule $t \rightarrow t'$ is called a *reduction rule* if $var(t') \subseteq var(t)$. A set of reduction rules is called a *term rewriting system*.

Notice that the set of rules in a term rewriting system need not be fully invariant, although the rules satisfy the variable property. This is not a serious problem, since for any set R of such reduction rules we can always form the fully invariant closure, which we denote by $\rightarrow_R$. Example 7.2.7 gave a term rewriting system for type $\tau = (2)$. For arbitrary type τ we can use some of the ideas developed in that example if we use the concept of an address u in a term t as a special term of type $\tau = (2)$. We define a partial order relation, called the *prefix order*, on the set of addresses by

$$u \leq v \quad \Leftrightarrow \quad v = uw \text{ for some address } w.$$

Then we have $t \rightarrow_R t'$ iff there exists a substitution $s : X \rightarrow W_\tau(X)$, a term $t'' \in W_\tau(X)$ and an address u of t'' such that (1) and (2) are satisfied.

Proposition 7.3.3 *Let* $R \subseteq W_\tau(X) \times W_\tau(X)$ *be a term rewriting system, and let* $t, t' \in W_\tau(X)$. *Then* $t \rightarrow_R t'$ *iff there exists a substitution* $s : X \rightarrow W_\tau(X)$, *an address* u *of* t *and a reduction rule* $t_1 \rightarrow t_2 \in R$ *such that* $t/u = \hat{s}[t_1]$ *and* $t' = t[u/\hat{s}[t_2]]$.

Proof: Let $t, t' \in W_\tau(X)$ be terms which satisfy the condition. If $\rightarrow_R$ is the term reduction system generated by R then $t_1 \rightarrow t_2 \in R$ implies $\hat{s}[t_1] \rightarrow_R \hat{s}[t_2]$ by rule (2). If t is a term with $t/u = \hat{s}[t_1]$ then $t[u/\hat{s}[t_1]] = t$, and by rule (1) we have $t = t[u/\hat{s}[t_1]] \rightarrow t[u/\hat{s}[t_2]] = t'$, and so $t \rightarrow t'$.

Assume conversely that $t \rightarrow_R t'$. Then there are terms t_1', t_2' with $t_1' \rightarrow_R t_2'$, an address u in t and a substitution $s : X \rightarrow W_\tau(X)$, such that $t/u = \hat{s}[t_1']$ and $t'/u = \hat{s}[t_2']$. Using (2) and (1) we have $t'[u/\hat{s}[t_2']] = t'$. Continuing in this way we come finally to a reduction rule $t_1 \rightarrow t_2 \in R$ with this property and our condition is satisfied.   ■

A special case occurs when t can be reduced to t' in one step, that is if there is a substitution $s : X \to W_\tau(X)$ such that $t' = \hat{s}[t]$. In this case this substitution s is called a *match* of t to t'.

All the properties of reduction systems described in Section 7.2 can be applied to term rewriting systems. Given a term rewriting system R, we have to find an equivalent complete terminating rewriting system. In this section we assume that our system is terminating. The termination of term rewriting systems will be considered in the next section. As we did in the example given in Section 7.2, we proceed by transforming critical pairs into new reduction rules.

To check the local confluence of our term rewriting system R, any two diverging one-step reductions $t \to t'$ and $t \to t''$ must be inspected for a common reduction $\bar{t}$ such that t' and t'' converge to $\bar{t}$. The easiest way is to find a substitution s under which t' and t'' are equal, making $\bar{t} = \hat{s}[t'] = \hat{s}[t'']$.

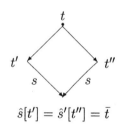

$$\hat{s}[t'] = \hat{s}'[t''] = \bar{t}$$

Two terms t and t' of $W_\tau(X)$ are called *unifiable* if there is a substitution $s : X \to W_\tau(X)$ such that $\hat{s}[t] = \hat{s}[t']$. More generally, a set $M \subseteq W_\tau(X)$ of terms is called unifiable if there is a substitution s such that $\hat{s}[t] = \hat{s}[t']$ for all $t, t' \in M$; such an s is called a *unifier* for M. A unifier s is called a *most general unifier* for M if for all unifiers s_1 for M there is a substitution s_2 such that $s_1 = s \circ s_2$. In this case we also write $s_1 \le s$. Clearly, the relation $\le$ is reflexive and transitive; so is a quasiorder on the set of all substitutions. We can then define an equivalence relation on the set of all substitutions, by

$$s_1 \sim s_2 :\Leftrightarrow s_1 \le s_2 \text{ and } s_2 \le s_1.$$

Let us denote by $Subst_\tau$ the set of all substitutions $s : X \to W_\tau(X)$. Then the relation $\le$ induces a partial order on the quotient set $Subst_\tau / \sim$. One can show that the relation $<$ defined by $s_1 \leq s_2$, $s_1 \neq s_2$ is noetherian on $Subst_\tau$ (see E. Eder, [42]). As a consequence we get the following result.

Proposition 7.3.4 *Let $M \subseteq W_\tau(X)$ be a finite unifiable set of terms. Then there exists a most general unifier for M.*

Proof: We consider the set U_M of all unifiers for M and its quotient set $U_M/ \sim$ when we factorise by the relation $\sim$. Clearly a most general unifier for M is a maximal element with respect to the partial order $\leq$ on this quotient set. Assume that s is not maximal in $U_M/ \sim$. Then there is an element s_1 in $U_M/ \sim$ such that $s < s_1$. If s_1 is maximal, we are finished. Otherwise, by iteration of this procedure (using the axiom of choice) we get a sequence $s_1, s_2, s_3, \ldots$ such that

$$s_1 < s_2 < s_3 < \cdots .$$

Since $<$ is noetherian this sequence must terminate, at some s_m for which there is no element s_{m+1} with $s_m < s_{m+1}$. Therefore s_m is a most general unifier. ∎

Remark 7.3.5 1. J. A. Robinson showed in [101] that for every nonempty finite set $M \subseteq W_\tau(X)$ there is an algorithm which decides whether M is unifiable or not. If M is unifiable, then the algorithm generates a most general unifier.

2. Another kind of unification problem can be defined for equational theories Σ. Two terms t and t' in Σ are called Σ-unifiable if there exists a substitution $s : X \to W_\tau(X)$ such that $\hat{s}[t] \approx \hat{s}[t'] \in \Sigma$. (See Section 15.1.)

Example 7.3.6 1. Consider the type $\tau = (2, 1, 1)$ with operation symbols f, g, h, and the terms $t = f(x_1, g(x_2))$ and $t' = f(h(x_3), x_4)$. Let $s_1 : X \to W_\tau(X)$ be a substitution with $x_1 \mapsto h(x_3)$, $x_2 \mapsto x_2$, $x_3 \mapsto x_3$ and $x_4 \mapsto g(x_2)$. Then s_1 is a unifier for t and t', since $\hat{s}_1[t] = \hat{s}_1[f(x_1, g(x_2))] = f(h(x_3), g(x_2)) = \hat{s}_1[f(h(x_3), x_4)] = \hat{s}_1[t']$.

2. Let $\tau = (2)$ and let Σ be the equational theory of the variety of all commutative semigroups, that is, the equational theory generated by

$$E = \{f(x_1, x_2) \approx f(x_2, x_1), f(x_1, f(x_2, x_3)) \approx f(f(x_1, x_2), x_3)\}.$$

Consider the terms $t = f(a, x_1)$ and $t' = f(b, x_2)$, where a and b are variables different from x_1 and x_2. Then the substitution s_1 defined by $s_1(x_1) = b$, $s_1(x_2) = a$ and $s_1(x_j) = x_j$ for all $j \geq 3$ is a Σ-unifier for t and t', since $\hat{s}_1[t] = f(a, b) \approx f(b, a) = \hat{s}_1[t']$ is an identity in Σ.

The next example illustrates the process of replacing critical pairs by new reduction rules, in our procedure for finding the completion of a given term rewriting system.

Example 7.3.7 Consider the type $\tau = (2, 1, 0)$ with the binary operation symbol $\cdot$, the unary operation symbol $^{-1}$ and the nullary operation symbol e. We start with three rules r_0, r_1 and r_2.

$$r_0 : \quad (x_1 \cdot x_2) \cdot x_3 \rightarrow x_1 \cdot (x_2 \cdot x_3),$$
$$r_1 : \quad x^{-1} \cdot x \rightarrow e, \quad \text{and}$$
$$r_2 : \quad e \cdot x \rightarrow x.$$

These rules clearly arise from the axioms of a group, with the imposition of a definite (one-way) orientation. We will show later, in Example 7.4.5, that the term rewriting system R defined by these three rules is terminating. But it is not confluent, since we can find some critical pairs. For instance, we have (using r_1) that $(x_1^{-1} \cdot x_1) \cdot x_3 \rightarrow e \cdot x_3$, and (using r_0) that $(x_1^{-1} \cdot x_1) \cdot x_3 \rightarrow x_1^{-1} \cdot (x_1 \cdot x_3)$. But the term $x_1^{-1} \cdot (x_1 \cdot x_3)$ cannot be further reduced with our three rules; so the pair $(e \cdot x_3, x_1^{-1} \cdot (x_1 \cdot x_3))$ is a critical pair. We can reduce the term $e \cdot x_3$ to x_3, using r_2, and we see from this that $x_1^{-1} \cdot (x_1 \cdot x_3) \approx x_3$ is an identity in the variety of all groups. But the corresponding rule $x_1^{-1} \cdot (x_1 \cdot x_3) \rightarrow x_3$ cannot be derived from the rules r_0, r_1 and r_2. Therefore, we add a new reduction rule:

$$r_3 : \quad x_1^{-1} \cdot (x_1 \cdot x_3) \rightarrow x_3.$$

For notational convenience, we shall not distinguish here between rules from the original R and new rules in the successive enlargements of R. Now from r_3 and r_1 we can get $(x_2^{-1})^{-1} \cdot (x_2^{-1} \cdot x_2) \rightarrow x_2$ and $(x_2^{-1})^{-1} \cdot (x_2^{-1} \cdot x_2) \rightarrow (x_2^{-1})^{-1} \cdot e$. This gives a new critical pair $(x_2, (x_2^{-1})^{-1} \cdot e)$ which cannot be further reduced. We add another rule, r_4:

$$r_4 : \quad (x_2^{-1})^{-1} \cdot e \rightarrow x_2.$$

From rules r_3 and r_2 we obtain $e^{-1} \cdot (e \cdot x_3) \to x_3$ and $e^{-1} \cdot (e \cdot x_3) \to e^{-1} \cdot x_3$. This gives the critical pair $(e^{-1} \cdot x_3, x_3)$, and the new rule r_5:

$$r_5 : \quad e^{-1} \cdot x_3 \to x_3.$$

The process continues in this way, as shown in Table 1 below, where we outline which derivations we get and which critical pairs and reduction rules we have to add. At the point where we have fourteen rules there are no more critical pairs, and the new term rewriting system is complete. We note that basically we needed to add, along with rule r_{14}, the rule $x_2^{-1} \cdot x_1^{-1} \to (x_1 \cdot x_2)^{-1}$. But as Knuth and Bendix showed in [66], this would have led to some strong complications.

Now we want to use our observations in this example to find a general method to construct "critical pairs" in a systematic way. First we need a more rigorous definition of a critical pair.

Lemma 7.3.8 *Let t_1 and t_2 be terms with $var(t_1) \cap var(t_2) = \emptyset$. Let v be an address of t_1 such that t_1/v is a subterm which is not a variable. If s_1 and s_2 are substitutions with $\hat{s}_1[t_1/v] = \hat{s}_2[t_2]$ then there is a substitution s with the property that $\hat{s}[t_1/v] = \hat{s}[t_2]$.*

Proof: Since $var(t_1) \cap var(t_2) = \emptyset$ we can define a substitution s by

$$s(x) = \begin{cases} s_1(x) \text{ if } x \in var(t_1) \\ s_2(x) \text{ if } x \in var(t_2). \end{cases}$$

Then we have $\hat{s}[t_1/v] = \hat{s}_1[t_1/v] = \hat{s}_2[t_2] = \hat{s}[t_2]$. ∎

We now use the concept of a most general unifier for a pair of terms, to make our definition of a critical pair.

Rules	Derivations	"Critical Pair"	New Reduction Rule
r_0, r_4	$((x_2^{-1})^{-1} \cdot e) \cdot x_1 \to$ $(x_2^{-1})^{-1} \cdot (e \cdot x_1),$ $((x_2^{-1})^{-1} \cdot e) \cdot x_1 \to$ $x_2 \cdot x_1$	$((x_2^{-1})^{-1} \cdot$ $(e \cdot x_1),$ $x_2 \cdot x_1)$	$r_6 =$ $(x_2^{-1})^{-1} \cdot x_1 \to$ $x_2 \cdot x_1$
r_4, r_6	$(x_2^{-1})^{-1} \cdot e \to x_2,$ $(x_2^{-1})^{-1} \cdot e \to x_2 \cdot e$	$(x_2 \cdot e, x_2)$	$r_7 = x_2 \cdot e$ $\to x_2$
r_4, r_7	$((x_2)^{-1})^{-1} \cdot e \to x_2,$ $((x_2)^{-1})^{-1} \cdot e \to$ $((x_2)^{-1})^{-1}$	$(((x_2)^{-1})^{-1}, x_2)$	$r_8 =$ $(x_2^{-1})^{-1} \to x_2$
r_5, r_7	$e^{-1} \cdot e \to e,$ $e^{-1} \cdot e \to e^{-1}$	(e^{-1}, e)	$r_9 = e^{-1} \to e$
r_1, r_8	$(x_2^{-1})^{-1} \cdot x_2^{-1} \to e,$ $((x_2)^{-1})^{-1} \cdot x_2^{-1} \to$ $x_2 \cdot x_2^{-1}$	$(x_2 \cdot x_2^{-1}, e)$	$r_{10} =$ $x_2 \cdot x_2^{-1} \to e$
r_0, r_{10}	$(x_2 \cdot x_2^{-1}) \cdot x_1 \to$ $x_2(x_2^{-1} \cdot x_1),$ $(x_2 \cdot x_2^{-1}) \cdot x_1 \to e \cdot x_1$	$(x_2 \cdot (x_2^{-1} \cdot x_1),$ $e \cdot x_1)$	$r_{11} = x_2 \cdot$ $(x_2^{-1} \cdot x_1) \to x_1$
r_0, r_{10}	$(x_1 \cdot x_2) \cdot (x_1 \cdot x_2)^{-1} \to$ $x_1 \cdot (x_2 \cdot (x_1 \cdot x_2)^{-1}),$ $(x_1 \cdot x_2) \cdot (x_1 \cdot x_2)^{-1}$ $\to e$	$(x_1 \cdot (x_2 \cdot$ $(x_1 \cdot x_2)^{-1}), e)$	$r_{12} = x_1(x_2 \cdot$ $(x_1 \cdot x_2)^{-1})$ $\to e$
r_{12}, r_3	$x_1^{-1} \cdot (x_1^{-1} \cdot (x_2$ $(x_1 \cdot x_2)^{-1}))$ $\to x_1^{-1} \cdot e,$ $x_1^{-1} \cdot (x_1(x_2 \cdot$ $(x_1 \cdot x_2)^{-1}))$ $\to x_2 \cdot (x_1 \cdot x_2)^{-1}$	$(x_2 \cdot (x_1 \cdot x_2)^{-1},$ $x_1^{-1} \cdot e)$	$r_{13} = x_2 \cdot$ $(x_1 \cdot x_2)^{-1} \to$ x_1^{-1}
r_3, r_{13}	$x_2^{-1}(x_2 \cdot (x_1 \cdot x_2)^{-1})$ $\to (x_1 \cdot x_2)^{-1},$ $x_2^{-1} \cdot (x_2 \cdot (x_1 \cdot x_2)^{-1})$ $\to x_2^{-1} \cdot x_1^{-1}$	$((x_1 \cdot x_2)^{-1},$ $x_2^{-1}, x_1^{-1})$	$r_{14} =$ $(x_1 \cdot x_2)^{-1}$ $\to x_2^{-1} \cdot x_1^{-1}$

Table 1

Definition 7.3.9 Let R be a term rewriting system and let $t_1 \to t_1'$ and $t_2 \to t_2'$ be two reduction rules of R. We may assume that $\text{var}(t_1) \cap \text{var}(t_2) = \emptyset$, since otherwise the variables can be renamed in an appropriate way. Let v be an address of t_1 such that t_1/v is a subterm which is not a variable. If there is a most general unifier s of t_1/v and t_2 (so that $\hat{s}[t_1/v] = \hat{s}[t_2]$), then the pair $(\hat{s}[t_1'], \hat{s}[t_1][v/\hat{s}[t_2']])$ is called a *critical pair* in R.

Remark 7.3.10 If we reduce $\hat{s}[t_1]$ using the rules $t_1 \to t_1'$ and $t_2 \to t_2'$, then we obtain $\hat{s}[t_1][v/\hat{s}[t_2']]$.

Example 7.3.11 Consider the reduction rules r_3 and r_{12} from Example 7.3.7:

$$x_1^{-1} \cdot (x_1 \cdot x_3) \to x_3 \quad \text{and} \quad x_1 \cdot (x_2 \cdot (x_1 \cdot x_2)^{-1}) \to e.$$

These rules have a common variable x_1; so we replace x_1 in the first rule by x_1', and consider instead the rules $r_3' = (x_1')^{-1} \cdot (x_1' \cdot x_3) \to x_3$ and r_{12}. We want to use our definition of critical pairs, with $t_1 = (x_1')^{-1} \cdot (x_1' \cdot x_3)$, $t_1' = x_3$, $t_2 = x_1 \cdot (x_2 \cdot (x_1 \cdot x_2)^{-1})$ and $t_2' = e$. The address $v = 2$ gives the subterm $x_1' \cdot x_3$ of t_1, not a variable, and these terms t_1/v and t_2 are unifiable with a most general unifier s mapping x_1' to x_1 and x_3 to $x_2 \cdot (x_1 \cdot x_2)^{-1}$. We have $\hat{s}[t_1'] = \hat{s}[x_3] = x_2 \cdot (x_1 \cdot x_2)^{-1}$ and $\hat{s}[t_1][v/\hat{s}[t_2']] = (x_1^{-1} \cdot (x_1 \cdot x_3))[v/e] = x_1^{-1} \cdot e$. Therefore $(x_2 \cdot (x_1 \cdot x_2)^{-1}, \ x_1^{-1} \cdot e)$ is a critical pair.

The next lemma shows that all other "critical situations" of the kind we have been considering are pairs which can be obtained by substitutions from a critical pair.

Lemma 7.3.12 *Let R be a term rewriting system and let $t_1 \to t_1'$ and $t_2 \to t_2'$ be two reduction rules of R. Let v be an address of t_1 such that t_1/v is a subterm which is not a variable. If there exist substitutions s_1 and s_2 such that $\hat{s}_1[t_1/v] = \hat{s}_2[t_2]$, then there exist a critical pair (t', t'') and a substitution s such that*

$$\hat{s}[t'] = \hat{s}_1[t_1'] \text{ and } \hat{s}[t''] = \hat{s}_1[t_1][v/\hat{s}_2[t_2']].$$

Proof: We can always find a bijection $r : X \to X$ which renames the variables in t_2 if necessary, so that we may assume that $\text{var}(t_1) \cap \text{var}(\hat{r}[t_2]) = \emptyset$. Then by Lemma 7.3.8 there is a substitution $\bar{s}$ such that $\hat{\bar{s}}[t_1/v] = \hat{\bar{s}}[\hat{r}[t_2]]$;

and from the proof of Lemma 7.3.8 this substitution is given by

$$\bar{s}(x) = \begin{cases} s_1(x), & \text{if } x \in var(t_1) \\ s_2(r^{-1}(x)), & \text{if } x \in var(\hat{r}[t_2]). \end{cases}$$

By assumption we have $var(t_1') \subseteq var(t_1)$ and $var(t_2') \subseteq var(t_2)$. By Proposition 7.3.4 there exists a most general unifier m of t_1/v and $\hat{r}[t_2]$. Since $\bar{s}$ is also a unifier of these terms, we have $\bar{s} \leq m$, and there is a substitution s with $\bar{s} = s \circ m$. Then $\bar{s}[t] = \hat{s}_1[\hat{m}[t]]$ for all terms t.

Now we define a critical pair (t', t'') by $t' = \hat{m}[t_1']$ and $t'' = \hat{m}[t_1[v/\hat{r}[t]]]$. Then we have $\hat{s}_1[t_1'] = \hat{\bar{s}}[t_1'] = \hat{s}[\hat{m}[t_1']] = \hat{s}[t']$, and $\hat{s}_1[t_1][v/\hat{s}_2[t_2']] = \hat{\bar{s}}[t_1][v/\hat{s}[\hat{r}[t_2']]] = \hat{\bar{s}}[t_1[v/\hat{r}[t_2']]] = \hat{s}[\hat{m}[t_1[v/\hat{r}[t_2']]]] = \hat{s}[t'']$. ∎

Using the previous results we obtain the following.

Proposition 7.3.13 *A term rewriting system is locally confluent iff all its critical pairs are convergent.* ∎

As we saw after Definition 7.2.1, in a terminating term rewriting system every term t can be reduced to a normal form, while confluence insures that each term has a unique normal form. We get the following result.

Theorem 7.3.14 *(Knuth-Bendix Theorem)* *A terminating term rewriting system R is confluent if and only if for all critical pairs (t', t''), the terms t' and t'' have a common (unique) normal form.*

Proof: Let $R \subseteq W_\tau(X) \times W_\tau(X)$ be a terminating term rewriting system. If (t', t'') is a critical pair in R then there is a term t with $t \underset{R}{\to} t'$ and $t \underset{R}{\to} t''$. If R is confluent then by Theorem 7.1.10 it has the Church-Rosser property, and as we saw in the remark after Definition 7.2.1, every term t has a unique normal form $NF(t)$. Then by Theorem 7.1.9 we have $NF(t') = NF(t'')$.

If conversely t' and t'' have a common normal form, for all critical pairs (t', t''), then R is locally confluent by Proposition 7.3.13. Since R is terminating, it is confluent by Theorem 7.2.4. Moreover in this case the normal form of any term is unique. ∎

Theorem 7.3.14 may be used to develop an algorithm to complete a given
term rewriting system. This algorithm is called the *Knuth-Bendix completion procedure*. This procedure produces a sequence $(R_n \mid n \in I\!N)$ of finite
sets of reduction rules satisfying the following conditions:

- Each R_n is terminating and equivalent to R,
- Critical pairs in R_n are convergent in R_{n+1}, for each n.

The main problem here is that termination has to be preserved. For this the
following Lemma is needed.

Lemma 7.3.15 *Let R be a term rewriting system. If $>$ is a terminating
fully invariant relation on $W_\tau(X)$, then R is terminating whenever $t > t'$
for each rule $t \to t'$ of R.*

Proof: Let $R \subseteq W_\tau(X)$ be a term rewriting system. Let $\underset{R}{\to}$ be the fully
invariant relation on $W_\tau(X)$ generated by R. If we have R contained in $>$,
with $>$ fully invariant, it follows that $\underset{R}{\to}$ is also contained in $>$. But as a
subrelation of a terminating relation, $\underset{R}{\to}$ is also terminating.   ■

To use this result in our algorithm, it must be decidable for each pair (t, t')
of terms whether $t > t'$ holds or not. If a critical pair (t, t') is produced
during the completing process for which the two components t and t' are
incomparable with respect to $>$, then the Knuth-Bendix algorithm gives an
error-message.

The inputs of the Knuth-Bendix algorithm are a finite term rewriting system
$R = \{(t_1, t'_1), \ldots, (t_m, t'_m)\}$ and a terminating fully invariant relation $>$. The
output is a complete term rewriting system equivalent to R, or an error
message.
The sequence $(R_n \mid n \in I\!N)$ of finite term rewriting systems is produced, if
possible, as follows:

begin $R_{-1} := \emptyset; n := 0;$
      for $i = 1, \ldots, m$ **do**
             if t_i, t'_i are incomparable **then**   stop with error message
                    else if $t_i > t'_i$ **then**   add $t_i \to t'_i$ to R_0
                              else   add $t'_i \to t_i$ to R_0;
      while $R_n \neq R_{n-1}$ **do**

begin $R_{n+1} := R_n$; compute $NF(t)$ and $NF(t')$ for any
          critical pair (t, t') in R_n;
if $NF(t), NF(t')$ are incomparable
then stop with error message
else if $NF(t) > NF(t')$ **then** add $NF(t) \rightarrow NF(t')$ to R_{n+1}
                   else add $NF(t') \rightarrow NF(t)$ to R_{n+1}
$n := n + 1$
end

end.

If this algorithm calculates a sequence $(R_n \mid n \in \mathbb{N})$ of finite term rewriting systems, without ever generating an error message, then the union of all these R_n is a completion of R.

7.4 Termination of Term Rewriting Systems

As we have seen, termination is an important property of term rewriting systems. In general termination is an undecidable property. Since a relation $\rightarrow$ on a set A is terminating if and only if its transitive closure $\rightarrow^T$ is terminating, we may assume without loss of generality that the relation is transitive. It is also necessary that a terminating relation be irreflexive. Relations which are both irreflexive and transitive are called *proper* order relations. To verify termination using the Knuth-Bendix completion algorithm, we need methods which are based on proper fully invariant orders.

Definition 7.4.1 A terminating proper order on $W_\tau(X)$ is called a *reduction order* if it is fully invariant.

Then Lemma 7.3.15 can be reformulated as follows.

Lemma 7.4.2 *Let R be a term rewriting system. Then R is terminating if there is a reduction order $>$ on $W_\tau(X)$ such that $t > t'$ for all reduction rules $t \rightarrow t'$ in R.*

There are several ways to define a reduction order for a term rewriting system. We must choose the best one, meaning that we want a reduction order which allows a proof of termination and for which membership can be effectively tested. The latter point is important since the Knuth-Bendix completion method requires comparisons between terms at every step of the process.

Here we will describe the Knuth-Bendix ordering ([66]). We assume that there is a total order $\leq$ defined on the set $\{f_i \mid i \in I\}$ of operation symbols of our type, and that we have a function $\varphi : \{f_i \mid i \in I\} \to I\!\!N$ called a *weight function*, which has the following properties:

(i) If f_0 is a nullary operation symbol then $\varphi(f_0) > 0$;

(ii) If f_1 is a unary operation symbol and $\varphi(f_1) = 0$, then f_1 is the greatest element with respect to $\leq$.

This weight function can then be extended inductively to a mapping $\hat{\varphi} : W_\tau(X) \to I\!\!N$ on the set of all terms, as follows:

(i) $\hat{\varphi}(x) = \min\{\varphi(f_0) \mid f_0 \text{ is nullary}\}$, for all $x \in X$,

(ii) $\hat{\varphi}(f_i(t_1, \ldots, t_{n_i})) = \varphi(f_i) + \hat{\varphi}(t_1) + \cdots + \hat{\varphi}(t_{n_i})$, for any term $f_i(t_1, \ldots, t_{n_i})$ where f_i is an n_i-ary operation symbol and $t_1, \ldots, t_{n_i}$ are terms in $W_\tau(X)$.

We will denote by $occ_x(t)$ the number of occurrences of the variable x in the term t.

Definition 7.4.3 Let $\leq$ be a total order on the set $\{f_i \mid i \in I\}$ of operation symbols of type τ, and let $\varphi : \{f_i \mid i \in I\} \to I\!\!N$ be a weight function. The *induced Knuth-Bendix order* $>_{KB}$ on $W_\tau(X)$ is defined as follows. We set $t >_{KB} t'$ iff either

(i) $\hat{\varphi}(t) > \hat{\varphi}(t')$ and $occ_x(t) > occ_x(t')$ for all $x \in X$, or

(ii) $\hat{\varphi}(t) > \hat{\varphi}(t')$ and $occ_x(t) = occ_x(t')$ for all $x \in X$ and either

(ii)(a) $t = f(\cdots f(t') \cdots)$ and f is the greatest element in $\{f_i \mid i \in I\}$ or

(ii)(b) $t = f_i(t_1, \ldots, t_{n_i})$, $t' = f'_j(t'_1, \ldots, t'_{n_j})$ and either $f_i > f'_j$ or $f_i = f'_j$ and $t_1 = t'_1, \ldots, t_{i-1} = t'_{i-1}$ and $t_i >_{KB} t'_i$ for some $i = 1, \ldots, n_i$ where $n_i = n_j$.

Then we can verify that this induced order is indeed a reduction order.

Proposition 7.4.4 *Every Knuth-Bendix order is a reduction order.*

Proof: Let τ be a fixed type, and let $\leq$ be a total order on the set $\{f_i \mid i \in I\}$ of operation symbols and $\varphi : \{f_i \mid i \in I\} \to I\!\!N$ be a weight function. We consider the induced Knuth-Bendix order $>_{KB}$ on $W_\tau(X)$. We have to show that $>_{KB}$ is fully invariant and terminating.

Assume that t_1 and t_2 are terms from $W_\tau(X)$ with $t_1 >_{KB} t_2$, and let $t \in W_\tau(X)$ be an arbitrary term. We have to prove that $t[u/t_1] >_{KB} t[u/t_2]$ for every address u of t. By the transitivity of $>_{KB}$ it suffices to verify this for all addresses u of t with $u \in I\!\!N \setminus \{0\}$.

Since $t_1 >_{KB} t_2$, either case (i) or case (ii) of Definition 7.4.3 holds. If case (i) holds, and $\hat{\varphi}(t) > \hat{\varphi}(t')$ and $occ_x(t) > occ_x(t')$ for all $x \in X$, then also $\hat{\varphi}(t[u/t_1]) > \hat{\varphi}(t[u/t_2])$ and $occ_x(t[u/t_1]) > occ_x(t[u/t_2])$ for all $x \in X$. Then again by case (i) of Definition 7.4.3 we have $t[u/t_1] >_{KB} t[u/t_2]$. If instead case (ii) holds, and $\hat{\varphi}[t_1] = \hat{\varphi}[t_2]$ and $occ_x(t[u/t_1]) = occ_x(t[u/t_2])$ for all $x \in X$, then by the same case we have $t[u/t_1] >_{KB} t[u/t_2]$. This shows that $>_{KB}$ is closed under the replacement rule. In a similar way it can be shown that $>_{KB}$ is closed under all substitutions, so that $>_{KB}$ is fully invariant.

Next we show by contradiction that $>_{KB}$ is terminating. Suppose that there is an infinite chain $t_0 >_{KB} t_1 >_{KB} t_2 >_{KB} \cdots >_{KB} \cdots$ of terms. By the definition of the Knuth-Bendix order, such a sequence is only possible if our type contains a nullary operation symbol. Also by definition we have $var(t_n) \subseteq var(t_0)$, for all $n \geq 0$. Therefore there is a substitution $s : X \to W_\tau(X)$ such that all the variables occurring in any t_i are mapped to nullary terms. Then $\hat{s}(t_0) >_{KB} \hat{s}(t_1) >_{KB} \cdots >_{KB} \cdots$ is an infinite chain of nullary terms. For any nullary term, let us denote by k_n the number of operation symbols of arity n in t. Then we can prove by induction that $k_0 = 1 + k_2 + 2k_3 + 3k_4 + \cdots$.

Since each nullary operation symbol has positive weight, the weight w of any nullary term is greater than or equal to k_0. Therefore for a fixed weight w there is only a finite number of choices for $k_0, k_2, k_3, \cdots$. If each unary operation symbol has a positive weight we have $w \geq k_1$ and therefore there are only finitely many nullary terms of weight w. That means an infinite chain

$$t_0 >_{KB} t_1 >_{KB} t_2 >_{KB} \cdots >_{KB} \cdots$$

is impossible unless there is a unary operation symbol of weight zero.

Thus we now assume that there is a unary operation symbol f_0 of weight zero. We define a mapping h from the set of all unary terms into itself such that $h(t)$ is the unary term obtained from t by replacing all occurrences of f_0 by another unary term. Clearly the mapping h preserves the weight and there are only finitely many nullary terms of weight w (by the previous remark). Now we show that there is no infinite chain $t_0 >_{KB} t_1 >_{KB} t_2 >_{KB} \cdots$ of nullary terms such that $h(t_0) = h(t_1) = h(t_2) = \cdots$. Each nullary term t can be regarded as a word over the set of all nullary operation symbols; that means t can be written in the form $t = f_0^{r_1} \alpha_1 f_0^{r_2} \alpha_2 \cdots f_0^{r_1}$ or $t = \alpha_1 f_0^{r_1} \alpha_2 f_0^{r_2} \cdots \alpha_n f_0^{r_n} \alpha_n$, where $r_1, \ldots, r_n$ are natural numbers and $\alpha_1, \ldots, \alpha_n$ are words built up by nullary operation symbols except f_0. Let $r(t) = (r_1, \ldots, r_n)$ be the n-tuples consisting of the exponents of the occurrences of f_0.

It can be shown that if $h(t) = h(t')$ then $t >_{KB} t'$ iff $r(t) >_{lex} r(t')$, where $>_{lex}$ is the lexicographic order on n-tuples. It can easily be shown that $>_{lex}$ is terminating (on words of equal length). This completes the proof of Proposition 7.4.4.   ■

Example 7.4.5 As an example of the Knuth-Bendix order we consider the type $\tau = (2, 1, 0)$, with operation symbols $\cdot$, $^{-1}$ and e (corresponding to groups). On this set of operation symbols we define the following order

$$e \leq \cdot \leq ^{-1}.$$

We also use the weight function $\varphi : \{\cdot, ^{-1}, e\} \to \mathbf{N}$ with $\varphi(\cdot) = 1$, $\varphi(e) = 1$ and $\varphi(^{-1}) = 0$. For the extension $\hat{\varphi}$ to terms, this gives for instance
$\hat{\varphi}(x) = 1$ for all variables $x \in X$,
$\hat{\varphi}((x \cdot y) \cdot z) = \varphi(\cdot) + (\varphi(\cdot) + \varphi(x) + \varphi(y)) + \varphi(z) = 5$,
$\hat{\varphi}(x \cdot (y \cdot z)) = \varphi(\cdot) + \varphi(x) + (\varphi(\cdot) + \varphi(y) + \varphi(z)) = 5$.

Then by Definition 7.4.3 (ii)(b), second case, we have

$$(x \cdot y) \cdot z \quad >_{KB} \quad x \cdot (y \cdot z).$$

Also, $(x \cdot y)^{-1} >_{KB} y^{-1} \cdot x^{-1}$ by Definition 7.4.3 (ii)(b), first case, and $(y^{-1})^{-1} > y$ by Definition 7.4.3 (ii)(a). This order can be used to prove the termination of the term rewriting system for groups from Example 7.3.7 (see [66]).

The following proposition generalizes Lemma 7.3.15.

Proposition 7.4.6 *A term rewriting system R is terminating if there is a terminating and invariant (under substitution) proper order $>$ on $W_\tau(X)$ such that $\hat{s}(t) > \hat{s}(t')$ for all reduction rules $t \to t'$ in R and all substitutions $s : X \to W_\tau(X)$.*

Proof: For a given term rewriting system R, we will denote by $\underset{R}{\to}$ the fully invariant relation generated by R. Then $\underset{R}{\to}$ is the invariant closure of the set

$$S = \{(\hat{s}(t), \hat{s}(t')) \mid t \to t' \in R \text{ and } s : X \to W_\tau(X) \text{ is a substitution}\}.$$

Assume that $>$ is a terminating invariant proper order on $W_\tau(X)$, such that $\hat{s}(t) > \hat{s}(t')$ for all $t \to t' \in R$ and all substitutions $s : X \to W_\tau(X)$. Then S is a subset of $>$, and since $>$ is invariant we have $\underset{R}{\to}$ a subset of $>$ too. Thus $\underset{R}{\to}$ is terminating. $\blacksquare$

Unfortunately, there are many cases for which the Knuth-Bendix ordering is not suitable. Suppose we want to show the termination of $t = x \cdot (y^{-1} \cdot y) \longrightarrow (y \cdot y)^{-1} \cdot x$, using the same order on the fundamental operation symbols and the same weight function as in Example 7.4.5. The only possible case is Definition 7.4.3 (ii)(b), second case, but then we must have $x >_{KB} (y \cdot y)^{-1}$, and this is a contradiction.

Several other orderings have therefore been developed; see for instance the Handbook of Formal Languages, Vol. 3 ([54]).

7.5 Exercises

7.5.1. Let ρ be a relation on a set A. Prove that

$$\rho^{SRT} = (\rho \cup \rho^{-1} \cup \triangle_A)^T.$$

7.5.2. Let ρ_1 and ρ_2 be relations on a set A. Prove that for all $n \in \mathbb{N}$,

$$\rho_1 \circ \rho_2^{RT} \subseteq \rho_2^{RT} \circ \rho_1^{RT} \Rightarrow (\rho_1 \circ \rho_2^{RT})^{(n)} \subseteq \rho_2^{RT} \circ \rho_1^{RT}.$$

7.5.3. Let $Subst_\tau$ be the set of all substitutions of type τ. Let $\leq$ be the relation defined on $Subst_\tau$ by $s_1 \leq s$ iff there is a substitution s such that $s_1 = s \circ s_2$. Prove that $\leq$ is a reflexive and transitive order on $Subst_\tau$, and that the relation $\sim$ defined by $s_1 \sim s_2$ iff $s_1 \leq s_2$ and $s_2 \leq s_1$ is an equivalence relation on $Subst_\tau$.

7.5.4. Prove that the relation $<$ defined on $Subst_\tau$ by $s_1 < s_2 :\Leftrightarrow s_1 \leq s_2$ and $s_1 \not\sim s_2$ is noetherian.

7.5.5. Let H be the commutative semigroup generated by the four elements a, b, c and s, and by the defining equations

$$(E) \quad as = c^2 s \text{ and } bs = cs.$$

Let $\rightarrow$ be the smallest relation on H such that $cs \rightarrow bs$, $b^2 s \rightarrow as$, and if $x \rightarrow y$ then $ux \rightarrow uy$ for all x, y and u in H. Prove that $\rightarrow$ is noetherian and has the Church-Rosser property.

Chapter 8

Algebraic Machines

An important topic in theoretical Computer Science is the study of which problems can be solved by machines such as automata or Turing Machines which arise as abstractions of "real-life" computers. As we saw in Example 1.2.16, automata are in fact multi-based algebras, and so we can think of them as "algebraic machines." The most important automata are those in which the set of states is finite, and we consider *finite automata* or *finite state machines*; these model the realistic situation of a computer with a finite amount of memory and other resources.

For any class of automata of a fixed type we obtain a family of languages, the languages recognized by the automata in the class. The relation of *recognition* between automata and languages thus defines a Galois-connection between the class of all finite automata of a given type and the family of all languages. The class of all regular languages is described in Section 8.1, while Section 8.2 introduces finite automata and the languages they recognize. In Section 8.2 we present Kleene's Theorem, which shows that the languages recognizable by finite automata are precisely the regular languages.

Finite automata are machines which recognize languages made up of "words," that is, of terms on the free monoid on a finite alphabet. In this way they are in fact a special case of more general machines which recognize terms of any fixed type. Referring to terms as trees, we study machines called *tree-recognizers*. Sections 8.4 to 8.8 present the theory, including a Kleene-type theorem, of such machines. Finally in Section 8.9 we study Turing machines, which we use in the next section to describe the solution to

147

various decidability problems in Universal Algebra.

8.1   Regular Languages

In order to discuss the languages accepted by finite automata, we must first recall the definition of words from Example 6.4.5. Let $X_n = \{x_1, \ldots, x_n\}$ be a finite alphabet of size $n \geq 0$. We denote by X_n^* the universe set of the free monoid generated by X_n. When we denote the binary operation of the monoid by juxtaposition, and use the associativity of this operation in a monoid to justify omitting all brackets, we see that any monoid term from X_n^* may be expressed as a *word* composed of the letters from X_n. Note that the letters may be repeated; for instance, each of $x_1 x_2$, $x_2 x_1 x_1$ and $x_1 x_2 x_3 x_3$ is a word on the alphabet X_3. Formally, any word w may be expressed as $w = x_{i_1} x_{i_2} \cdots x_{i_m}$, for some $m \geq 0$ and $x_{i_1}, \ldots, x_{i_m} \in X$. The number m is called the *length* of the word w, and is denoted by $|w|$. The special case that $m = 0$ corresponds to the *empty word*, denoted by e. The set X_n^+ is defined to be the set of all non-empty words on the alphabet X_n. This set is the universe of the free semigroup generated by the set X_n. Of course, X_n^+ and X_n^* are (the universes of) a semigroup and a monoid respectively, with the binary operation of juxtaposition or *concatenation* of words.

A *language*, or more precisely a *language over the alphabet* X_n, is simply any subset of X_n^*. The family of all possible languages on X_n is then the power set of X_n^*. We can introduce an algebraic structure on this power set, by defining three operations on sets of languages. The first operation is the binary one of union (where the union of two languages is just their set-theoretic union). The second binary operation we need is called the *product*, and is usually denoted by juxtaposition: for any two languages U and V, we set

$$UV := \{uv \mid u \in U, v \in V\}.$$

Note that this binary product is associative, meaning that $U(VW) = (UV)W$ for all languages U, V and W on X_n, and satisfies in addition the properties that $U\emptyset = \emptyset U = \emptyset$ and $U\{e\} = \{e\}U = U$ for every language U.

For any language U, we inductively define powers U^m for all $m \geq 0$, by

(i) $U^0 = \{e\}$,   and

(ii) $U^m = U^{m-1}U$, for $m \geq 0$.

Then we define $U^* = \bigcup_{m \in \mathbb{N}} U^m$ and $U^+ = \bigcup_{m \geq 1} U^m$.

A word $w \in X_n^*$ belongs to U^* if and only if it is the empty word or it can be expressed in the form $u_1 u_2 \cdots u_m$ for some $m \geq 1$ and $u_1, \ldots, u_m \in U$. Clearly X_n^m is the set of all words of length m on the alphabet X_n, and $X_n^* = \bigcup_{m \in \mathbb{N}} X_n^m$.

The unary operation taking a language U to the language U^* is called *iteration*. The three operations just discussed, the union, product and iteration, are called the *regular language operations*. A regular language is any language which can be built out of languages consisting of a single length-one word, using these three operations only.

Definition 8.1.1 The set $RegX$ of all regular languages over an alphabet X is the smallest set R such that

(i) $\emptyset \in R$ and $\{x\} \in R$ for each $x \in X$,   and

(ii) for any U and V in R, all of $U \cup V$, UV and U^* are in R.

It is easy to see from this definition that all finite languages are regular. The set $RegX$ is the smallest set of languages over X which contains all the finite languages and is closed under the three regular language operations.

Algebraically, we have defined three operations on the base set $\mathcal{P}(X_n^*)$. If we define two binary operation symbols $+$ and $\cdot$ to represent the union and product operations, and a unary operation symbol $*$ for the iteration operation, then using variables and nullary operation symbols for $\emptyset$ and e we can define terms in this new language. These terms are called *regular expressions*, and they correspond precisely to regular languages. As usual, we omit the binary operation symbol and just use juxtaposition for the product operation.

Example 8.1.2 Let $X_2 = \{x_1, x_2\}$. Some elements of $RegX_2$ are $\{x_1\}$, $\{x_2\}$, $\{x_1 x_2\}$, $\{x_1 x_2\} \cup \{x_2 x_2\} = \{x_1 x_2, x_2 x_2\}$.

8.2   Finite Automata

In Section 1.2 we introduced a finite automaton as a multi-based algebra of
the form $\mathcal{H} = (Z, X; \delta, z_0)$, where

(1) Z is a finite non-empty set of states;
(2) X is the (finite) input alphabet;
(3) $\delta : Z \times X \to Z$ is the state transition function;   and
(4) $z_0 \in Z$ is an initial state.

We sometimes also want to have some states designated as final states. In
this case we have a special subset Z' of Z, called the set of final states, and
we write our automaton as a quintuple $\mathcal{H} = (Z, X; \delta, z_0, Z')$. In this chapter
we shall consider only the case that the input alphabet is a finite set X_n,
but automata can also be defined on an infinite alphabet.

The "action" of a finite state automaton is given by the state transition
function δ. We think of the equation $\delta(z, x) = y$ as telling us that when the
machine receives input x when it is in state z, it moves or changes to state
y. Any state transition function δ can be uniquely extended to a mapping

$$\hat{\delta} : Z \times X^* \to Z,$$

where X^* is the set of all words over X, in the following inductive way:

(i) $\hat{\delta}(z, e) = z$, for each state $z \in Z$, where e is the empty word,   and
(ii) $\hat{\delta}(z, wx) = \delta(\hat{\delta}(z, w), x)$, for each state $z \in Z$, each letter $x \in X$ and each
word $w \in X^*$.

The inductive step (ii) of this definition says that to compute which state
the machine is in, when it has started in state z with input word wx, we
first find what state it is in after input w, then move from that state with
input x. That is, words are read in one letter at a time, with a change of
state each time.

Now that we know how an automaton reads in words, we can describe how
an automaton acts as a language recognizer, over the alphabet X. We input
a word w to the machine, in the initial state z_0, and compute the resulting
state $\hat{\delta}(z_0, w)$. If this state is a final state from the set Z', the input word w is
recognized or *accepted* by the automaton; otherwise, it is said to be rejected

by the automaton. The *language recognized by* an automaton $\mathcal{H}$ is the set $L(\mathcal{H})$ of all the words from X^* which are recognized by the automaton. Thus

$$L(\mathcal{H}) = \{w \in X^* \mid \hat{\delta}(z_0, w) \in Z'\}.$$

A language L over an alphabet X is said to be *recognizable* if there exists an automaton $\mathcal{H}$ with alphabet X such that $L = L(\mathcal{H})$. We will denote by $RecX$ the set of all recognizable languages over X, and by Rec the family of all recognizable languages.

Notice that the finiteness of the state set is important here; otherwise, every language over X would be recognizable. We now have two classes of languages over an alphabet: the regular languages of the previous section, and the new recognizable languages. The first important result of finite automata theory is *Kleene's Theorem* (see S. C. Kleene, [63]), which says that these two classes are in fact the same. We can easily prove one direction of this theorem now, that any recognizable language is regular, but the other direction will take us more work.

Theorem 8.2.1 *Let X be a finite alphabet. Any recognizable language over X is regular, so that $RecX \subseteq RegX$.*

Proof: Let $L = L(\mathcal{H})$ be a language recognized by an automaton $\mathcal{H}$. We will show that L is a regular language. We may assume that the states of H are enumerated as $\{z_0, z_1, \ldots, z_n\}$, where z_0 is an initial state. For each $0 \leq k \leq n + 1$ and each $0 \leq i, j \leq n$, let L_{ij}^k be the set of words w such that $\hat{\delta}(z_i, w) = z_j$ and such that in the computation of $\hat{\delta}(z_i, w)$ no intermediate state z_m with $m \geq k$ is used. It is clear that $L(\mathcal{H})$ is the union of the sets L_{0j}^{n+1} over those indices j for which z_j is a final state. This means that it will suffice to prove, by induction on k, that the sets L_{ij}^k are regular, that is, that they can be represented by regular expressions.

If $k = 0$, then no intermediate steps are allowed in our computations, and the set L_{ij}^k is either $\{x \in X \mid \delta(z_i, x) = z_j\}$, if $i \neq j$, or the union of this set with the set $\{e\}$, if $i = j$. In either case the set L_{ij}^k is regular.

Now assume that all the sets L_{ab}^k are regular. Then L_{ij}^{k+1} is the union of L_{ij}^k with the set of all words w such that $\hat{\delta}(z_i, w) = z_j$ and the state z_k is used at least once in the computation. We show that this latter set is also regular.

Now words in this set use state z_k at least once, and we can describe them in terms of the occurrences of z_k. Up to the first occurrence of z_k, no state z_m with $m \geq k$ is used, and the same is true between any two consecutive uses of z_k and after the last use of z_k. Thus we can express any such word w in the form $uw_1 \cdots w_l v$, where $u \in L_{ik}^k$, $v \in L_{kj}^k$ and $w_r \in L_{kk}^k$ for each r. This means that we can express $L_{ij}^{k+1} = L_{ij}^k \cup L_{ik}^k (L_{kk}^k)^* L_{kj}^k$. By the induction hypothesis the sets used in this expression are each regular, and thus so is L_{ij}^{k+1}. ∎

We present the following example of the language recognized by a finite automaton.

Example 8.2.2 Let $\mathcal{H} = (Z, X; \delta, z_0, Z')$, where
$Z = \{z_0, z_1, z_2, z_3, z_4\}$, $X = \{0, 1\}$, $Z' = \{z_3\}$ and the state transition function δ is given by

δ	0	1
z_0	z_1	z_3
z_1	z_2	z_3
z_2	z_3	z_0
z_3	z_4	z_4
z_4	z_4	z_4

The graph below describes the work of this automaton. Notice that an arrow from state z_i to state z_j is labelled with a letter x when $\delta(z_i, x) = z_j$.

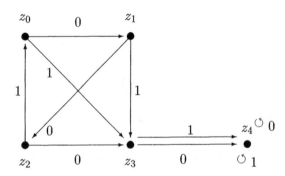

Consider the word $w = 0001101$. The machine reads this word in one letter at a time, from left to right, starting with input 0 in the initial state z_0. This gives us $\delta(z_0, 0) = z_1$, $\delta(z_1, 0) = z_2$, $\delta(z_2, 0) = z_3$, $\delta(z_3, 1) = z_4$, $\delta(z_4, 1) = z_4$, $\delta(z_4, 0) = z_4$ and finally $\delta(z_4, 1) = z_4$. When the word is completely read in, the machine ends in state z_4, which is not one of the final states from Z'. Therefore the word 0001101 is rejected. Inputting the word 00101 gives us $\delta(z_0, 0) = z_1$, $\delta(z_1, 0) = z_2$, $\delta(z_2, 1) = z_0$, $\delta(z_0, 0) = z_1$ and then $\delta(z_1, 1) = z_3$. This word results in a final state; so it is accepted by this machine.

We want to show now that the language accepted by this automaton is regular, by finding a regular expression for it. We notice that the final state z_3 can be reached only after a final step from states z_0, z_1 or z_2, while if state z_4 is reached there is no way to leave it for another state. Thus we may travel through the states z_0, z_1 and z_2 as many times as we wish, but must then finish with a transition to z_3. For instance, if the last transition used goes from z_0 to z_3, the word accepted has the form $(001)^n 1$, for some $n \geq 1$. This gives a set of words represented by the regular expression $(001)^* 1$. If the last transition used goes from z_1 or z_2 to z_3, we get words of the form $(001)^n 01$ or $(001)^n 000$, respectively.

Thus we guess that the language $L(\mathcal{H})$ is represented by the regular expression $(001)^* (1 + 01 + 000)$. Now we must prove that this is indeed the case, by induction.

As a first step, we show that for all $n \geq 0$ we have $\hat{\delta}(z_0, (001)^n) = z_0$. For the base case of $n = 0$ our word $(001)^0 = e$, the empty word, and here $\hat{\delta}(z_0, e) = z_0$. Inductively, suppose that the claim is true for a value n. Then $\hat{\delta}(z_0, (001)^{n+1}) = \hat{\delta}(z_0, (001)^n 001) = \delta(\hat{\delta}(z_0, (001)^n 00), 1) = \delta(\delta(\hat{\delta}(z_0, (001)^n 0), 0), 1) = \delta(\delta(\delta(\hat{\delta}(z_0, (001)^n), 0), 0), 1) = \delta(\delta(\delta(z_0, 0), 0), 1) = \delta(\delta(z_1, 0), 1) = \delta(z_2, 1) = z_0$.

Next we prove the converse of the first claim, that if $\hat{\delta}(z_0, w) = z_0$, then $w = (001)^n$ for some $n \geq 0$. This is obvious if w is the empty word e; so we will assume that $w = w_1 r$. Then since $\delta(\hat{\delta}(z_0, w_1), r) = \hat{\delta}(z_0, w) = z_0$, we see that $r = 1$ and $\hat{\delta}(z_0, w_1) = z_2$. In a similar way we get $w = w_2 001$, where $\hat{\delta}(z_0, w_2) = z_0$. Since w_2 is a shorter word than w, we conclude by the induction hypothesis that $w_2 = (001)^n$ for some n, so that $w = (001)^{n+1}$.

Now we let w be any word belonging to the language represented by the expression $(001)^* (1 + 01 + 000)$. Then w has the form $(001)^n v$ for some $n \geq 0$, where v is one of the words 1, 01 or 000. If $v = 1$, then $\hat{\delta}(z_0, w) =$

$\hat{\delta}(z_0, (001)^n 1) = \delta(\hat{\delta}(z_0, (001)^n), 1) = \delta(z_0, 1) = z_3$, and so the word w is in $L(\mathcal{H})$. The other two cases for v are very similar, and we omit the details.

Conversely, we must show that any word w in $L(\mathcal{H})$ has a representation of the required form. We must have $\hat{\delta}(z_0, w) = z_3$, and so $w \neq e$. Let $w = w_1 r$ where r is either 0 or 1. For $r = 0$ we have $\delta(\hat{\delta}(z_0, w_1), 0) = z_3$, which forces $\hat{\delta}(z_0, w_1) = z_2$. Similarly, if we write $w_1 = w_2 0$ we have $\hat{\delta}(z_0, w_2) = z_1$, and again $w_2 = w_3 0$ with $\hat{\delta}(z_0, w_3) = z_0$. Using the previous result, we obtain $w_3 = (001)^n$ for some n, so that $w = (001)^n 000$ and belongs to the language represented by our regular expression. The case that $r = 1$ is handled similarly.

Our next aim is to prove the other direction of Kleene's Theorem, by showing that for any given regular expression there is a finite automaton which recognizes the corresponding language. To do this we need the concept of a *derived language*. Let $L \subset X^*$ be a language over an alphabet X. For any letter $a \in X$ we define the derived language L_a of L with respect to a by

$$L_a := \{w \in X^* \mid aw \in L\}.$$

We want to find a finite automaton to recognize a given language L. Suppose that for each $a \in X$ we have found an automaton $\mathcal{H}_a$ which recognizes the derived language L_a. We form a new automaton $\mathcal{H}$ as follows. We may assume (by relabelling if necessary) that the automata $\mathcal{H}_a$ have no states in common. Our new set of states consists of the union of the state sets of the $\mathcal{H}_a$, along with a new initial state labelled z_0. We define a transition function δ so that for each letter a the state $\delta(z_0, a)$ is equal to the initial state of $\mathcal{H}_a$, and otherwise δ copies the state transitions of the $\mathcal{H}_a$. The final states of $\mathcal{H}$ are all states which are final in any of the $\mathcal{H}_a$. It is easy to see that $\mathcal{H}$ recognizes the language L.

However, we cannot assume inductively that the $\mathcal{H}_a$ have been constructed, since unfortunately it is possible for the derived language L_a to equal L. This happens for instance when $L = \{a, b\}^*$. This problem can be overcome by the use of loops. More serious is the possibility that there might be infinitely many different languages derived from the initial one, in which case the automaton resulting from our construction would not be finite. The next Lemma deals with this problem.

We extend our definition of derived languages, to allow words as well as single letters. That is, for any language L and any word $w \in X^*$, we set

$$L_w := \{v \in X^* \mid wv \in L\}.$$

Notice that such languages L_w can be obtained by repeatedly forming derived languages. For example, $L_{aab} = (((L_a)_a)_b)$.

Lemma 8.2.3 *If L is a language represented by a regular expression on a finite alphabet X, then there are only finitely many distinct languages of the form L_w for $w \in X^*$.*

Proof: Let E be another language represented by a regular expression, and let $L_E = \bigcup\{L_w \mid w \in E\}$. The language $L_w = L_{\{w\}}$ is represented by a regular expression, since any single word has this property. We show by induction on the construction of the regular expression for L that there are only finitely many distinct languages of the form L_E, and from this we get the required result.

We consider first the three base cases, that $L = \emptyset$, $L = \{e\}$ or $L = \{a\}$ for some letter $a \in X$. If $L = \emptyset$ then L_E can only be the empty set too. If $L = \{e\}$, then L_E is either $\{e\}$ if $e \in E$ or the empty set again if not. If $L = \{a\}$ for a letter a, then L_E is one of the following:

$$
\begin{array}{ll}
\{e, a\}, & \text{if } a, e \in E; \\
\{e\}, & \text{if } a \in E \text{ and } e \notin E; \\
\{a\}, & \text{if } e \in E \text{ and } a \notin E; \\
\emptyset, & \text{if } a, e \notin E.
\end{array}
$$

This gives only finitely many possibilities for L_E.

For the inductive step, we first consider languages $L_1 \cup L_2$ formed by taking unions. It is evident that $(L_1 \cup L_2)_w = (L_1)_w \cup (L_2)_w$, and if we extend to sets E by taking unions we have $(L_1 \cup L_2)_E = (L_1)_E \cup (L_2)_E$. Thus if two sets E and E' produce distinct languages $(L_1 \cup L_2)_E \neq (L_1 \cup L_2)_{E'}$, then either $(L_1)_E \neq (L_1)_{E'}$ or $(L_2)_E \neq (L_2)_{E'}$. If by induction there are only finitely many, say n, distinct languages of the form $(L_1)_E$ and finitely many, say m, distinct languages of the form $(L_2)_E$, then there are no more than mn different languages of the form $(L_1 \cup L_2)_E$.

Next we consider products $L_1 L_2$. Here we will prove that

$$(L_1 L_2)_E = (L_1)_E L_2 \cup (L_2)_{E_{L_1}}. \qquad (*)$$

From this it will follow, as in the case for unions, that there are no more than mn distinct languages of the form $(L_1 L_2)_E$ (with m and n as before).

For the first direction of $(*)$ let $v \in (L_1 L_2)_E$. Then $wv = w_1 w_2$ for some $w_1 \in L_1$, $w_2 \in L_2$ and $w \in E$. There are two cases possible, depending on the length of w:

Case 1: If the length of w is less than or equal to the length of w_1, then w_1 $= wv_1$ and $v = v_1 w_2$ for some v_1, and v is in $(L_1)_E L_2$.
Case 2: If the length of w is greater than that of w_1, then $w = w_1 v_1$ and w_2 $= v_1 v$, which implies that $v_1 \in E_{L_1}$, and thus $v \in (L_2)_{E_{L_1}}$.

This establishes one inclusion for $(*)$. Conversely, if $v \in (L_1)_E L_2$, then v $= v_1 v_2$ for some v_1 in $(L_1)_E$ and $v_2 \in L_2$. Thus for some $w \in E$ we have $ww_1 \in L_1$ and $wv_1 v_2 \in L_1 L_2$, which shows that $v = v_1 v_2$ is in $(L_1 L_2)_E$. If $v \in (L_2)_{E_{L_1}}$, then $w_2 v \in L_2$ for some $w_2 \in E_{L_1}$, and $w_1 w_2 \in E$ for some $w_1 \in L_1$, and therefore $w_1 w_2 v \in L_1 L_2$. This shows that $v \in (L_1 L_2)_E$, and completes the proof of $(*)$.

For our final step we consider languages L^*, assuming that there are only finitely many different languages of the form L_E. We show that $(L^*)_E$ is either $L_{E_{L^*}} . L^*$, if the empty word e is not in E, or the union of this set with $\{e\}$, if $e \in E$.

Suppose that $v \in (L^*)_E$, and $v \neq e$. Then $wv \in L^*$ for some $w \in E$, and we can write $wv = w_1 w_2 \cdots w_n$ for some words $w_i \in L$. Since $v \neq e$ we also have $n > 0$, and there is a value r such that $w = w_1 \cdots w_{r-1} w_r'$ and $v = w_r'' w_{r+1} \cdots w_n$, where $w_r = w_r' w_r''$. Since $w_1 w_2 \cdots w_{r-1} \in L^*$ and $w_1 \cdots w_{r-1} w_r' \in E$, we get $w_r' \in E_{L^*}$. It follows that $w_r'' \in L_{E_{L^*}}$. Therefore $v = w_r'' w_{r+1} \cdots w_n \in L_{E_{L^*}} L^*$.

Conversely, if $v \in L_{E_{L^*}} . L^*$, then $v = u w_1 w_2 \cdots w_n$ for some $u \in L_{E_{L^*}}$ and $w_i \in L$. Then $u'u \in L$ for some $u' \in E_{L^*}$, and $u_1 u_2 \cdots u_m u' \in E$ for some $u_i \in L$. Therefore $u_1 u_2 \cdots u_m u' u w_1 w_2 \cdots w_n \in L^*$, and since $u_1 u_2 \cdots u_m u' \in E$ we have $b = u w_1 w_2 \cdots w_n \in (L^*)_E$. Finally, we note that if $e \in E$ then also $e \in (L^*)_E$.

Now we note that there are at most twice as many different languages of the form $(L^*)_E$ as there are of the form L_E, and hence only finitely many. ∎

This result can now be used to finish our proof of Kleene's Theorem.

Theorem 8.2.4 *(Kleene's Theorem) Let X be a finite alphabet. Then any language L over X is recognizable iff it is regular, so $RecX = RegX$.*

Proof: Since one direction of this was proved in Theorem 8.2.1, we now have to prove that any regular language L is recognizable. By Lemma 8.2.3, there are only finitely many different languages of the form L_w; we will label these as $L_1, \ldots, L_k$. We construct our new automaton $\mathcal{H}$ with states indexed by these L_i. We have a transition $\delta(L_i, a) = L_j$ if $L_j = (L_i)_a$. This gives a complete (deterministic) definition of our transition function δ. Our initial state is that indexed by L itself, and the final states are those L_i for which the language L_i contains the empty word e. We now claim that the language accepted by this automaton $\mathcal{H}$ is precisely our given language L. This is because for any word w, L_w is the language which labels the state $\hat{\delta}(z_0, w)$. Therefore

$$w \in L(\mathcal{H}) \quad \text{iff} \quad \hat{\delta}(z_0, w) \text{ is final}$$
$$\text{iff } e \text{ belongs to the language which labels } \hat{\delta}(z_0, w)$$
$$\text{iff } w \in L. \qquad \blacksquare$$

The automata we have considered thus far, of the form $\mathcal{H} = (Z, X; \delta, z_0, Z')$, are called *initial deterministic automata*. Initial refers to the fact that we have a unique initial state z_0. It is also possible to have a set $Z_0 \subseteq Z$ of initial states, with $|Z_0| \geq 2$; in this case we have a *weak initial automaton*.

The deterministic property refers to the fact that we have required that our transition information be given by a function δ from $Z \times X$ to Z; this ensures that any input combination of a state and letter leads to exactly one resulting state. A *non-deterministic automaton* is one in which some combinations of state-letter inputs may lead to no state, or to more than one choice of resulting state. That is, the transition δ may be a partial mapping, from $Z \times X$ to the power set $\mathcal{P}(Z)$. It can be proven that a language is recognizable by a non-deterministic automaton if and only if it is recognizable by a deterministic one.

The remaining variant of our basic definition that we need to consider is a finite automaton with output, of the form $\mathcal{H} = (Z, X, B; \delta, \lambda)$. Here we have an additional output alphabet B and an output function $\lambda : Z \times X \to B$, which maps each state-letter pair (z, a) to an output element in B. Just as was the case for δ, this output function λ can be uniquely extended to a map $\hat{\lambda}$ from $Z \times X^*$ to B^*. This is done inductively, by the following rules:

(i) $\hat{\lambda}(z, e) = z$,
(ii) $\hat{\lambda}(z, x) = \lambda(z, x)$ for any letter x,   and
(iii) $\hat{\lambda}(z, xw) = \lambda(z, x)\hat{\lambda}(\hat{\delta}(z, x), w)$.

Just as in the non-output case, the definition of an automaton with output can be extended to an *initial automaton with output*, when one initial state is selected, or to a *weak initial automaton with output*, when a larger set of initial states is chosen.

Example 8.2.5 We consider the following finite deterministic automaton with output: $\mathcal{H} = (Z, X, B; \delta, \lambda)$, where $X = \{0, 1\}$, $B = \{0, 1, 2, 3, 4\}$, $Z = \{z_0, z_1, z_2, z_3, z_4\}$, and the state transition function δ and the output function λ are given by the following tables.

δ	0	1
z_0	z_0	z_1
z_1	z_2	z_3
z_2	z_4	z_0
z_3	z_1	z_2
z_4	z_3	z_4

λ	0	1
z_0	0	1
z_1	2	3
z_2	4	0
z_3	1	2
z_4	3	4

The reader should draw the graph of this automaton. We will illustrate the behaviour of the extended output function $\hat{\lambda}$ by calculating the output of this automaton when it is started in state z_0 with input word $w = 110$. We have

$$\hat{\lambda}(z_0, 110) = \lambda(z_0, 1)\hat{\lambda}(\hat{\delta}(z_0, 1), 10)$$
$$= 1\ \hat{\lambda}(z_1, 10)$$
$$= 1\ \lambda(z_1, 1)\hat{\lambda}(\delta(z_1, 1), 0)$$
$$= 13\ \hat{\lambda}(z_3, 0) = 131.$$

8.3 Algebraic Operations on Finite Automata

As we have remarked earlier, finite automata may be regarded as heterogeneous or multi-based algebras, and all the theory developed in Chapters 1, 3 and 4 regarding subalgebras, quotients and homomorphic images may be extended to this setting. In this section we develop this theory for the particular case of finite automata, beginning with the concept of a subautomaton.

Definition 8.3.1 Let $\mathcal{H}_1 = (Z_1, X_1, B_1; \delta_1, \lambda_1)$ and $\mathcal{H}_2 = (Z_2, X_2, B_2; \delta_2, \lambda_2)$ be two deterministic automata with output. Then $\mathcal{H}_1$ is called a *subautomaton* of $\mathcal{H}_2$ if the following conditions are satisfied:

(i) $Z_1 \subseteq Z_2$, $X_1 \subseteq X_2$ and $B_1 \subseteq B_2$, and
(ii) For all $x \in X_1$ and $z \in Z_1$, the functions δ_i and λ_i satisfy
   $\delta_2(z, x) = \delta_1(z, x)$ and $\lambda_2(z, x) = \lambda_1(z, x)$.

Other concepts regarding subalgebras from Section 1.3 can similarly be generalized to the multi-based setting. Let $\mathcal{H} = (Z, X, B; \delta, \lambda)$ be a finite deterministic automaton with output, and let $\{\mathcal{H}_j = (Z_j, X_j, B_j; \delta_j, \lambda_j) \mid j \in J\}$ be an indexed family of subautomata of $\mathcal{H}$. Assume that the intersections $\bigcap_{j \in J} Z_j$, $\bigcap_{j \in J} X_j$ and $\bigcap_{j \in J} B_j$ are non-empty. Then we define a new subautomaton called the *intersection* $\bigcap_{j \in J} \mathcal{H}_j$, defined by

$$\bigcap_{j \in J} \mathcal{H}_j := (\bigcap_{j \in J} Z_j, \bigcap_{j \in J} X_j, \bigcap_{j \in J} B_j; \delta', \lambda'),$$

where δ' and λ' are the restrictions of δ and λ respectively to the corresponding intersections. That this is indeed a subautomaton of each of the $\mathcal{H}_j$ can be proved in the same way that we proved Corollary 1.3.5.

Let $\mathcal{H}$ be an automaton. When $Z_0 \subseteq Z$, $X_0 \subseteq X$ and $B_0 \subseteq B$, we can consider the intersection

$$\bigcap \{\mathcal{H}' = (Z', X', B'; \delta', \lambda') \mid Z_0 \subseteq Z', X_0 \subseteq X', B_0 \subseteq B'\},$$

and this gives a new subautomaton of $\mathcal{H}$ which is called the subautomaton generated by the triple (Z_0, X_0, B_0). We will denote this by $\langle (Z_0, X_0, B_0) \rangle$.

As in the homogeneous algebra case, the union of an indexed set $\{\mathcal{H}_j \mid j \in J\}$ of subautomata of a given automaton $\mathcal{H}$ is not usually a subautomaton. But we can take the subautomaton generated by the unions, $\langle(\bigcup_{j \in J} Z_j, \bigcup_{j \in J} X_j, \bigcup_{j \in J} B_J)\rangle$; this will be the least subautomaton of $\mathcal{H}$ to contain all the $\mathcal{H}_j$, and we denote it by $\bigvee_{j \in J} \mathcal{H}_j$. In this way we construct a meet operation (intersection) and a join operation on the set of all subautomata of $\mathcal{H}$. As before, these operations give us a complete lattice of substructures.

Proposition 8.3.2 *Let $\mathcal{H}$ be a deterministic finite automaton. With meet operation of intersection and join operation as defined above, the set of all subautomata of $\mathcal{H}$ forms a complete lattice $Sub(\mathcal{H})$.*   ∎

To define homomorphisms between automata, we need triples of mappings which preserve the structure.

Definition 8.3.3 Let $\mathcal{H}_1 = (Z_1, X_1, B_1; \delta_1, \lambda_1)$ and $\mathcal{H}_2 = (Z_2, X_2, B_2; \delta_2, \lambda_2)$ be two deterministic automata with output. Let $f = (f_I, f_S, f_O)$ be a triple of mappings, with $f_I : X_1 \to X_2$, $f_S : Z_1 \to Z_2$ and $f_O : B_1 \to B_2$. (The subscripts I, S and O refer to inputs, states and outputs, respectively.) Then f is called a *homomorphism* of $\mathcal{H}_1$ into $\mathcal{H}_2$ when for every $z \in Z_1$ and every $x \in X_1$, we have

$$f_S(\delta_1(z, x)) = \delta_2(f_S(z), f_I(x)) \text{ and } f_O(\lambda_1(z, x)) = \lambda_2(f_S(z), f_I(x)).$$

If all three functions in the triple $f = (f_I, f_S, f_O)$ are injective (surjective, bijective) then the homomorphism f is said to be injective (surjective, bijective). A bijective homomorphism is called an isomorphism.

Notice that the fact that $\mathcal{H}_1$ is a subautomaton of $\mathcal{H}_2$ can also be expressed by an injective homomorphism $f : \mathcal{H}_1 \to \mathcal{H}_2$. We define the composition of two homomorphisms $f = (f_I, f_S, f_O)$ and $g = (g_I, g_S, g_O)$ by $f \circ g = (f_I \circ g_I, f_S \circ g_S, f_O \circ g_O)$. It is easy to show that the composition of two homomorphisms is again a homomorphism. (See Exercise 8.11.2.)

Our goal now is to formulate a version of the Homomorphism Theorem for automata. To do this we must first define the appropriate analogues of congruences and kernels.

Definition 8.3.4 Let $\mathcal{H} = (Z, X, B; \delta, \lambda)$ be a finite deterministic automaton with output. The triple $R = (R_I, R_S, R_O)$ is called a *congruence* on $\mathcal{H}$ if R_I, R_S and R_O are equivalence relations on the sets X, Z and B respectively, and each is compatible with respect to the operations of δ and λ. This means that for all $(z_1, z_2) \in R_S$ and all $(x_1, x_2) \in R_I$ we have $(\delta(z_1, x_1), \delta(z_2, x_2)) \in R_S$ and $(\lambda(z_1, x_1), \lambda(z_2, x_2)) \in R_O$.

Let $\mathcal{H} = (Z, X, B; \delta, \lambda)$ be an automaton with output, and let $R = (R_I, R_S, R_O)$ be a congruence on $\mathcal{H}$. The *quotient automaton* of $\mathcal{H}$ with respect to the congruence R is defined as the automaton

$$\mathcal{H}/R := (Z/R_S, X/R_I, B/R_O; \delta', \lambda'),$$

where δ' and λ' are defined by

$$\delta'([z]_{R_S}, [x]_{R_I}) := [\delta(z, x)]_{R_S}, \text{ and } \lambda'([z]_{R_S}, [x]_{R_I}) := [\lambda(z, x)]_{R_O}.$$

The reader should check that our congruence property means precisely that these two mappings are well defined. As before, congruences occur as kernels of homomorphisms.

Definition 8.3.5 Let $f : \mathcal{H}_1 \to \mathcal{H}_2$ be a homomorphism of $\mathcal{H}_1$ into $\mathcal{H}_2$. The triple $ker f := (ker f_I, ker f_S, ker f_O)$ is called the *kernel* of the homomorphism f.

It is a straightforward exercise (see Exercise 8.11.3) to show that kernels of homomorphisms have the nice properties from Chapter 3: the kernel of any homomorphism $f : \mathcal{H}_1 \to \mathcal{H}_2$ is a congruence relation; and for any congruence R on an automaton $\mathcal{H}$ the natural mapping $f : \mathcal{H} \to \mathcal{H}/R$ is a homomorphism, whose kernel is the congruence R.

Theorem 8.3.6 *(The Homomorphism Theorem for Automata) Let $f : \mathcal{H} \to \mathcal{H}'$ be a surjective homomorphism of automata. Then there is a unique isomorphism $h : \mathcal{H}/ker f \to \mathcal{H}'$ such that the diagram below commutes.* ∎

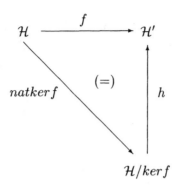

All of these results for automata with output can be extended to the variations of initial and weak initial automata. In the initial case, we require that a homomorphism $f : \mathcal{H}_1 \to \mathcal{H}_2$ satisfy the additional requirement that f_S maps the initial state of $\mathcal{H}_1$ to the initial state of $\mathcal{H}_2$, and maps final states of $\mathcal{H}_1$ to final states of $\mathcal{H}_2$.

The definition of algebraic constructions such as subautomata and homomorphic images gives us an algebraic way to compare different automata. We can also compare automata by comparing how they behave with respect to the same input words. We will call two automata *equivalent* if they behave the same way when given the same inputs. We will define this more formally by means of another type of equivalence, that of equivalence of two states in the same automaton.

Definition 8.3.7 Let $\mathcal{H} = (Z, X, B; \delta, \lambda)$ be an automaton with output. Two states z_1 and z_2 in Z are said to be *equivalent* if for all words $w \in X^*$ we have $\hat{\lambda}(z_1, w) = \hat{\lambda}(z_2, w)$. In this case we write $z_1 \sim z_2$.

Now we can define equivalence of automata.

Definition 8.3.8 Let $\mathcal{H}_1 = (Z_1, X_1, B_1; Z_{01}, \delta_1, \lambda_1)$ and $\mathcal{H}_2 = (Z_2, X_2, B_2;$ $Z_{02}, \delta_2, \lambda_2)$ be two weak initial finite deterministic automata with output. Then $\mathcal{H}_1$ and $\mathcal{H}_2$ are called *equivalent*, written as $\mathcal{H}_1 \sim \mathcal{H}_2$, if there are functions $\varphi_1 : Z_{01} \to Z_{02}$ and $\varphi_2 : Z_{02} \to Z_{01}$, such that for all $z_1 \in Z_{01}$ and all $z_2 \in Z_{02}$ we have $z_1 \sim \varphi_1(z_1)$ and $z_2 \sim \varphi_2(z_2)$.

In the case that $\mathcal{H}_1$ and $\mathcal{H}_2$ are initial, so $Z_{0i} = \{z_{0i}\}$, this condition reduces to $\mathcal{H}_1 \sim \mathcal{H}_2$ iff $z_{01} \sim z_{02}$. We also leave it as an exercise for the reader to show that $\sim$ defines an equivalence relation on the class of all finite automata.

The intention here is to use equivalence of states in a machine to reduce the machine, in the sense of producing a smaller (having fewer states) machine that accepts the same language. If two states are equivalent, they act the same on all inputs, and hence we can keep one of them and omit the other; that is, we form a quotient in which all equivalent states are identified, without affecting the language accepted. An automaton in which no such reductions are possible is called *reduced*.

Definition 8.3.9 A finite deterministic weak initial automaton with output $\mathcal{H} = (Z, X, B; \delta, \lambda)$ is called *reduced* if for all states z_1 and z_2 in Z, whenever $z_1 \sim z_2$ we have $z_1 = z_2$.

Theorem 8.3.10 *For each weakly initial deterministic finite automaton, there exists an equivalent reduced automaton.*

Proof: Let $\mathcal{H} = (Z, X, B; \delta, \lambda)$ be a weakly initial deterministic finite automaton with output. We define a new automaton $\mathcal{H}' = (Z', X', B'; \delta', \lambda')$ as follows. We use the same alphabet and output sets, so $X' = X$ and $B' = B$. The states in Z' are the equivalence classes of states from Z with respect to the equivalence $\sim$ of states, so

$$Z' = \{[z]_\sim \mid z \in Z\}.$$

We take $Z'_0 = \{[z_0] \mid z_0 \in Z_0\}$. We define the transition and output functions $\delta' : Z' \times X \to Z'$ and $\lambda' : Z' \times X \to B$ by

$$\delta'([z]_\sim, x) := [\delta(z, x)]_\sim \text{ and } \lambda'([z]_\sim, x) := \lambda(z, x).$$

We show first that these functions are well defined, then check that $\mathcal{H}'$ does recognize the same language as $\mathcal{H}$. Suppose that $([z_1]_\sim, x_1) = ([z_2]_\sim, x_2)$, so that $x_1 = x_2$ and $z_1 \sim z_2$. By the definition of equivalent states this means that for any input $w \in X^*$ we have

$$\hat{\lambda}(z_1, w) = \lambda(z_2, w). \quad (*)$$

In particular, we have $\lambda(z_1, x_1) = \lambda(z_2, x_2)$ when $x_1 = x_2$. It follows that $\lambda'([z_1]_\sim, x_1) = \lambda(z_1, x_1) = \lambda(z_2, x_2) = \lambda'([z_2]_\sim, x_2)$, showing that λ' is well defined.

For δ', we have to show (using $x = x_1 = x_2$) that when $[z_1]_\sim = [z_2]_\sim$, then $[\delta(z_1, x)]_\sim = [\delta(z_2, x)]_\sim$. This is equivalent to showing that when $(*)$ holds, then also the two states $\delta(z_1, x)$ and $\delta(z_2, x)$ always produce the same output, that is, that $\hat{\lambda}(\delta(z_1, x), p) = \hat{\lambda}(\delta(z_2, x), p)$ for any input p. We apply the equation $(*)$ with input $w = xp$. Then we have

$$\hat{\lambda}(z_1, xp) = \lambda(z_2, xp).$$

From the inductive definition of $\hat{\lambda}$, this gives

$$\lambda(z_1, x)\hat{\lambda}(\hat{\delta}(z_1, x), p) = \lambda(z_2, x)\hat{\lambda}(\hat{\delta}(z_2, x), p).$$

By $(*)$ with $w = x$, the first part in each of these concatenations is the same; so the remaining parts must also be equal:

$$\hat{\lambda}(\hat{\delta}(z_1, x), p) = \hat{\lambda}(\hat{\delta}(z_2, x), p).$$

This is enough to show that $\delta(z_1, x)$ is equivalent to $\delta(z_2, x)$, as required to show that δ' is well defined.

To see that $\mathcal{H}'$ is equivalent to $\mathcal{H}$, we define mappings $\varphi_1 : Z_0 \to Z_0'$ and $\varphi_2 : Z_0' \to Z_0$. We let $\varphi_1(z_0) = [z_0]_\sim$ for all $z_0 \in Z_0$. For φ_2 we need to first choose a system of representatives with respect to $\sim$, one for each equivalence class; then we let $\varphi_2([z_0]_\sim)$ be this representative. Then using the definition of λ' we can show by induction that $\lambda(z_0, p) = \lambda'([z_0]_\sim, p)$. This gives the equivalence of the two automata.   ∎

It can also be shown that the equivalent reduced automaton $\mathcal{H}'$ constructed in this proof is uniquely determined by $\mathcal{H}$, up to isomorphism. (See Exercise 8.11.4.)

The equivalence of two automata was defined only in terms of their behaviour on the same input words. But such equivalence also turns out to have an algebraic interpretation.

Theorem 8.3.11 Let $\mathcal{H} = (Z, X, B; \delta, \lambda)$ be a weakly initial deterministic finite automaton with output, and let $\mathcal{H}'$ be the corresponding equivalent reduced automaton, as constructed in the previous proof. Then $\mathcal{H}'$ is a homomorphic image of $\mathcal{H}$.

Proof: We use the triple of mappings $\varphi = (\varphi_I, \varphi_S, \varphi_O)$ given by $\varphi_I = id_X$, $\varphi_O = id_B$ and $\varphi_S : Z \to Z'$ defined by $\varphi_S(z) = [z]_\sim$ for all $z \in Z$. Then clearly φ is surjective, and it is a homomorphism since $\varphi_S(\delta(z, x)) = [\delta(z, x)]_\sim = \delta'([z]_\sim, x) = \delta'(\varphi_S(z), x) = \delta'(\varphi_S(z), \varphi_I(x))$ and $\varphi_O(\lambda(z, x)) = \lambda'([z]_\sim, x) = \lambda'(\varphi_S(z), \varphi_I(x))$. ∎

8.4 Tree Recognizers

In Section 8.2 we defined language recognizers which accept or reject words over a given finite alphabet. As we saw in Example 6.4.5, such words are a convenient way of representing the terms in the relatively free algebra in the variety of semigroups. In this section we want to define automata which are able to accept or reject terms of any type, that is, elements of the free algebra $\mathcal{F}_\tau(X)$ of type τ. Since terms can always be regarded as trees (see Section 5.1), we think of accepting or rejecting trees, and our automata will be called tree-recognizers.

We will require that our alphabet X be finite, and also that our type τ contain only finitely many operation symbols. We group the elements of the set $\{f_i \mid i \in I\}$ of operation symbols of type τ into subsets of different arities, to consider a so-called *ranked* alphabet $\Sigma = \Sigma_0 \cup \cdots \cup \Sigma_n$ of operation symbols; here Σ_j contains all the j-ary operation symbols of type τ. We will now write our algebras $(A; (f_i^A)_{i \in I})$ as $(A; \Sigma^A)$, where Σ^A denotes a set of operations defined on the set A and induced by the operation symbols from Σ. We refer to Σ as a ranked alphabet of symbols, and to X as an input alphabet or variable set. We will also use the name $W_\Sigma(X)$ for the set of all terms on the alphabet X using the operation symbols from Σ.

Definition 8.4.1 A $\Sigma - X$-*tree-recognizer* is a sequence

$$\mathbf{A} = (\mathcal{A}, \Sigma, X, \alpha, A'),$$

where
$\mathcal{A} := (A; \Sigma^A)$ is a finite algebra,
Σ is a *ranked alphabet of operation symbols*,
X is a set of *individual variables*,
$\alpha : X \cup \Sigma_0 \longrightarrow A \cup \Sigma_0^A$ is a mapping, called the *evaluation mapping*, and
A' is a subset of the (finite) set A.

The evaluation mapping α maps each variable from X to an element of A, and each nullary operation symbol from Σ (if any) to the corresponding nullary operation, that is, element of A. Let us denote by $\mathcal{F}_\Sigma(X)$ the absolutely free algebra $\mathcal{F}_\tau(X)$ of the type τ defined by the operation symbols from Σ. Then any evaluation map α can be uniquely extended to a homomorphism $\hat{\alpha} : \mathcal{F}_\Sigma(X) \longrightarrow \mathcal{A}$ (see Chapter 5).

A *language* (over X and Σ) is any subset T of terms from $W_\Sigma(X)$. We want of course to use tree-recognizers to recognize or accept a language. To think of a tree-recognizer as an automaton, we call the elements of the finite set A *states*, while states in the special subset $A' \subseteq A$ are called *final states*. The extension evaluation $\hat{\alpha}$ maps any term to an element of A, and those terms which are mapped to a final state are said to be accepted or recognized by the automaton **A**. Thus we call the set

$$T(\mathbf{A}) := \{t \mid t \in W_\Sigma(X) \text{ and } \hat{\alpha}(t) \in A'\}$$

the *language recognized by* **A**. By definition, $T(\mathbf{A})$ is the preimage of A' under $\hat{\alpha}$, that is, $T(\mathbf{A}) = \hat{\alpha}^{-1}(A')$.

A language $T \subseteq W_\Sigma(X)$ is called *recognizable* if there is a $\Sigma - X$-tree-recognizer **A** such that $T = T(\mathbf{A})$.

Tree-recognizers defined in this way work in a deterministic top-down fashion on terms regarded as trees. Our evaluation map α specifies the action of the tree-recognizer at the top of a tree: variables or nullary operation symbols which label the leaves of a given tree are replaced by elements of A and by the corresponding nullary operations of the algebra $\mathcal{A}$, respectively. Then inductively the m-ary operation symbols which label the vertices of the given tree are replaced by the corresponding m-ary fundamental operations of $\mathcal{A}$. Thus step-by-step the value $\hat{\alpha}(t)$ will be calculated from the top down, starting from the leaves and ending in the root of the tree. We illustrate this with an example.

Example 8.4.2 We take $\Sigma = \Sigma_1 \cup \Sigma_2$, where $\Sigma_1 = \{h\}$ and $\Sigma_2 = \{f, g\}$, and alphabet $X = \{x_1, x_2\}$. We consider the finite algebra $\mathcal{A} = (\{0,1\}; \wedge, \vee, \neg)$, where $h^A = \neg$, $f^A = \wedge$ and $g^A = \vee$. We define an evaluation α by $\alpha(x_1) = 1$ and $\alpha(x_2) = 0$. We also designate $A' = \{1\}$. Consider the term $t = f(h(f(x_2, x_1)), g(h(x_2), x_1))$. The tree corresponding to this term is accepted by this tree-recognizer, since $\hat{\alpha}(t) = (\neg(0 \wedge 1)) \wedge (\neg(0) \vee 1) = 1 \in A'$.

We remark that tree-recognizers can also be defined in a "bottom-up" style, to work from the root of a tree to its leaves.

The following example shows that not every language is recognizable.

Example 8.4.3 Let $\Sigma = \Sigma_2 = \{f\}$ and let X be an arbitrary nonempty finite alphabet. We will show that the language $T = \{f(t,t) \mid t \in W_\Sigma(X)\}$ is not recognizable by any $\Sigma - X$-tree-recognizer. For suppose that there was a $\Sigma - X$-tree-recognizer $\mathbf{A}$ with $T = T(\mathbf{A})$. Since A is finite, there must be two different $\Sigma - X$-trees t and t' such that $\hat{\alpha}(t) = \hat{\alpha}(t')$. But then

$$\hat{\alpha}(f(t,t')) = f^A(\hat{\alpha}(t), \hat{\alpha}(t')) = f^A(\hat{\alpha}(t), \hat{\alpha}(t)) = \hat{\alpha}(f(t,t)) \in A'.$$

This gives the contradiction $f(t,t') \in T$.

Our $\Sigma - X$-tree-recognizers are designed to recognize terms from the absolutely free algebra $\mathcal{F}_\Sigma(X)$. One can also define recognizers for the relatively free algebra $\mathcal{F}_V(X)$ with respect to any variety V of type Σ. The reader should verify that the usual finite automata from Section 8.2, recognizing words (free semigroup terms) on a finite alphabet, are a special case of this more general definition.

Another variation in the basic definition of a $\Sigma - X$-tree-recognizer involves the property of determinism. Our definition so far gives a deterministic machine, but in some applications and in the theoretical development we shall also need a non-deterministic version. To define this we must first define the concepts of a non-deterministic operation and a non-deterministic algebra.

Let $\mathcal{P}(A)$ be the power set of the set A. For $n \geq 1$, an n-ary mapping

$$f^A : A^n \longrightarrow \mathcal{P}(A)$$

is called an n-ary non-deterministic operation on A. Such a mapping assigns to each element of A a set (possibly empty) of elements of A; so instead of a function with exactly one output for each element of A we have a situation where one input may be assigned no output or several different outputs. Nullary non-deterministic operations are defined as mappings

$$f^A : \{\emptyset\} \longrightarrow \mathcal{P}(A).$$

A *non-deterministic algebra* $\mathcal{A} = (A; \Sigma^A)$ is a pair consisting of a set A and a set Σ^A of non-deterministic operations on A.

Of course any deterministic algebra may be viewed as a non-deterministic one, with the convention that elements of the universe are identified with the corresponding singletons. Conversely, given any non-deterministic algebra $\mathcal{A} = (A; \Sigma^A)$ we can form a related ordinary algebra. This is the power set algebra $\mathcal{P}(\mathcal{A}) = (\mathcal{P}(A); \Sigma^{\mathcal{P}(A)})$, where the fundamental operations are defined by

$$f^{\mathcal{P}(A)}(A_1, \ldots, A_n) := \bigcup \{ f^A(a_1, \ldots, a_n) \mid a_1 \in A_1, \ldots, a_n \in A_n \},$$

for $f^A \in \Sigma^A$ and $A_1, \ldots, A_n \in \mathcal{P}(A)$.

Definition 8.4.4 A *non-deterministic $\Sigma - X$-tree-recognizer* is a sequence

$$\mathbf{A} = (\mathcal{A}, \Sigma, X, \alpha, A'),$$

where
$\mathcal{A} := (A; \Sigma^A)$ is a finite non-deterministic algebra,
Σ is a ranked alphabet of operation symbols,
X is a set of individual variables,
$\alpha : X \cup \Sigma_0 \longrightarrow \mathcal{P}(A) \cup \Sigma_0^A$ is a mapping (called the *evaluation mapping*), and
A' is a subset of the (finite) set A.

The set
$$T(\mathbf{A}) := \{ t \mid t \in W_\Sigma(X) \text{ and } \hat{\alpha}(t) \cap A' \neq \emptyset \}$$

is called the *language recognized by* $\mathbf{A}$.
It turns out that deterministic and non-deterministic tree-recognizers are equivalent, in the sense that the families of languages which they recognize are equal. This can be useful, since non-deterministic tree-recognizers are sometimes easier to work with than deterministic ones.

Proposition 8.4.5 *A language is recognized by a deterministic Σ-X-tree-recognizer iff it is recognized by a non-deterministic one.*

Proof: Since any deterministic $\Sigma - X$-tree-recognizer can be viewed as a non-deterministic one, one direction of our proof is immediate. For the opposite direction, let $\mathbf{A} = (\mathcal{A}, \Sigma, X, \alpha, A')$ be a non-deterministic $\Sigma - X$-tree-recognizer, based on the non-deterministic algebra $\mathcal{A} = (A; \Sigma^A)$. We construct a deterministic tree-recognizer which recognizes the same language. For our algebra, we use the deterministic power set algebra $\mathcal{P}(\mathcal{A})$ $= (\mathcal{P}(A); \Sigma^{\mathcal{P}(A)})$ discussed above, and for our set of final states we use $A'' :=$ $\{A_1 \in \mathcal{P}(A) \mid A_1 \cap A' \neq \emptyset\}$. For our tree-recognizer then we have $\mathbf{P}(\mathcal{A}) =$ $(\mathcal{P}(\mathcal{A}), \Sigma, X, \alpha, A'')$. Then for any term t,

$$t \in T(\mathbf{P}(\mathcal{A})) \Leftrightarrow \hat{\alpha}(t) \in A'' \Leftrightarrow \hat{\alpha}(t) \cap A' \neq \emptyset \Leftrightarrow t \in T(\mathbf{A}). \qquad \blacksquare$$

8.5 Regular Tree Grammars

In Section 8.2 we proved that any language which is recognizable by a finite automaton without output is a regular language, and conversely that every regular language is recognizable by some finite automaton. To prove a similar result for tree-languages, we introduce the concept of a regular tree grammar.

Definition 8.5.1 A *regular $\Sigma - X$-tree grammar* is a sequence

$$G = (N, \Sigma, X, P, a_0),$$

where

N is a finite non-empty set, called the set of *non-terminal symbols*,

Σ is a ranked alphabet of operation symbols,

X is an alphabet of *individual variables*,

P is a finite set of *productions* (or *rules of derivation*) which have the form $a \to r$ for some $a \in N$ and $r \in W_{\Sigma}(N \cup X)$, and $a_0 \in N$ is called an *initial symbol*.

We also specify that $N \cap (\Sigma \cup X) = \emptyset$.

Let G be a regular $\Sigma - X$-tree grammar. We think of a production $a \to r$ from G as a rule allowing us to replace the non-terminal symbol a by the term (tree) r. Such a replacement is done within terms, as follows. Let s be a term from $W_{\Sigma}(X \cup N)$ in which the non-terminal symbol a occurs as a subterm. If we have a production $a \to r$ in G, we can change s to a new term t by replacing the subterm a by the term r. In this case we write

$$s \Rightarrow_G t,$$

to indicate that s can be transformed into t by one application of a production rule from G.

We write
$$s \Rightarrow_G^* t$$

if either $s = t$ or there is a sequence $(t_0, \ldots, t_n)$ of terms from $W_\Sigma(X \cup N)$ with $t_0 = s$, $t_n = t$ and

$$t_0 \Rightarrow_G t_1 \Rightarrow_G \cdots \Rightarrow_G t_{n-1} \Rightarrow_G t_n.$$

Such a sequence is called a *derivation* of t from s. It is common to omit the G-subscript, if the grammar is clear from the context. Note that $\Rightarrow_G^*$ is the reflexive and transitive closure of the relation $\Rightarrow_G$, if we consider $\Rightarrow_G$ as a relation on $W_\Sigma(X \cup N)$.

Definition 8.5.2 Let $G = (N, \Sigma, X, P, a_0)$ be a regular $\Sigma - X$-tree grammar. Then the set
$$T(G) = \{t \mid t \in W_\Sigma(X) \text{ and } a_0 \Rightarrow_G^* t\}$$

is called the *language generated by* the grammar G. Two regular $\Sigma - X$-tree grammars G_1 and G_2 are said to be *equivalent* if $T(G_1) = T(G_2)$.

Example 8.5.3 Consider the regular tree grammar
$$G = (\{a, b\}, \Sigma, X, P, a),$$

where $\Sigma = \Sigma_0 \cup \Sigma_2$, $\Sigma_0 = \{h\}$, $\Sigma_2 = \{f\}$, $X = \{x\}$, and P contains the three productions

$$a \to f(x, f(x, b)), \quad a \to f(h, a) \text{ and } b \to f(x, x).$$

Then the tree $f(h, f(x, f(x, f(x, x))))$ has the derivation

$$a \Rightarrow_G f(h, a) \Rightarrow_G f(h, f(x, f(x, b))) \Rightarrow_G f(h, f(x, f(x, f(x, x)))),$$

and the tree $f(x, f(x, f(x, x)))$ has the derivation .

$$a \Rightarrow_G f(x, f(x, b)) \Rightarrow_G f(x, f(x, f(x, x))).$$

As usual when we define an equivalence between objects, we want to be able to replace a given grammar by a "simpler" one equivalent to it. To make this more precise, we define the concept of a normal form of a tree grammar, by selecting particular forms of productions. We will measure the depth of a production $a \to r$ by the depth of the term r.

Definition 8.5.4 A regular tree grammar is said to be in *normal form* if each of its productions $a \to r$ is of one of the following types:

(i) if r has depth 0, then r is either a variable $x \in X$ or a nullary symbol $f \in \Sigma_0$;

(ii) otherwise r has the form $f_i(a_1, \ldots, a_{n_i})$ for some $n_i \geq 1$ and $f_i \in \Sigma_{n_i}$ and some $a_1, \ldots, a_{n_i} \in N$.

Then we can prove the following.

Lemma 8.5.5 *Every regular tree grammar is equivalent to a regular tree grammar in normal form.*

Proof: Let G be a regular $\Sigma - X$-tree grammar. We describe a procedure for producing a normal form grammar equivalent to G. As a first step, we delete from the production set P any production of the form $a \to b$ where both a and b are non-terminal symbols in N, replacing it instead by all productions of the form $a \to r$ for which we have $a \Rightarrow_G^* b$ and $b \to r$ in P, for some $r \in W_\Sigma(X \cup N) \setminus N$.

Next let $a \to r$ be a production with depth greater than one. Then r has the form $f_i(r_1, \ldots, r_{n_i})$, where $n_i \geq 1$, $f_i \in \Sigma_{n_i}$ and $depth(r_j) < depth(r)$ for all $j = 1, \ldots, n_i$. In this case we delete $a \to r$, and replace it with productions of the form

$$a \to f_i(a_1, \ldots, a_{n_i}) \qquad (*)$$

where $a_1, \ldots, a_{n_i}$ are new non-terminal symbols, and

$$a_i \to r_j, \qquad (**)$$

for each $1 \leq j \leq n_i$.

Clearly any application of the rule $a \to r$ can be replaced by an application of $(*)$ followed by applications of productions of the form $(**)$. Conversely, any application of a production $(**)$ occurs only after an application of $(*)$, and if $(*)$ has been used then $(**)$ has to follow. Thus these steps have the same result as a single application of $a \to r$. If one of the symbols r_j in $a \to f_i(r_1, \ldots, r_{n_i})$ is a variable, a nullary operation symbol or a non-terminal symbol, then we substitute a new non-terminal symbol for it and introduce a rule $d \to r_j$ of depth 0.

In this way, step by step, any production with a depth greater than 1 can be replaced by productions of lower depth, until we reach a normal form. None of these steps changes the language generated by the grammar; so we obtain an equivalent normal-form grammar.   ∎

For technical reasons we introduce the following slight generalization of a regular $\Sigma - X$ grammar. Instead of the selected initial symbol a_0, we use a set $A' \subseteq N$ of initial symbols. This gives us the concept of an *extended regular $\Sigma - X$ grammar* $G = (N, \Sigma, X, P, A')$. The language generated by an extended regular grammar is defined as

$$T(G) := \{t \in W_\Sigma(X) \mid \exists a_0 \in A' (a_0 \Rightarrow_G^* t)\}.$$

Clearly, every language generated by an extended regular tree grammar can be generated by an ordinary tree grammar; see Exercise 8. 11.7.

Regular grammars and recognizable languages are connected by the following Kleene-type theorem:

Theorem 8.5.6 *A language is recognizable by a $\Sigma - X$-tree-recognizer iff it can be generated by a regular tree grammar.*

Proof: Let $\mathbf{A} = (\mathcal{A}, \Sigma, X, \alpha, A')$ be a non-deterministic $\Sigma - X$-recognizer. We use $\mathbf{A}$ to construct an extended $\Sigma - X$-grammar G as follows. For the set N of non-terminal symbols of G we take the base set A of the algebra from $\mathbf{A}$. Our grammar G will have the form (A, Σ, X, P, A'), where P is a set of normal form productions defined as follows. We put into P productions of three kinds:

(i) For any $x \in X$ and any $a \in \alpha(x)$, we have a production $a \to x$;
(ii) For any nullary symbol $f \in \Sigma_0$ and any $a \in f^A$, we have a production $a \to f$;
(iii) For any symbol $f_i \in \Sigma_{n_i}$ with $n_i \geq 1$ and any $a, a_1, \ldots, a_{n_i} \in A$, $a \in f_i^A(a_1, \ldots, a_{n_i})$, we have a production $a \to f_i(a_1, \ldots, a_{n_i})$.

These productions have the form required to ensure that the grammar G is in normal form. Conversely, given any grammar G, we may construct an equivalent normal form grammar and then use the productions to define a non-deterministic $\Sigma - X$-recognizer. Now we claim that this grammar and recognizer produce the same language, that is, that $T(\mathbf{A}) = T(G)$. To prove this it will suffice to show that the following equivalence is satisfied:

$$a \in \hat{\alpha}(t) \text{ iff } a \Rightarrow^*_G t \text{ for all } a \in A \text{ and } t \in W_\Sigma(X). \quad (*)$$

We proceed by induction on the depth of the tree t. For the base case, if t has depth zero it is either a variable x or a nullary operation symbol f. If $t = x \in X$ and $a \in \hat{\alpha}(x)$ then $a \Rightarrow^*_G x$ since $a \to x$ is a production. If conversely $a \Rightarrow^*_G x$ then to start this derivation we can only use a production of the form $a \to x$. But then $a \in \alpha(x)$. A similar argument holds in the case that $t = f$ for some $f \in \Sigma_0$, using the second kind of productions in P.

For the inductive step, let $t = f_i(t_1, \ldots, t_{n_i})$ and suppose that

$$a \in \hat{\alpha}(t_j) \text{ iff } a \Rightarrow^*_G t_j \text{ for all } a \in A \text{ and all } j \in \{1, \ldots, n_i\}.$$

If $a \in \hat{\alpha}(t)$, then there exist elements $a_1, \ldots, a_{n_i} \in A$ such that $a_j \in \hat{\alpha}(t_j)$ for $1 \le j \le n_i$ and such that $a \in f_i^A(a_1, \ldots, a_{n_i})$. Then the induction hypothesis implies $a_j \Rightarrow^*_G t_j$, for all $j \in \{1, \ldots, n_i\}$. Using the production $a \to f_i(a_1, \ldots, a_{n_i})$ we obtain the derivation

$$a \Rightarrow_G f_i(a_1, \ldots, a_{n_i}) \Rightarrow^*_G f_i(t_1, \ldots, t_{n_i}) = t.$$

Conversely, if $a \Rightarrow^*_G t$, then there is a derivation of the form

$$a \Rightarrow_G f_i(a_1, \ldots, a_{n_i}) \Rightarrow^*_G f_i(t_1, \ldots, t_{n_i}),$$

where $a_1, \ldots, a_{n_i} \in A$ and $a_j \Rightarrow^*_G t_j$ for $j = 1, \ldots, n_i$. Since the first step of this derivation applies a rule of the form $a \to f_i(a_1, \ldots, a_{n_i})$, we get $a \in f_i^A(a_1, \ldots, a_{n_i})$, and the induction hypothesis implies $a_j \in \hat{\alpha}(t_j)$ for $j = 1, \ldots, n_i$. But this means that $a \in f_i^{\mathcal{P}(\mathbf{A})}(\hat{\alpha}(t_1), \ldots, \hat{\alpha}(t_{n_i})) = \hat{\alpha}(t)$.

This proves that the equivalence $(*)$ holds. Now for every tree t we have $t \in T(\mathbf{A}) \Leftrightarrow \hat{\alpha}(t) \cap A' \ne \emptyset \Leftrightarrow \exists a \in A'(a \in \hat{\alpha}(t)) \Leftrightarrow \exists a \in A'(a \Rightarrow^*_G t) \Leftrightarrow t \in T(G)$.

This finally gives $T(\mathbf{A}) = T(G)$. ∎

8.6   Operations on Tree Languages

Let us denote by $Rec(\Sigma, X)$ the set of all recognizable $\Sigma - X$-languages.
We will study some algebraic properties of this set, beginning with some
operations under which it is closed.

Proposition 8.6.1 *If S and T are languages in $Rec(\Sigma, X)$, then $S \cap T$,
$S \cup T$ and $S \setminus T$ are in $Rec(\Sigma, X)$.*

Proof: Suppose that S and T are recognized by the $\Sigma - X$-tree-recognizers
A and **B**, respectively. For our new tree-recognizers, we use as underlying
Σ-algebra the direct product $\mathcal{C} := \mathcal{A} \times \mathcal{B}$. We use the evaluation mapping
$\gamma : X \to \mathcal{C} := A \times B$, defined by $x \mapsto (\alpha(x), \beta(x))$, for all $x \in X$, where
α and β are the evaluation mappings from **A** and **B**, respectively. Clearly,
$\hat{\gamma}(t) = (\hat{\alpha}(t), \hat{\beta}(t))$ for all $t \in F_\Sigma(X)$. If we now take

$$\mathbf{C_1} = (\mathcal{C}, \Sigma, X, \gamma, A' \times B'),$$
$$\mathbf{C_2} = (\mathcal{C}, \Sigma, X, \gamma, A' \times B \cup A \times B'), \text{ and}$$
$$\mathbf{C_3} = (\mathcal{C}, \Sigma, X, \gamma, A' \times (B \setminus B')),$$

we obtain $\Sigma - X$-recognizers for the languages $S \cap T$, $S \cup T$ and $S \setminus T$ respec-
tively. For the intersection, we have $t \in T(\mathbf{C_1})$ iff $\hat{\gamma}(t) = (\hat{\alpha}(t), \hat{\beta}(t)) \in A' \times B'$
iff $t \in T(\mathbf{A}) \cap T(\mathbf{B})$. This shows $T(\mathbf{C_1}) = S \cap T$. The verifications for the
union and set difference are similar, and we leave them as exercises (see Ex-
ercise 8.11.8).   ∎

Now we consider mappings which transform trees from one language into
trees from another one. Let $\Sigma := \{f_i \mid i \in I\}$ be a set of operation symbols
of type $\tau_1 = (n_i)_{i \in I}$, where f_i is n_i-ary, $n_i \in I\!N$ and let $\Omega = \{g_j \mid j \in J\}$
be a set of operation symbols of type $\tau_2 = (n_j)_{j \in J}$ where g_j is n_j-ary. We
denote by $W_\Sigma(X)$ and by $W_\Omega(Y)$ the sets of all terms of type τ_1 and τ_2,
respectively, where X and Y are alphabets of variables.

Definition 8.6.2 Let $\chi_{n_i} := \{\xi_1, \ldots, \xi_{n_i}\}$ be an auxiliary alphabet. Let
$h_X : X \to W_\Omega(Y)$ and for each $n_i \in I\!N$ let $h_{n_i} : \Sigma_{n_i} \to W_\Omega(Y \cup \chi_{n_i})$ be
mappings. Then the *tree homomorphism* determined by these mappings is
the mapping
$$h : W_\Sigma(X) \to W_\Omega(Y)$$

defined inductively by

(i) $h(x) := h_X(x)$ for every $x \in X$.

(ii) $h(f_i(t_1, \ldots, t_{n_i})) = h_{n_i}(f_i)(\xi_1 \leftarrow h(t_1), \ldots, \xi_{n_i} \leftarrow h(t_{n_i}))$, where $\xi_j \leftarrow h(t_j)$, means to substitute $h(t_j)$ for ξ_j.

The tree homomorphism h is said to be *linear* if for any $n_i \geq 0$ and $f_i \in \Sigma_{n_i}$ no variable ξ_j appears more than once in (ii).

Example 8.6.3 Consider the type $\tau = (2, 2)$ with operation symbols $\Sigma := \{f, g\}$ and the alphabet $X = Y = \{x, y\}$. Then the mappings $h_X(x) := x$ for every $x \in X$ and $h_2 : \{f, g\} \to W_\Sigma(X)$ defined by $f \mapsto f(x, g(x, y))$ and $g \mapsto f(y, x)$ define a tree homomorphism h. In fact this is a special kind of tree homomorphism called a *hypersubstitution*, which will be studied in Chapter 14.

We remark that tree homomorphisms do not always preserve recognizability, as the following example shows:

Example 8.6.4 We put $\Sigma = \Sigma_1 = \{f\}, \Omega = \Omega_2 = \{g\}, X = Y = \{x\}$. We define h_X and h_1 by

$$h_X(x) = x, \text{ and } h_1(f) = g(\xi_1, \xi_1).$$

The $\Sigma - X$-trees have the form $t_k = f(f(\cdots f(x) \cdots)) = f^k(x)$, for $k \geq 0$. Clearly, $h(W_\Sigma(X))$ consists of the trees

$$s_0 := x, s_1 := g(x, x), \ldots, s_{k+1} := g(s_k, s_k), \ldots.$$

Suppose there is an $\Omega - Y$-recognizer $\mathbf{A} = (\mathcal{A}, \Omega, Y, \alpha, A')$ such that $T(\mathbf{A}) = h(W_\Sigma(X))$. Then there must exist two integers $i, j \geq 0$ with $i \neq j$ such that $\hat{\alpha}(s_i) = \hat{\alpha}(s_j)$. But then $\hat{\alpha}(g(s_i, s_j)) = g^{\mathcal{A}}(\hat{\alpha}(s_i), \hat{\alpha}(s_j)) = g^{\mathcal{A}}(\hat{\alpha}(s_i), \hat{\alpha}(s_i)) = \hat{\alpha}(s_{i+1}) \in A'$. This means that $g(s_i, s_j) \in h(W_\Sigma(X))$. Therefore $h(W_\Sigma(X))$ cannot be recognizable.

If we have the additional condition that the tree homomorphism h is linear, then the image $h(T)$ of a recognizable language T is recognizable.

Theorem 8.6.5 *If $h : W_\Sigma(X) \to W_\Omega(Y)$ is a linear tree homomorphism and if T is a recognizable language, then $h(T)$ is a recognizable language.*

Proof: Since T is recognizable, there is a regular tree grammar $G = (N, \Sigma, X, P, a_0)$ which generates T. We may assume that G is in normal form and that G has no non-terminal symbols from which no $\Sigma - X$-tree from T can be generated. Adding all non-terminal symbols from N as nullary operation symbols to Σ and to Ω we obtain the sets Σ' and Ω' of operation symbols. Now we extend the given linear tree homomorphism to a tree homomorphism $h' : W_{\Sigma'}(X) \to W_{\Omega'}(Y)$, by extending h_0 to a mapping $h_0' : \Sigma_0 \cup N \to W_{\Omega'}(Y)$ so that $h_0'(a) = a$ for all $a \in N$. Consider now the regular $\Omega - Y$-tree grammar $G' = (N, \Omega, Y, P', a_0)$ with

$$P' := \{a \to h'(t) \mid a \to t \in P\}.$$

The theorem will be proved if we show that $T(G') = h(T)$. This in turn can be proved by showing that $a \Rightarrow_{G'}^* t$ iff $(\exists s \in W_\Sigma(X))$ $(h(s) = t$ and $a \Rightarrow_G^* s$ for all $a \in N$, $t \in W_\Omega(Y))$. We leave the remaining details to the reader (see for instance F. Gécseg and M. Steinby, [48]).   ∎

8.7   Minimal Tree Recognizers

In this section we generalize our considerations of Section 8.3 to the case of tree-recognizers. We compare tree-recognizers by comparing how they behave with respect to the same input words, and use this to define equivalence of tree-recognizers. Then we look for tree-recognizers which are minimal, in a class of equivalent tree-recognizers, with respect to the size of state sets. Then minimal-tree recognizers will be characterized in terms of homomorphisms of tree-recognizers and quotient recognizers.

As usual, a homomorphism of tree-recognizers should be a mapping which preserves the structure.

Definition 8.7.1 A *homomorphism* from a $\Sigma - X$-tree-recognizer $\mathbf{A} = (\mathcal{A}, \Sigma, X, \alpha, A')$ to a $\Sigma - X$-tree-recognizer $\mathbf{B} = (\mathcal{B}, \Sigma, X, \beta, B')$ is a mapping $\varphi : A \to B$ such that

  (i) φ is a homomorphism from the algebra $\mathcal{A}$ to the algebra $\mathcal{B}$,

 (ii) $\varphi(\alpha(x)) = \beta(x)$ for all $x \in X$,   and

(iii) $\varphi^{-1}(B') = A'$.

If φ is injective (surjective, bijective) then the homomorphism φ is called injective (surjective, bijective). A bijective homomorphism is called an isomorphism.

Lemma 8.7.2 *Let* **A** *and* **B** *be two* $\Sigma - X$*-tree-recognizers. If there exists a homomorphism* $\varphi : \mathbf{A} \to \mathbf{B}$*, then* $T(\mathbf{A}) = T(\mathbf{B})$*.*

Proof: We will show by induction on the depth of the term t that $\varphi(\hat{\alpha}(t)) = \hat{\beta}(t)$, for all $t \in W_\Sigma(X)$. The base case is given by the definition of a homomorphism. Now inductively assume that $\varphi(\hat{\alpha}(t_j)) = \hat{\beta}(t_j)$, for $1 \le j \le n_i$. Then $\varphi(\hat{\alpha}(f_i(t_1, \ldots, t_{n_i}))) = \varphi(f_i^A(\hat{\alpha}(t_1), \ldots, \hat{\alpha}(t_{n_i}))) = f_i^B(\varphi(\hat{\alpha}(t_1)), \ldots, \varphi(\hat{\alpha}(t_{n_i})))$
$= f_i^B(\hat{\beta}(t_1), \ldots, \hat{\beta}(t_{n_i})) = \hat{\beta}(f_i(t_1, \ldots, t_{n_i}))$.
Using this equation we have

$$t \in T(\mathbf{B}) \quad \Leftrightarrow \quad \hat{\beta}(t) \in B' \quad \Leftrightarrow \varphi(\hat{\alpha}(t)) \in B' \quad \Leftrightarrow \quad \hat{\alpha}(t) \in \varphi^{-1}(B') = A'$$
$$\Leftrightarrow \quad t \in T(\mathbf{A}). \qquad \blacksquare$$

An equivalence relation ϱ on a set A is said to *saturate* a subset A' of A if A' is the union of ϱ-equivalence classes. In this case we write $A' = A'\varrho$.

It turns out the following concept of congruence relation is suitable to express the expected relationship between congruence relations and homomorphisms.

Definition 8.7.3 *A congruence* of a $\Sigma - X$-tree-recognizer **A** is a congruence on the algebra $\mathcal{A}$ which saturates A', so that $A'\varrho = A'$. We will denote by $C(\mathbf{A})$ the set of all congruence relations defined on **A**.

Notice that by definition the set $C(\mathbf{A})$ is a subset of the set $Con\mathcal{A}$ of all congruences on the algebra $\mathcal{A}$.

Proposition 8.7.4 $C(\mathbf{A})$ *is a principal ideal of the complete lattice* $Con\mathcal{A}$*, and therefore* $C(\mathbf{A})$ *is also a complete lattice.*

Proof: From the definition of a principal ideal in a lattice, we have to verify the following facts:

(i) $\Delta_A \in C(\mathbf{A})$ and therefore $C(\mathbf{A}) \ne \emptyset$;

(ii) $\theta \subseteq \varrho \in C(\mathbf{A})$ and $\theta \in Con(\mathcal{A})$ imply $\theta \in C(\mathbf{A})$;

(iii) $\bigcup_{C(\mathbf{A})} \{\varrho \mid \varrho \in C(\mathbf{A})\} \in C(\mathbf{A})$.

The details of this proof are left to the reader; see Exercise 8.11.5.   ∎

The join in (iii) is the greatest and thus the generating element of $C(\mathbf{A})$. Later on we will give a more useful description of the greatest element of $C(\mathbf{A})$.

Quotient recognizers are defined in the following way:

Definition 8.7.5 Let $\varrho \in C(\mathbf{A})$. Then the quotient recognizer of $\mathbf{A}$ with respect to the congruence ϱ is defined by

$$\mathbf{A}/\varrho = (A/\varrho, \Sigma, X, \alpha_\varrho, A'/\varrho),$$

where $\alpha_\varrho(x) = [\alpha(x)]_\varrho$ for each $x \in X$.

It is easy to check that the kernel $ker\varphi := \varphi^{-1} \circ \varphi$ of a homomorphism $\varphi : \mathbf{A} \to \mathbf{B}$ is a congruence relation on $\mathbf{A}$. Conversely, if $\varrho \in C(\mathbf{A})$, then the natural mapping

$$nat\varrho : \mathbf{A} \to \mathbf{A}/\varrho, \quad \text{given by} \quad a \mapsto [a]_\varrho \text{ for } a \in A,$$

is a surjective homomorphism. With these facts, we have the analogue of the Homomorphic Image Theorem.

Theorem 8.7.6 *(Homomorphic Image Theorem for Tree-Recog-nizers) Let $\varphi : \mathbf{A} \to \mathbf{B}$ be a surjective homomorphism of tree-recognizers. Then there exists a unique isomorphism f from $\mathbf{A}/ker\,\varphi$ onto $\mathbf{B}$ with $f \circ nat(ker\,\varphi) = \varphi$.*   ∎

Using the natural homomorphism and Lemma 8.7.2 we have the following conclusion.

Corollary 8.7.7 *If $\varrho \in C(\mathbf{A})$, then $T(\mathbf{A}/\varrho) = T(\mathbf{A})$.*   ∎

Now we can define what it means for two tree-recognizers to be equivalent.

Definition 8.7.8 Two states a and b of a $\Sigma - X$-tree-recognizer **A** are said to be *equivalent*, and we write $a \sim_{\mathbf{A}} b$ (or simply $a \sim b$ if the context is clear), if

$$f(a) \in A' \Leftrightarrow f(b) \in A' \text{ for all unary polynomial operations } f \text{ of } \mathcal{A}.$$

A tree-recognizer is called *reduced* if no two distinct states are equivalent. We will also define the concept of a minimal tree-recognizer.

Definition 8.7.9 The $\Sigma - X$- tree-recognizer **A** is called

(i) *reduced* if $\sim_{\mathbf{A}} = \Delta_A$;

(ii) *connected* if every state is reachable, in the sense that for every $a \in A$ there exists a tree $t \in W_\Sigma(X)$ such that $\hat{\alpha}(t) = a$;

(iii) *minimal* if it is both connected and reduced.

In a tree-recognizer which is connected, the set $\alpha(X)$ generates the algebra $\mathcal{A}$. Non-reachable states can be deleted without changing the language which is recognized. In the finite case, the greatest congruence on **A** gives the smallest quotient recognizer.

Theorem 8.7.10 *For any $\Sigma - X$-tree-recognizer **A** the relation $\sim_{\mathbf{A}}$ is the greatest congruence of **A**, and $\mathbf{A}/\sim_{\mathbf{A}}$ is a reduced $\Sigma - X$-tree-recognizer which is equivalent to **A**.*

Proof: To prove that $\sim_{\mathbf{A}}$ is a congruence, it suffices to prove the congruence property for all unary polynomial operations of $\mathcal{A}$. If f, g are unary polynomial operations of $\mathcal{A}$ then the composition $g \circ f$ is a unary polynomial operation of $\mathcal{A}$. Then $g(f(a)) \in A'$ iff $g(f(b)) \in A'$ and this implies $f(a) \sim_{\mathbf{A}} f(b)$. If $a \sim_{\mathbf{A}} b$ and $a \in A'$, then $b = id_A(b) \in A'$. Thus $A' \sim_{\mathbf{A}} = A'$ and therefore $\sim_{\mathbf{A}}$ is a congruence relation on **A**.

Now let ϱ be a congruence of **A**. If $(a, b) \in \varrho$ and if f is a unary polynomial operation of $\mathcal{A}$, then $(f(a), f(b)) \in \varrho$. From $A'\varrho = A'$ we obtain $f(a) \in A'$ iff $f(b) \in A'$ and thus $a \sim_{\mathbf{A}} b$. This shows that $\sim_{\mathbf{A}}$ is the greatest congruence on **A**. By Corollary 8.7.7, $T(\mathbf{A}) = T(\mathbf{A}/\sim_{\mathbf{A}})$. Now we have to show that $\mathbf{A}/\sim_{\mathbf{A}}$ is reduced. It follows from $([a]_{\sim_{\mathbf{A}}}, [b]_{\sim_{\mathbf{A}}}) \in \sim_{\mathbf{A}/\sim_{\mathbf{A}}}$ that for all unary polynomial operations f of $\mathbf{A}/\sim_{\mathbf{A}}$,

$f([a]_{\sim_\mathbf{A}}) \in A'/\mathbf{A}$ iff $f([b]_{\sim_\mathbf{A}}) \in A'/\mathbf{A}$
$\Leftrightarrow f(a) \in A'$ iff $f(b) \in A'$
$\Leftrightarrow a \sim_\mathbf{A} b \quad \Leftrightarrow [a]_{\sim_\mathbf{A}} = [b]_{\sim_\mathbf{A}}.$

The other inclusion is clear since $\sim_\mathbf{A}$ is an equivalence relation. ∎

The quotient recognizer $\mathbf{A}/\sim_\mathbf{A}$ is called the *reduced form* of $\mathbf{A}$. Tree recognizers which have isomorphic reduced forms are equivalent. For connected tree-recognizers the converse is also true.

Theorem 8.7.11 *Let* $\mathbf{A}$ *and* $\mathbf{B}$ *be two minimal tree-recognizers. If they are equivalent then they are also isomorphic.*

Proof: We show that the mapping $\varphi : A \to B$ defined by

$$\varphi(\hat{\alpha}(t)) = \hat{\beta}(t) \text{ for all } t \in W_\Sigma(X)$$

is an isomorphism from $\mathbf{A}$ onto $\mathbf{B}$. We prove this in several steps:

1. φ maps every state of $\mathbf{A}$ to a state of $\mathbf{B}$ since $\mathbf{A}$ is connected. Since $\mathbf{B}$ is connected φ is surjective.

2. We show that φ is well defined. Assume that $\hat{\alpha}(s) = \hat{\alpha}(t)$ and that $\hat{\beta}(s) \neq \hat{\beta}(t)$. Then $\hat{\beta}(s)$ and $\hat{\beta}(t)$ are non-equivalent since $\mathbf{A}$ and $\mathbf{B}$ are reduced. Then there exists a unary polynomial operation f of the algebra $\mathcal{A}$ such that $f(\hat{\beta}(s)) \in B'$ and $f(\hat{\beta}(t)) \notin B'$, or vice versa. Furthermore, there is a tree $p \in W_\Sigma(B \cup \{\xi\})$, where ξ is an auxiliary variable and $\xi \notin B \cup X$, such that for all $b \in B$ we have $f(b) = p^B(\beta_b)$ where $\beta_b : B \cup \{\xi\} \to B$ is defined by $\beta_b|B = 1_B$ and $\beta_b(\xi) = b$. Since $\mathbf{B}$ is connected, for each $b \in B$ there exists a $\Sigma - X$-tree p_b such that $\hat{\beta}[p_b] = b$. Now let

$$q = p(b \leftarrow p_b \mid b \in B) \quad (\in W_\Sigma(X \cup \{\xi\})).$$

Consider the $\Sigma - X$-trees q_s and q_t which arise from q by substitution of the trees s and t, respectively, for ξ. Then

$$\hat{\beta}(q_s) = p^B(\beta_{\hat{\beta}(s)}) = f(\hat{\beta}(s)) \in B'$$

and

$$\hat{\beta}(q_t) = p^B(\beta_{\hat{\beta}(t)}) = f(\hat{\beta}(t)) \notin B'.$$

Now we assign to every letter $x \in X$ in q the value $\alpha(x)$, to get a unary polynomial operation g of $\mathcal{A}$ such that $g(a) = q^{\mathcal{A}}(\alpha_a)$ for each $a \in A$, where α_a is defined by $\alpha_a|X = \alpha$ and $\alpha_a(\xi) = a$. Then we have $\varphi(g(\hat{\alpha}(s))) = \varphi(\alpha_{\hat{\alpha}(s)}(q^{\mathcal{A}})) = \varphi(\hat{\alpha}(q_s)) = \hat{\beta}(q_s) \in B'$ and $\varphi(g(\hat{\alpha}(t))) = \varphi(\alpha_{\hat{\alpha}(t)}(q^{\mathcal{A}})) = \varphi(\hat{\alpha}(q_t)) = \hat{\beta}(q_t) \notin B'$. This contradicts the original assumption that $\hat{\alpha}(s) = \hat{\alpha}(t)$. Hence $q_s \in T(\mathbf{B})$, but $q_t \notin T(\mathbf{B})$. On the other hand, $\hat{\alpha}(s) = \hat{\beta}(t)$ implies $\hat{\alpha}(q_s) = \hat{\alpha}(q_t)$, and this is a contradiction of the assumption that $T(\mathbf{A}) = T(\mathbf{B})$.

3. In a similar way, reversing the roles of $\mathbf{A}$ and $\mathbf{B}$, it can be shown that $\hat{\beta}(s) = \hat{\beta}(t)$ implies $\hat{\alpha}(s) = \hat{\alpha}(t)$ for all $\Sigma - X$-trees s and t. This means that φ is injective.

4. We show that φ is compatible with the operations of $\mathcal{A}$. Since $\mathbf{A}$ is connected, for any $a_1, \ldots, a_{n_i}$ there exist terms $t_1, \ldots, t_{n_i} \in W_\Sigma(X)$ such that $\hat{\alpha}(t_1) = a_1, \ldots, \hat{\alpha}(t_{n_i}) = a_{n_i}$. If $f_i \in \Sigma_{n_i}$ for $n_i \geq 0$, then

$$\varphi(f_i^{\mathcal{A}}(a_1, \ldots, a_{n_i})) = \varphi(f_i^{\mathcal{A}}(\hat{\alpha}(t_1), \ldots, \hat{\alpha}(t_{n_i})))$$
$$= \varphi(\hat{\alpha}(f_i(t_1, \ldots, t_{n_i}))) = \hat{\beta}(f_i(t_1, \ldots, t_{n_i}))$$
$$= f_i^{\mathcal{B}}(\hat{\beta}(t_1), \ldots, \hat{\beta}(t_{n_i})) = f_i^{\mathcal{B}}(\varphi(\hat{\alpha}(t_1)), \ldots, \varphi(\hat{\alpha}(t_{n_i})))$$
$$= f_i^{\mathcal{B}}(\varphi(a_1), \ldots, \varphi(a_{n_i})).$$

5. For each $x \in X$ we have $\varphi(\alpha(x)) = \hat{\beta}(x)$, and thus $\varphi \circ \alpha = \hat{\beta}$.

6. If $\hat{\alpha}(t) \in A'$ and $t \in W_\Sigma(X)$, then $\varphi(\hat{\alpha}(t)) = \hat{\beta}(t) \in B'$ since $t \in T(\mathbf{A}) = T(\mathbf{B})$. Similarly, $\varphi(\hat{\alpha}(t)) \in B'$ implies $\hat{\alpha}(t) \in A'$. Hence $\varphi^{-1}(B') = A'$. ∎

As a corollary we have the following result.

Corollary 8.7.12 *If $\mathbf{A}$ and $\mathbf{B}$ are connected $\Sigma - X$-tree recognizers such that $T(\mathbf{A}) = T(\mathbf{B})$, then $\mathbf{A}/ \sim_{\mathbf{A}}$ is isomorphic to $\mathbf{B}/ \sim_{\mathbf{B}}$.* ∎

For every $\Sigma - X$-language T there is at least the (infinite) $\Sigma - X$- recognizer $\mathbf{F}_T = (\mathcal{F}_\Sigma(X), \Sigma, X, id_X, T)$, where $\mathcal{F}_\Sigma(X)$ is the absolutely free algebra. Clearly, for each term $t \in W_\Sigma(X)$ we have

$$id_X(t) = t \in T(\mathbf{F}_T) \text{ iff } t \in T.$$

The tree-recognizer $\mathbf{F}_T$ is connected. To show that $\mathbf{F}_T/\sim_{\mathbf{F}_T}$ is also connected, we show more generally that the homomorphic image of a connected recognizer is connected. Let $\varphi : \mathbf{A} \to \mathbf{B}$ be a surjective homomorphism of $\Sigma - X$-recognizers. Let b be an arbitrary state of $\mathbf{B}$. Then there exists an $a \in A$ such that $\varphi(a) = b$. Since $\mathbf{A}$ is connected, there is a tree such that $\hat{\alpha}(t) = a$. Then we have

$$\hat{\beta}(t) = \varphi(\hat{\alpha}(t)) = \varphi(a) = b.$$

Altogether we have proved the following theorem.

Theorem 8.7.13 *For every language T there is a minimal (possibly infinite) tree-recognizer, and it is unique up to isomorphism. If $\mathbf{A}$ is a connected recognizer of the language T, then the minimal recognizer of T is a homomorphic image of $\mathbf{A}$ (under a surjective homomorphism). The quotient recognizer $\mathbf{A}/\sim_{\mathbf{A}}$ is minimal.* ■

We remark that the previous results can be used to prove that an arbitrary $\Sigma - X$-language T is recognizable iff there exist a finite Σ-algebra $\mathcal{A}$, a homomorphism $\varphi : \mathcal{F}_\Sigma(X) \to \mathcal{A}$ and a subset $A' \subseteq A$ such that $\varphi^{-1}(A) = T$. This proposition allows us to give a new definition of recognizability for subsets of the universes of arbitrary algebras, not just free algebras. Let $\mathcal{A}$ be any algebra. Then a subset $T \subseteq A$ is called *recognizable* if there exist a finite algebra $\mathcal{B}$ of the same type, a homomorphism $\varphi : \mathcal{A} \to \mathcal{B}$ and a subset $B' \subseteq B$ such that $\varphi^{-1}(B') = T$. This definition gives us the recognizability of $\Sigma - X$-languages when $\mathcal{A}$ is the free algebra of its type, and the recognizable languages of Section 8.1 when $\mathcal{A}$ is the free monoid X^* generated by X.

8.8   Tree Transducers

In this section we consider sets of terms or trees of two different types, Σ and Ω. A tree transducer is a system which transforms trees of one type into trees of the other (just as automata transform strings into strings). Such systems also give us tree transformations, which are subsets of $W_\Sigma(X) \times W_\Omega(X)$. The concept of a tree transducer used here is due to J. W. Thatcher ([113]).

More precisely, let $\Sigma := \{f_i \mid i \in I\}$ be a set of operation symbols of type $\tau_1 = (n_i)_{i \in I}$, where f_i is n_i-ary for $n_i \in \mathbb{N}$, and let $\Omega = \{g_j \mid j \in J\}$ be a set of operation symbols of type $\tau_2 = (n_j)_{j \in J}$ where g_j is n_j-ary. As in Section

8.6 we denote by $W_\Sigma(X)$ and $W_\Omega(X)$ the sets of all terms of type τ_1 and τ_2, respectively. Then we define a *tree transformation* as a binary relation

$$T_{\tau_1,\tau_2} \subseteq W_\Sigma(X) \times W_\Omega(X).$$

The most important tree transformations are those which can be given in an effective way.

Definition 8.8.1 Let τ_1 and τ_2 be two types. A $(\tau_1 - \tau_2)$-*tree transducer* is a sequence

$$\underline{A} = (\Sigma, X, A, \Omega, P, A'),$$

where
$\Sigma = \{f_i \mid i \in I\}$ is a set of operation symbols of type τ_1,
$\Omega = \{g_j \mid j \in J\}$ is a finite set of operation symbols of type τ_2,
$A = \{a_1, \ldots, a_m\}$ is a finite set of unary operation symbols,
$A' \subseteq A$, and
P is a finite set of *productions* or *rules of derivation* of the forms

(i) $x \to a(t)$, for $x \in X$, $a \in A$ and $t \in W_\Omega(X)$,

(ii) $f_i(a_1(\xi_1), \ldots, a_{n_i}(\xi_{n_i})) \to a\, t(\xi_1, \ldots, \xi_{n_i})$, for $f_i \in \Sigma_{n_i}$, $a_1, \ldots, a_m \in A$, $a \in A$ and $\xi_1, \ldots, \xi_{n_i} \in \chi_m$, where $\chi_m = \{\xi_1, \ldots, \xi_m\}$ is an auxiliary alphabet, and $t(\xi_1, \ldots, \xi_m) \in W_\Omega(X \cup \chi_m)$.

For two trees s and t, we will say that s *directly derives* t *in* $\underline{A}$, if t can be obtained from s by the following steps:

(1) replacement of an occurrence of a variable $x \in X$ in s by the right hand side of a production from (i) or

(2) replacement of an occurrence of a subtree
$f_i(a_1(q_1), \ldots, a_{n_i}(q_{n_i}))$ in s, for $a_1, \ldots, a_{n_i} \in A$ and $q_1, \ldots, q_{n_i} \in W_\Omega(X \cup \chi_m)$, by $a\, t(q_1, \ldots, q_m)$, if $f_i(a_1(\xi_1), \ldots, a_{n_i}(\xi_{n_i})) \to a\, t(\xi_1, \ldots, \xi_m)$ is a production.

If s directly derives t in $\underline{A}$, we write $s \to_{\underline{A}} t$. Forming the reflexive and transitive closure of this relation $\to_{\underline{A}}$, we say that s *derives* t *in* $\underline{A}$ if there is a sequence $s \to_{\underline{A}} s_1 \to_{\underline{A}} s_2 \to_{\underline{A}} \cdots \to_{\underline{A}} s_n = t$ of direct derivations of t from s or if $s = t$. In this case we write $s \Rightarrow^*_{\underline{A}} t$.

Every tree transducer induces a tree transformation, in a natural way.

Definition 8.8.2 If $\underline{A}$ is a $(\tau_1 - \tau_2)$-tree transducer, then the *tree transformation induced by $\underline{A}$* is the set

$$T_{\underline{A}} := \{(s,t) \mid s \in W_{\tau_1}(X) \text{ and } s \Rightarrow^*_{\underline{A}} a_0 t \text{ for some } a_0 \in A'\}.$$

This means that tree transformations of the form $T_{\underline{A}}$, induced from a tree transducer, can be described in an effective (algorithmic) way. Now we consider the following example of a tree transducer.

Example 8.8.3 Let $\underline{A} = (\Sigma, \{x\}, \{a_0, a_1\}, \Omega, P, \{a_1\})$ be the tree transducer with $\Sigma = \Sigma_2 = \{f\}$, $\Omega = \Omega_2 = \{g\}$, and where P consists of the productions $x \rightarrow a_0 x$, $f(u_0, a_0) \;\;\rightarrow\;\; a_1 g(\xi_1, \xi_2)$, $f(a_0, a_1) \rightarrow a_0 g(\xi_1, \xi_2)$, $f(a_1, a_0) \rightarrow a_1 g(\xi_1, \xi_2)$ and $f(a_1, a_1) \rightarrow a_1 g(\xi_1, \xi_2)$.
Then the term $t = f(f(x, x), x)$ has the following derivation:

$$f(f(x,x),x) \quad \Rightarrow^*_{\underline{A}} \quad f(f(a_0 x, a_0 x), a_0 x) \quad \Rightarrow^*_{\underline{A}} \quad f(a_1 g(x,x), a_0 x) \quad \Rightarrow^*_{\underline{A}}$$
$$a_1 g(g(x,x),x).$$

Therefore $(f(f(x,x),x), g(g(x,x),x)) \in T_{\underline{A}}$.

The tree homomorphisms $h : W_\Sigma(X) \rightarrow W_\Omega(X)$ defined in Section 8.6 also induce tree transformations $T_h := \{(t, h(t)) \mid t \in W_\Sigma(X)\}$. This raises the question of whether for each tree homomorphism $h : W_\Sigma(X) \rightarrow W_\Omega(X)$ there is a tree transducer $\underline{A}$ such that $T_h = T_{\underline{A}}$.

Definition 8.8.4 A tree transducer is called an *H-transducer* if it has the form $\underline{A} = (\Sigma, X, \{a\}, \Omega, P, a)$, with

$$P = \{x \rightarrow a h_X(x) \mid x \in X\} \cup$$
$$\{f_i(a(\xi_1), \ldots, a(\xi_{n_i})) \rightarrow a h_{n_i}(f_i)(\xi_1, \ldots, \xi_{n_i}) \mid f_i \in \Sigma_{n_i}\},$$

for some tree homomorphism h.

Theorem 8.8.5 *Let $h : W_\Sigma(X) \rightarrow W_\Omega(X)$ be a tree homomorphism and let T_h be the tree transformation defined by h. Then T_h can be induced by an H-transducer. Conversely, tree transformations induced by H-transducers are defined by tree homomorphisms.*

Proof: We show that any pair $(t, h(t)) \in T_h$ is also in $T_{\underline{A}}$, proceeding by induction on the complexity of the tree t. If $t \in X$, then $t = x$ derives $a h_X(x)$

$= h(x)$ and therefore $(x, h(x)) \in T_A$. Inductively, let $t = f_i(t_1, \ldots, t_{n_i})$ and suppose that $(t_j, h(t_j)) \in T_h$ implies $t_j \Rightarrow^*_{\underline{A}} ah(t_j)$ for all $j = 1, \ldots, n_j$. Then t has the following derivation in $\underline{A}$:

$$t = f_i(t_1, \ldots, t_{n_i}) \Rightarrow^*_{\underline{A}} ah_{n_i}(f_i)(h(t_1), \ldots h(t_{n_i})) = ah(t);$$

and this means $(t, h(t)) \in T_{\underline{A}}$. This shows $T_h \subseteq T_{\underline{A}}$. If conversely $t \Rightarrow^*_{\underline{A}} as$, and $t = x$ is a variable, then the only possible start for a derivation is to use a rule $x \to ah_X(x) = h(x)$, and then $(t, s) \in T_h$. Now we assume that $t = f_i(t_1, \ldots t_{n_i})$ and that $t_j \Rightarrow^*_{\underline{A}} as_j$ implies $(t_j, s_j) \in T_h$, so that $s_j = h(t_j)$ for $j = 1, \ldots, n_i$. A derivation of t has to start with a rule of the second kind. Therefore t has the form $t = f_i(as_1, \ldots, as_{n_i}) \to ah_{n_i}(f_i)(s_1, \ldots, s_{n_i}) = ah_{n_i}(f_i)(h(t_1), \ldots, h(t_{n_i})) = h(t)$. Therefore $(t, s) \in T_h$. The second proposition can be proved in a similar way. ∎

8.9 Turing Machines

The concept of a *Turing machine* was introduced by A. Turing to define the class of "computable" functions. The Turing machine is a way to make precise the concept of an algorithm, and we will use Turing machines in the next section to show that some algebraic properties are undecidable. The definition of a Turing machine is similar to that of a finite automaton, in that we have an input alphabet, an initial state, final states and a transition function. The main difference is that a Turing machine also has an infinite storage capacity. This is provided by a doubly infinite tape, marked off into squares, and a read-write head which when positioned on a square can read the letter currently on the square or write a letter in the square.

To define Turing machines more precisely, we use the following notation. We use $X = \{0, 1\}$ for the input alphabet, and also refer to 0 and 1 as *tape symbols*. The set $\Gamma = \{L, R\}$ is the set of *motion symbols*, indicating whether the read-write head will move one square to the left or to the right. We have a doubly infinite tape, on which each square is printed with exactly one of the tape symbols. Thus the tape can be regarded as a function $t : \mathbb{Z} \to \{0, 1\}$, with $t(n)$ equal to the symbol printed on the n-th square of the tape. The *blank tape* corresponds to the constant function t_0 with value 0. We also have an infinite set Z of states, indexed by the non-negative integers:

$$Z = \{\mu_n \mid n \geq 0\}.$$

The state μ_1 is called the *initial state*, and we set $Z_1 = \{\mu_1\}$, while the state μ_0 is called the *final state*.

An *instruction* of the machine is a quintuple $\mu_i r s T \gamma$, where $r, s \in \{0, 1\}$, T is one of L or R, μ_i is a state symbol other than μ_0, and γ is any state symbol (possibly μ_0). Such a quintuple is interpreted as follows: if the machine is in state μ_i and reads the symbol r on the current square, it performs the following steps:

1. The machine replaces r by s on the current square.
2. If $T = L$, the read-write head moves one square to the left;
   if $T = R$, the read-write head moves one square to the right.
3. The state changes to state γ.

It is clear that the work of a Turing machine could be described by a partial transition function $\delta : \mathbb{Z} \times X \to \mathbb{Z} \times \Gamma \times X$. (Notice that δ is not defined on pairs using state μ_0, so it is only a partial function.) An input is accepted if the state μ_0 is reached; otherwise it is rejected. The language accepted by the Turing machine $\mathcal{T}$ is the set

$$L(\mathcal{T}) := \{w \in X^* \mid w \text{ is accepted by } \mathcal{T}\}.$$

This version of a Turing machine is modelled on the language recognizers of the previous sections. We can also view Turing machines as a set of machine instructions.

Definition 8.9.1 A Turing machine is a finite set $\mathcal{T}$ of machine instructions, for which there is some natural number k such that

(i) For each $1 \leq i \leq k$ and for each $r \in \{0, 1\}$, $\mathcal{T}$ contains exactly one instruction $\mu_i r s T \gamma$.
(ii) No state symbol μ_j for which $j > k$ occurs in the instructions in $\mathcal{T}$.

A *configuration* of a Turing machine $\mathcal{T}$ is a triple $Q = (t, n, \gamma)$, where t is a tape, n is an integer and γ is a state symbol. This encodes the information that the machine is in state γ, reading the n-th square on tape t. If there is an instruction $\gamma t(n) s T \gamma'$, then the machine will move into state γ', and the read-write head moves either right or left according to T. In addition, the tape t is converted to a tape t', for which $t'(n) = s$ and $t'(k) = t(k)$ for all $k \neq n$. The result is described by a new configuration, either $(t', n-1, \gamma')$

when $T = L$ or $(t', n+1, \gamma')$ when $T = R$. We write $Q' = \mathcal{T}(Q)$ to indicate the transition from the configuration Q to the resulting one. The initial configuration $Q_0 = (t_0, 0, \mu_1)$ starts the machine in its initial state, reading at the 0 position on a blank tape t_0.

In this way, starting with Q_0 and applying the transition to new configurations, we obtain a sequence Q_0, Q_1, Q_2, ... of configurations. This process will only stop if we reach a configuration Q_m which uses the final state μ_0, since there is no instruction for this state. In this case the sequence of configurations stops at some finite stage m; otherwise it will continue infinitely. We say that the Turing machine $\mathcal{T}$ *halts* iff its sequence of configurations is finite. In the next section we will use the well-known fact that the problem of deciding whether a Turing machine will halt, the so-called *Halting Problem*, is recursively undecidable. More details may be found in S. C. Kleene [63].

8.10 Undecidable Problems

In 1993 R. McKenzie resolved several longstanding and challenging problems concerning varieties generated by finite algebras. One of these was the problem known as *Tarski's Finite Basis Problem*, which asked whether it is algorithmically decidable whether any finite algebra is finitely axiomatizable. Since many properties of varieties generated by finite algebras do turn out to be algorithmically decidable, the following results of McKenzie were surprising:

1. There is no algorithm to decide whether any given finite algebra is finitely axiomatizable.

2. There is no algorithm to decide whether any given finite algebra generates a residually finite variety.

The main idea of McKenzie's proofs involves the construction of a finite algebra $\mathcal{A}(\mathcal{T})$ which encodes the computation of a Turing machine $\mathcal{T}$. We will outline here the construction of this algebra, and its use in the proof, but the full proofs are beyond the scope of this book. Complete details may be found in [79], [80] and [81].

We begin by describing the algebra $\mathcal{A}(\mathcal{T})$ constructed from a given Turing machine $\mathcal{T}$. The universe of this algebra is rather complicated, and will be described in pieces. We let $\mu_0, \ldots, \mu_k$ be the states of the Turing machine. We define the following sets:

$U = \{1, 2, H\}$,
$W = \{C, D, \overline{C}, \overline{D}\}$,
$A = \{0\} \cup U \cup W$,
$V = \bigcup_{i=0}^{k} V_i$, where $V_i = V_{i0} \cup V_{i1}$ and $V_{ir} = V_{ir}^0 \cup V_{ir}^s$,

with $V_{ir}^s = \{C_{ir}^s, D_{ir}^s, M_i^r, \overline{C}_{ir}^s, \overline{D}_{ir}^s, \overline{M}_i^r\}$, for $0 \le i \le k$ and $\{r, s\} \subseteq \{0, 1\}$.
The unbarred symbols are used to encode configurations of the Turing machine, while the barred versions control the finite subdirectly irreducible algebras. The universe of $\mathcal{A}(\mathcal{T})$ is defined to be the set $A \cup V$. We point out that the cardinality of this set is $20k + 28$, where k is the number of non-halting states of the Turing machine.

On this base set, we define the following operations. The first is a semilattice operation $\wedge$, defined by

$$x \wedge y = \begin{cases} x, & \text{if } x = y \\ 0, & \text{otherwise.} \end{cases}$$

We need a multiplication $\cdot$ defined by

$$\begin{aligned} 2 \cdot D = H \cdot C = D, & \quad 1 \cdot C = C, \\ 2 \cdot \overline{D} = H \cdot \overline{C} = \overline{D}, & \quad 1 \cdot \overline{C} = \overline{C}, \\ x \cdot y = 0 \text{ for all other pairs } & (x, y). \end{aligned}$$

We also need the following operations:

$$J(x, y, z) = \begin{cases} x, & \text{if } x = y \neq 0, \\ x \wedge z, & \text{if } x = \overline{y} \in V \cup W, \\ 0, & \text{otherwise.} \end{cases}$$

$$J'(x, y, z) = \begin{cases} x \wedge z, & \text{if } x = y \neq 0, \\ x, & \text{if } x = \overline{y} \in V \cup W, \\ 0, & \text{otherwise.} \end{cases}$$

$$S_0(u, v, x, y, z) = \begin{cases} (x \wedge y) \vee (x \wedge z), & \text{if } u \in V_0, \\ 0, & \text{otherwise.} \end{cases}$$

$$S_1(u, v, x, y, z) = \begin{cases} (x \wedge y) \vee (x \wedge z), & \text{if } u \in \{1, 2\}, \\ 0, & \text{otherwise.} \end{cases}$$

$$S_2(u, v, x, y, z) = \begin{cases} (x \wedge y) \vee (x \wedge z), & \text{if } u = \overline{v} \in V \cup W, \\ 0, & \text{otherwise.} \end{cases}$$

$$T(x, y, z, u) = \begin{cases} x \cdot y, & \text{if } x \cdot y = z \cdot u \neq 0 \text{ and } x = z \text{ and } y = w, \\ \overline{x \cdot y}, & \text{if } x \cdot y = z \cdot u \neq 0 \text{ and } x \neq z \text{ or } y \neq w, \\ 0, & \text{otherwise.} \end{cases}$$

The following unary operation I serves to set up the initial configuration:

$$I(x) = \begin{cases} C_{10}^0, & \text{if } x = 1, \\ M_1^0, & \text{if } x = H, \\ D_{10}^0, & \text{if } x = 2, \\ 0, & \text{otherwise.} \end{cases}$$

For each instruction $\mu_i r s L \mu_j$ of the Turing machine $\mathcal{T}$ and each $t \in \{0, 1\}$, we have an operation L_{irt} which will describe the operation of the machine when it is reading symbol r in state μ_i, and the square just to the left of the current one has symbol t on it.

$$L_{irt}(x, y, u) = \begin{cases} C_{jt}^{s'}, & \text{if } x = y = 1, u = C_{ir}^{s'} \text{ for some } s', \\ M_j^t, & \text{if } x = H, y = 1, u = C_{ir}^t, \\ D_{jt}^s, & \text{if } x = 2, y = H, u = M_i^r, \\ D_{jt}^{s'}, & \text{if } x = y = 2, u = D_{ir}^{s'} \text{ for some } s', \\ \overline{v}, & \text{if } u \in V \text{ and } L_{irt}(x, y, \overline{u}) = v \in V, \\ & \text{according to the previous lines} \\ 0, & \text{otherwise.} \end{cases}$$

For each instruction of the form $\mu_i r s R \mu_j$ in $\mathcal{T}$ and for each $t \in \{0, 1\}$, we have an operation R_{irt}, given by

$$R_{irt}(x,y,u) = \begin{cases} C_{jt}^{s'}, & \text{if } x = y = 1, u = C_{ir}^{s'} \text{ for some } s', \\ C_{jt}^{s}, & \text{if } x = H, y = 1, u = M_i^t, \\ M_j^t, & \text{if } x = 2, y = H, u = D_{ir}^t, \\ D_{jt}^{s'}, & \text{if } x = y = 2, u = D_{ir}^{s'} \text{ for some } s', \\ \overline{v}, & \text{if } u \in V \text{ and } R_{irt}(x,y,\overline{u}) = v \in V, \\ & \text{according to the previous lines} \\ 0, & \text{otherwise.} \end{cases}$$

Let $\mathcal{L}$ be the collection of all these operations L_{irt}, and dually for $\mathcal{R}$. We will assume that $F_1, \ldots, F_c$ is a complete list of all these operations. We define a binary relation $\preceq$ on U by

$$\preceq := \{(2,2),(2,H),(H,1),(1,1)\}.$$

Now we use this to define operations U_i^1 and U_i^2, for each $1 \le i \le c$, by

$$U_i^1(x,y,z,u) = \begin{cases} \overline{F_i(x,y,u)}, & \text{if } x \preceq z, \ y \ne z, \ F_i(x,y,u) \ne 0, \\ F_i(x,y,u) & \text{if } x \preceq z, \ y = z, \ F_i(x,y,u) \ne 0, \\ 0, & \text{otherwise.} \end{cases}$$

$$U_i^2(x,y,z,u) = \begin{cases} \overline{F_i(y,z,u)}, & \text{if } x \preceq z, \ x \ne y, \ F_i(y,z,u) \ne 0, \\ F_i(y,z,u) & \text{if } x \preceq z, \ x = y, \ F_i(y,z,u) \ne 0, \\ 0, & \text{otherwise.} \end{cases}$$

The *configuration algebra* of a Turing machine consists of all its configurations, together with a partial function which maps any configuration Q whose state is not the final or halting state to the configuration $Q' = \mathcal{T}(Q)$. Consider the direct product $\mathcal{B} = \mathcal{A}(\mathcal{T})^X$ for a non-empty set X. The operations in $\mathcal{A}(\mathcal{T})$ allow us to encode in $\mathcal{B}$ certain subsets of the configuration algebra, along with the production function restricted to a subset. If X is finite, any connected subset of the configuration algebra can be encoded in $\mathcal{B}$. For the details, see [80].

Now the two possibilities for $\mathcal{T}$, that it halts or not, are considered. McKenzie showed that if $\mathcal{T}$ does not halt, the variety generated by $\mathcal{A}(\mathcal{T})$ contains a denumerably infinite subdirectly irreducible algebra, and the residual bound of $\mathcal{A}(\mathcal{T})$ satisfies $\kappa(\mathcal{A}(\mathcal{T})) \ge \omega_1$. However, if $\mathcal{T}$ halts, then $\mathcal{A}(\mathcal{T})$ is residually finite, with a finite cardinal m such that $\kappa(\mathcal{A}(\mathcal{T})) \le m$. Since it is not decidable whether a Turing machine halts or not, there can be no algorithm to decide whether any given finite algebra is residually finite or not.

Now we can apply Willard's theorem, Theorem 6.7.5. The variety $V(\mathcal{A}(\mathcal{T}))$ can be shown to be congruence meet-semidistributive, and if $\mathcal{T}$ halts, the variety is also residually finite. Therefore, if $\mathcal{T}$ halts, the algebra $\mathcal{A}(\mathcal{T})$ is finitely axiomatizable. But it is algorithmically undecidable if $\mathcal{T}$ halts; so it is also undecidable whether $\mathcal{A}(\mathcal{T})$ is finitely axiomatizable or not.

8.11 Exercises

8.11.1. Draw a directed graph illustrating the action of the automaton from Example 8.2.5.

8.11.2. Prove that the composition of two homomorphisms $f = (f_I, f_S, f_O)$ and $g = (g_I, g_S, g_O)$, as defined in Definition 8.3.3, is again a homomorphism.

8.11.3. a) Prove that the kernel of a homomorphism $f : \mathcal{H}_1 \to \mathcal{H}_2$ between automata is a congruence relation.
b) Let R be a congruence on an automaton $\mathcal{H}$. Prove that the natural mapping $f : \mathcal{H} \to \mathcal{H}/R$ is a homomorphism, whose kernel is the congruence R.

8.11.4. Prove that for any weakly initial deterministic finite automaton $\mathcal{H}$, the reduced automaton $\mathcal{H}'$ constructed in the proof of Theorem 8.3.10 is unique up to isomorphism.

8.11.5. Prove that the set of all congruences of a $\Sigma - X$-tree-recognizer $C(\mathbf{A})$ forms a principal ideal of the complete lattice $Con\mathcal{A}$, and therefore that $C(\mathbf{A})$ is a complete lattice itself.

8.11.6 Prove the Homomorphic Image Theorem for $\Sigma - X$-tree-recognizers.

8.11.7. Prove that every language generated by an extended regular tree grammar can be generated by an ordinary tree grammar.

8.11.8. Let $\mathbf{A} = (A, \Sigma, X, \alpha, A')$ and $\mathbf{B} = (B, \Sigma, X, \beta, B')$ be two $\Sigma - X$-tree-recognizers, with $T(\mathbf{A}) = S$ and $T(\mathbf{B}) = T$. Let $\mathbf{C_2} = (A \times B, \Sigma, X, \alpha \times \beta, A' \times B \cup A \times B')$ and $\mathbf{C_3} = (A \times B, \Sigma, X, \alpha \times \beta, A' \times (B/B'))$, where

$\alpha \times \beta : W_\tau(X) \to W_\tau(X) \times W_\tau(X)$ is defined by $(\alpha \times \beta)(t) := (\alpha(t), \beta(t))$, for all $t \in W_\tau(X)$. Prove that $T(\mathbf{C_2}) = S \cup T$ and $T(\mathbf{C_3}) = S \setminus T$.

8.11.9. Let X_n be a fixed finite alphabet. Let R be the relation between the set of all languages on the alphabet X_n and the set of all finite automata on X_n, defined by $(L, \mathcal{H}) \in R$ iff $\mathcal{H}$ recognizes L. Describe the Galois-connection induced by this relation R.

Chapter 9

Mal'cev-Type Conditions

As we saw in Chapter 1, any algebra $\mathcal{A}$ has associated with it a lattice, the lattice $Con(\mathcal{A})$ of all congruence relations on $\mathcal{A}$. We can often use properties of the algebra $\mathcal{A}$ itself to deduce properties of the associated congruence lattice, and we can also sometimes use properties of $Con(\mathcal{A})$ to tell us about the algebra $\mathcal{A}$ as well. Thus we want to relate properties of lattices, such as permutability, distributivity, modularity, and so on, to properties of algebras and varieties. The first result in this direction was given by A. I. Mal'cev in 1954 ([74], [75]): he showed that all the congruence relations of any algebra in a variety are permutable with respect to the relational product iff the variety satisfies a certain identity (equality of terms). The special term used in this identity is called a *Mal'cev term*, and theorems like this one which relate properties of the congruence lattices of all the algebras in a variety to identities of the variety are usually called *Mal'cev-type conditions*. In this chapter we investigate a number of properties of congruence lattices, and the corresponding Mal'cev-type conditions. We also investigate these properties in detail for a particular example, the case of varieties generated by algebras of size two; our analysis here will be used in Chapter 10 to describe the lattice of all clones on a two-element set.

9.1 Congruence Permutability

The first Mal'cev-type condition is the original one given by Mal'cev in 1954 ([74]), characterizing congruence permutable varieties of algebras. We recall that two congruence relations θ and ψ on an algebra $\mathcal{A}$ are called *permutable*, or are said to *permute*, if $\theta \circ \psi = \psi \circ \theta$. An algebra $\mathcal{A}$ is called *congruence permutable* if any two congruences of $\mathcal{A}$ are permutable. A class K of algebras of type τ is called *congruence permutable* if each algebra from K is congruence permutable.

It is easy to show that for any two equivalence relations θ and ψ defined on a set A, the union $\theta \cup \psi$ is again an equivalence relation defined on A iff θ and ψ are permutable. In this case $\theta \circ \psi$ is the least equivalence relation containing θ and ψ, and we have $\theta \cup \psi = \theta \circ \psi$. When θ and ψ are congruences on an algebra $\mathcal{A}$, this makes the join $\theta \vee \psi$ in the congruence lattice equal to $\theta \circ \psi$. (We have used this argument already in the proof of Theorem 4.1.4.)

Mal'cev gave the following characterization of congruence permutable varieties.

Theorem 9.1.1 *A variety V of algebras of type τ is congruence permutable iff there is a ternary term $p \in W_\tau(X)/Id\,V$, called a Mal'cev term for V, such that*

$$p(x, x, y) \approx y, \quad p(x, y, y) \approx x \in Id\,V.$$

Proof: Assume that V is congruence permutable. To produce the term p, and verify the identities claimed, we will work with the V-free algebra on three generators. That is, we use X_3, or equivalently the three element generator set $Y = \{x, y, z\}$. From the absolutely free algebra $\mathcal{F}_\tau(Y)$ we form the quotient $\mathcal{F}_V(Y) = \mathcal{F}_\tau(Y)/Id\,V$, which we know is generated by the three equivalence classes $\overline{x} := [x]_{Id\,V}$, $\overline{y} := [y]_{Id\,V}$ and $\overline{z} := [z]_{Id\,V}$ and is in the variety V. Now recall that the congruence relation $\theta(a, b)$ generated by a pair (a, b) of elements of an algebra is the intersection of all congruences containing this pair. We consider the congruences $\theta(\overline{x}, \overline{y})$ and $\theta(\overline{y}, \overline{z})$ on the V-free algebra. We have $(\overline{x}, \overline{z}) \in \theta(\overline{y}, \overline{z}) \circ \theta(\overline{x}, \overline{y})$. But congruence permutability of V implies $(\overline{x}, \overline{z}) \in \theta(\overline{x}, \overline{y}) \circ \theta(\overline{y}, \overline{z})$, and therefore there is a term $p(\overline{x}, \overline{y}, \overline{z}) \in W_\tau(\{x, y, z\})/Id\,V$ with $(\overline{x}, p(\overline{x}, \overline{y}, \overline{z})) \in \theta(\overline{y}, \overline{z})$ and $(p(\overline{x}, \overline{y}, \overline{z}), \overline{z}) \in \theta(\overline{x}, \overline{y})$. This gives us our term p, whose properties we now need to verify. For this we consider a function from $Y = \{x, y, z\}$ to $\mathcal{F}_V(\{x, y\})$ which maps $\overline{x}$ to $\overline{x}$ and both of $\overline{y}$ and $\overline{z}$ to $\overline{y}$. Since $\mathcal{F}_V(\{x, y\})$ is an algebra in V, and $\mathcal{F}_V(Y)$ has the freeness property of the relatively free algebra, this function extends to a unique homomorphism φ from $\mathcal{F}_V(Y)$ to $\mathcal{F}_V(\{x, y\})$. Since $\overline{y}$ and $\overline{z}$ have the same image, $(\overline{y}, \overline{z}) \in ker\,\varphi$ and we must have $\theta(\overline{y}, \overline{z}) \subseteq ker\,\varphi$. Since $(\overline{x}, p(\overline{x}, \overline{y}, \overline{z})) \in \theta(\overline{y}, \overline{z}) \subseteq ker\,\varphi$, we also have $\varphi(\overline{x})$

$= \varphi(p(\overline{x}, \overline{y}, \overline{z}))$. Then, using the fact that φ is a homomorphism, we get

$$\overline{x} = \varphi(\overline{x}) = \varphi(p(\overline{x}, \overline{y}, \overline{z})) = p(\varphi(\overline{x}), \varphi(\overline{y}), \varphi(\overline{z})) = p(\overline{x}, \overline{y}, \overline{y}).$$

This means that in $\mathcal{F}_V(\{x, y\})$ we have $\overline{x} = p(\overline{x}, \overline{y}, \overline{y})$, and hence that $p(x, y, y) \approx x \in IdV$. A similar argument shows that $p(x, x, y) \approx y$ is also in IdV.

For the converse, assume now that there is a term p such that $p(x, x, y) \approx y$, $p(x, y, y) \approx x \in IdV$. To show that V is congruence permutable, we let $\mathcal{A}$ be any algebra in V, let θ and ψ be any congruences on $\mathcal{A}$, and let $(a, b) \in \psi \circ \theta$. By definition there is an element $c \in A$ such that $(a, c) \in \theta$ and $(c, b) \in \psi$. Using (a, a), (b, c), $(b, b) \in \psi$ and (a, a), (c, a), $(b, b) \in \theta$, and the compatibility property of terms from Theorem 5.2.4, we have

$$(p(a, b, b), \ p(a, c, b)) \in \psi \text{ and } (p(a, c, b), \ p(a, a, b)) \in \theta.$$

Thus $(a, b) = (p(a, b, b), \ p(a, a, b)) \in \theta \circ \psi$. This shows that $\psi \circ \theta \subseteq \theta \circ \psi$, and in the same way we show $\psi \circ \theta \supseteq \theta \circ \psi$, to get the desired equality. ∎

Example 9.1.2 Consider the class of all groups, viewed as algebras of type $(2,1,0)$. This is a variety, defined equationally by the identities

$$(xy)z \approx x(yz), \ ex \approx xe \approx x, \ xx^{-1} \approx x^{-1}x \approx e.$$

(As usual, we omit the binary multiplication symbol.) This variety has a Mal'cev term p, given by $p(x, y, z) := xy^{-1}z$. Clearly for this term we have $p(x, x, y) = xx^{-1}y \approx ey \approx y$ and $p(x, y, y) = xy^{-1}y \approx xe \approx x$, in any group. Theorem 9.1.1 then tells us that the variety of all groups is congruence permutable. Similarly, the class of all rings forms a variety with a Mal'cev term given by $p(x, y, z) = x - y + z$, since in any ring $p(x, x, y) = x - x + y \approx y$ and $p(x, y, y) = x - y + y \approx x$. So the variety of rings is also congruence permutable.

9.2 Congruence Distributivity

Another important property of lattices, and in particular of the congruence lattice of an algebra, is *distributivity*.

Definition 9.2.1 An algebra $\mathcal{A}$ is called *congruence distributive* if its congruence lattice $Con\,\mathcal{A}$ is distributive; that is, if for all congruences θ, ψ and $\Phi \in Con\,\mathcal{A}$, the following equations are satisfied:

$$\theta \wedge (\psi \vee \Phi) = (\theta \wedge \psi) \vee (\theta \wedge \Phi),$$
$$\theta \vee (\psi \wedge \Phi) = (\theta \vee \Psi) \wedge (\theta \vee \Phi).$$

(Note that these two equations are equivalent to each other, so that it is enough to check one of them.) A class $K \subseteq Alg(\tau)$ of algebras of type τ is called *congruence distributive* if each algebra of K is congruence distributive.

In order to show congruence distributivity of a class, we will need to consider joins of congruences. We recall from Chapter 1 that the join $\psi \vee \Phi$ of two congruences in a congruence lattice is the congruence relation generated by their union $\psi \cup \Phi$. We will need the following characterization of the elements in such a join.

Lemma 9.2.2 *Let ψ and Φ be two congruences on an algebra $\mathcal{A}$, and let a and b be elements of A. Then $(a, b) \in \Phi \vee \psi$ iff there exist finitely many elements $a_1, \ldots, a_n$ in A such that $(a, a_1) \in \psi$, $(a_1, a_2) \in \Phi$, \ldots, $(a_{n-1}, a_n) \in \psi$ and $(a_n, b) \in \Phi$.*

Proof: Let (a, b) be in $\Phi \vee \psi$. From Remark 1.4.9 and the fact that the union of two congruences is reflexive and symmetric, we see that the join is the transitive closure of the relation $\Phi \cup \psi$. This means that we can produce (a, b) in a finite number of instances of transitivity using pairs in the union. Thus we have elements with the desired property.

Conversely, suppose there are such elements $a_1, \ldots, a_n$. Then from $(a, a_1) \in \psi$ and $(a_1, a_2) \in \Phi$ we get $(a, a_2) \in \Phi \vee \psi$ since $\Phi \vee \psi$ is transitive. Similarly we get $(a_2, a_4) \in \Phi \vee \psi$; and now combining these two facts we have $(a, a_4) \in \Phi \vee \psi$. Continuing in this way we finally get $(a, b) \in \Phi \vee \psi$. ∎

The following theorem gives a Mal'cev-type condition for congruence distributivity. The ternary term m used here is usually called a *(two-thirds) majority term*.

Theorem 9.2.3 *Let V be a variety with a ternary term m such that*

$$m(x, x, y) \approx m(x, y, x) \approx m(y, x, x) \approx x \in Id\,V.$$

Then V is congruence distributive.

Proof: Consider any algebra $\mathcal{A} \in V$, and any three congruences θ, Φ and ψ on $\mathcal{A}$, and assume that $(a, b) \in \theta \wedge (\Phi \vee \psi)$. Then (a, b) is in both θ and $\Phi \vee \psi$. By Lemma 9.2.2, there exist some elements $a_1, \ldots, a_n$ in A such that $(a, a_1) \in \psi$, $(a_1, a_2) \in \Phi$, $\ldots$, $(a_{n-1}, a_n) \in \psi$ and $(a_n, b) \in \Phi$. We will use these elements, along with the converse direction of Lemma 9.2.2, to prove that (a, b) is in $(\theta \wedge \Phi) \vee (\theta \wedge \psi)$.

Taking $a_0 = a$ and $a_{n+1} = b$, we have (a_j, a_{j+1}) in either ψ or Φ, for each $0 \leq j \leq n + 1$. Applying the majority term m, and the fact that (a, a) and (b, b) are in any congruence, gives us $(m(a, a_j, b), m(a, a_{j+1}, b))$ also in the same congruence ψ or Φ. But we also have, for any a_j, that $m(a, a_j, b)$ is θ-related to $m(a, a_j, a)$ since (b, a) is in θ; $m(a, a_j, a)$ in turn is equal to a by the properties of the majority term m, and for the same reason also equal to $m(a, a_{j+1}, a)$, which is then related by θ to $m(a, a_{j+1}, b)$. Thus we have each pair $(m(a, a_j, b), m(a, a_{j+1}, b))$ in both θ and one of ψ or Φ. Therefore, $(m(a, a_0, b), m(a, a_{n+1}, b)) \in (\theta \wedge \Phi) \vee (\theta \wedge \psi)$. But this pair is just $(m(a, a, b), m(a, b, b)) = (a, b)$. Thus we have $\theta \wedge (\Phi \vee \psi) \subseteq (\theta \wedge \Phi) \vee (\theta \wedge \psi)$. The opposite inclusion is valid in all lattices, giving the equality needed for distributivity. ∎

Example 9.2.4 We will use Theorem 9.2.3 to show that the variety of all lattices is congruence distributive. Let m be the ternary lattice term

$$m(x, y, z) = (x \vee y) \wedge (x \vee z) \wedge (y \vee z).$$

Then in any lattice, using the idempotency and absorption laws, we have
$$m(x, x, y) = (x \vee x) \wedge (x \vee y) \wedge (x \vee y) \approx x \wedge (x \vee y) \approx x,$$
$$m(x, y, x) = (x \vee y) \wedge (x \vee x) \wedge (y \vee x) \approx (x \vee y) \wedge x \approx x, \quad \text{and}$$
$$m(y, x, x) = (y \vee x) \wedge (y \vee x) \wedge (x \vee x) \approx (y \vee x) \wedge x \approx x.$$

Thus m is a two-thirds majority term, and the variety of all lattices is congruence distributive.

Theorem 9.2.3 gives only a sufficient condition for congruence distributivity. A condition that is both necessary and sufficient was given by B. Jonsson in 1967 ([61]), in what is known as Jonsson's Condition.

Theorem 9.2.5 *(Jonsson's Condition) A variety V is congruence distributive iff there exist a natural number $n \geq 1$ and ternary terms $t_0, \ldots, t_n \in$*

$W_\tau(X)/Id\,V$ such that the following identities are satisfied in V:

$$
\left.
\begin{aligned}
t_0(x,y,z) &\approx x\,, \\
t_j(x,y,x) &\approx x \text{ for all } 0 \le j \le n\,, \\
t_j(x,x,y) &\approx t_{j+1}(x,x,y) \text{ for all } 0 \le j < n,\ j \text{ even}\,, \\
t_j(x,y,y) &\approx t_{j+1}(x,y,y) \text{ for all } 0 \le j < n,\ j \text{ odd}\,, \\
t_n(x,y,z) &\approx z\,.
\end{aligned}
\right\}\ (\Delta_n)
$$

We note that the previous characterization of congruence distributivity, in Theorem 9.2.2, is actually a special case of this theorem: when V has a majority term m, we may take $n = 2$, and terms $t_0 = x_1$, $t_1 = m$ and $t_2 = x_3$, to obtain the identities (Δ_2) of Jonsson's Condition.

When the conditions (Δ_n) from this theorem are satisfied for a variety V, the variety is said to be n-*distributive*. It is possible to show that an n_1-distributive variety is also n_2-distributive for any $n_2 \ge n_1$. But the converse direction is false, since n_2-distributivity does not imply n_1-distributivity (see K. Fichtner, [44]).

Proof: If V is congruence-distributive, then in particular the relatively free algebra $\mathcal{F}_V(\{x,y,z\})$ is congruence-distributive. We know from Chapter 6 that this algebra is generated by the equivalence classes $[x]_{Id\,V}$, $[y]_{Id\,V}$ and $[z]_{Id\,V}$. For notational convenience we shall write these generators simply as x, y and z respectively. To produce the necessary terms, we consider the congruences $\theta(x,y)$, $\theta(x,z)$ and $\theta(y,z)$ of $\mathcal{F}_V(\{x,y,z\})$ generated by the pairs (x,y), (x,z) and (y,z), respectively. These congruences must satisfy the distributive property, so we have

$$(\theta(x,y) \wedge \theta(x,z)) \vee (\theta(y,z) \wedge \theta(x,z)) = (\theta(x,y) \vee \theta(y,z)) \wedge \theta(x,z). \quad (D)$$

Since (x,y) and (y,z) are both in $\theta(x,y) \vee \theta(y,z)$, and this relation is a congruence, we have $(x,z) \in \theta(x,y) \vee \theta(y,z)$. Hence the pair (x,z) is in the relation on the right hand side of the equation (D) above, and because of the equality must be in the relation on the left hand side too. By the join criterion from Lemma 9.2.2, there are a natural number n and finitely many elements $d_0,\ldots,d_n$ of $\mathcal{F}_V(\{x,y,z\})$ such that $d_0 = x$, $d_n = z$, and $(d_0,d_1) \in \theta(y,z)$, $(d_1,d_2) \in \theta(x,y)$, and so on, with the last pair (d_{n-1},d_n) in either $\theta(x,y)$ or $\theta(y,z)$, depending on whether the number n is even or odd.

The elements $d_0, \ldots, d_n$ belong to $W_\tau(\{x, y, z\})/Id\,V$, which means that they are (equivalence classes of) terms in V: for each $0 \le j \le n$ we have $d_j := t_j(x, y, z)$ for some term t_j. When j is even, we have:

$$(t_j(x, y, z), \ t_{j+1}(x, y, z)) \in \theta(y, z)$$
$$\Rightarrow (t_j(x, x, z), \ t_{j+1}(x, x, z)) \in \theta(y, z).$$

If we restrict $\theta(y, z)$ to the subalgebra of $\mathcal{F}_V(\{x, y, z\})$ generated by x and z, we obtain $t_j(x, x, z) \approx t_{j+1}(x, x, z)$. The other identities needed are obtained similarly.

Assume now that all the identities (Δ_n) are satisfied for some number n and some terms $t_0, \ldots, t_n$. Let $\mathcal{A} \in V$ and let θ, ψ and Φ be congruences on $\mathcal{A}$. Since we always have $(\theta \wedge \Phi) \vee (\theta \wedge \psi)$ contained in $\theta \wedge (\Phi \vee \psi)$, it will suffice to prove the converse inclusion. We let $(x, y) \in \theta \wedge (\Phi \vee \psi)$. Then $(x, y) \in \theta$, and from Lemma 9.2.2 there are a number m and elements $d_0, \ldots, d_m \in A$ with $x = d_0$, $(d_0, d_1) \in \psi$, $(d_1, d_2) \in \Phi$, $\ldots$, $(d_{m-1}, d_m) \in \Phi$ and $d_m = y$. For each $j = 1, \ldots, n - 1$, applying the term t_j to these pairs and the pairs (x, x) and (y, y) gives us $(t_j(x, x, y), t_j(x, d_1, y)) \in \psi$, $(t_j(x, d_1, y), \ t_j(x, d_2, y)) \in \Phi$, $\ldots$, $(t_j(x, d_{m-1}, y), \ t_j(x, y, y)) \in \Phi$. We also have each pair $(t_j(x, d_k, y), \ t_j(x, d_{k+1}, x)) \in \theta$, by the same argument as in the proof of Theorem 9.2.3. Combining these facts we have $(t_j(x, x, y), \ t_j(x, d_1, y)) \in \theta \wedge \psi$, $(t_j(x, d_1, y), \ t_j(x, d_2, y)) \in \theta \wedge \Phi$, $\ldots$, $(t_j(x, d_{m-1}, y), \ t_j(x, y, y)) \in \theta \wedge \Phi$. Hence by Lemma 9.2.2 we have $(t_j(x, x, y), t_j(x, y, y)) \in (\theta \wedge \Phi) \vee (\theta \wedge \psi)$. Altogether we have

$$x = t_0(x, x, y) = t_1(x, x, y),$$
$$(t_1(x, x, y), t_1(x, y, y)) \in (\theta \wedge \Phi) \vee (\theta \wedge \psi),$$
$$(t_2(x, y, y), t_2(x, x, y)) \in (\theta \wedge \Phi) \vee (\theta \wedge \psi),$$
$$\cdots$$
$$t_{n-1}(x, d, y) = t_n(x, d, y) = y,$$

with $d = x$ or $d = y$. Thus, $(x, y) \in (\theta \wedge \Phi) \vee (\theta \wedge \psi)$, and we have congruence distributivity. ∎

The next famous theorem of Baker and Pixley ([6]) shows that the presence of a majority term operation has some far-reaching consequences.

Theorem 9.2.6 *(Baker-Pixley Theorem) Let $\mathcal{A}$ be a finite algebra with a majority function m among its term operations. Then for any $n \in \mathbb{N}$, an* *n-ary operation $f : A^n \to A$ on A is a term operation of $\mathcal{A}$ iff f preserves each subalgebra of $\mathcal{A}^2$.*

Proof: It is straightforward to show that any term operation on $\mathcal{A}$ preserves each subalgebra of $\mathcal{A}^2$, and we leave this as an exercise for the reader. For the converse, we assume that $f : A^n \to A$ is an operation of $\mathcal{A}$ which preserves every subalgebra of $\mathcal{A}^2$. We shall consider sets $D \subseteq A^n$ such that there exists a term operation p of $\mathcal{A}$ which agrees with f on D; so that

$$(x_1, \ldots, x_n) \in D \Rightarrow f(x_1, \ldots, x_n) = p(x_1, \ldots, x_n).$$

We will give an inductive proof that for all $k \geq 2$, any set $D \subseteq A^n$ of size k has this property. Since A is finite, this will suffice.

When $k = 2$ we can write $D = \{(x_1, \ldots, x_n), (y_1, \ldots, y_n)\}$. Since by assumption f preserves every subalgebra of $\mathcal{A}^2$, the pair $(f(x_1, \ldots, x_n), f(y_1, \ldots, y_n))$ belongs to the subalgebra of $\mathcal{A}^2$ which is generated by $\{(x_1, y_1), \ldots, (x_n, y_n)\}$. Then by Theorem 1.3.8, there is a term operation p of $\mathcal{A}$ for which $f(x_1, \ldots, x_n) = p(x_1, \ldots, x_n)$ and $f(y_1, \ldots, y_n) = p(y_1, \ldots, y_n)$. Thus D belongs to J.

Assume now that D has size $k+1 \geq 3$ and that $(x_1, \ldots, x_n), (y_1, \ldots, y_n), (z_1, \ldots, z_n)$ are pairwise different elements of D. By induction f agrees on $D_1 := D \setminus \{(x_1, \ldots, x_n)\}$ with some term operation p_1 of $\mathcal{A}$, f agrees on $D_2 := D \setminus \{(y_1, \ldots, y_n)\}$ with some term operation p_2 of $\mathcal{A}$, and f agrees on $D_3 := D \setminus \{(z_1, \ldots, z_n)\}$ with some term operation p_3 of $\mathcal{A}$.

Now we claim that f agrees with the term $m(p_1, p_2, p_3)$ on D. For any $(u_1, \ldots, u_n) \in D$, there are numbers $i \neq j \in \{1, 2, 3\}$ such that $(u_1, \ldots, u_n) \in D_i$ and $(u_1, \ldots, u_n) \in D_j$. Then $p_i(u_1, \ldots, u_n) = f(u_1, \ldots, u_n)$ and $p_j(u_1, \ldots, u_n) = f(u_1, \ldots, u_n)$, and we have $m(p_1(u_1, \ldots, u_n), p_2(u_1, \ldots, u_n), p_3(u_1, \ldots, u_n)) = f(u_1, \ldots, u_n)$, as required. ∎

Remark 9.2.7 Theorem 9.2.6 can be generalized in the following way. If the finite algebra $\mathcal{A}$ has a $(d+1)$-ary term operation $u(x_1, \ldots, u_{d+1})$ with $u(x, \ldots, x, y, x, \ldots, x) \approx x$ for any position of y, then $f : A^n \to A$ is a term operation of $\mathcal{A}$ iff f preserves any subalgebra of $\mathcal{A}^d$.

The following theorem shows how congruence distributivity influences the structural properties of a variety. For a congruence distributive variety

$V(\mathcal{A})$ there is a simpler structural characterization than the usual $V(\mathcal{A})$ = $\mathbf{HSP}(\mathcal{A})$, given by the famous Jonsson's Lemma ([61]). In the case that $\mathcal{A}$ is a finite algebra the Lemma takes the following form:

Theorem 9.2.8 *If $\mathcal{A}$ is a finite non-trivial algebra such that the variety $V(\mathcal{A})$ is congruence distributive, then every algebra from $V(\mathcal{A})$ is isomorphic to a subdirect product of homomorphic images of subalgebras of $\mathcal{A}$. That is, $V(\mathcal{A}) = \mathbf{IP}_s\mathbf{HS}(\mathcal{A})$.* ∎

If $\mathcal{A}$ is simple, and either has no proper subalgebras or has at most one-element subalgebras, then $V(\mathcal{A}) = \mathbf{IP}_s(\mathcal{A})$. This means that $\mathcal{A}$ is (up to isomorphism) the only subdirectly irreducible algebra in $V(\mathcal{A})$.

9.3 Arithmetical Varieties

A lattice which is both permutable and distributive is called arithmetical. Since we have Mal'cev-type conditions for both permutability and distributivity of congruence lattices, we should expect to find such conditions for arithmeticity of congruence lattices as well.

Definition 9.3.1 A variety K of algebras of type τ is called *arithmetical* if each algebra from K is both congruence permutable and congruence distributive.

Theorem 9.3.2 *For a variety V the following are equivalent:*

(i) V is arithmetical;

(ii) there are terms p and m with $p(x, x, y) \approx y \approx p(y, x, x) \in IdV$ and $m(x, x, y) \approx m(x, y, x) \approx m(y, x, x) \approx x \in IdV$;

(iii) there is a ternary term q such that $q(x, y, y) \approx q(x, y, x) \approx q(y, y, x) \approx x \in IdV$.

Proof: (i) $\Rightarrow$ (ii): Let V be arithmetical. Then V is congruence permutable, and by Theorem 9.1.1 there is a Mal'cev term p which fulfills the desired identities for p. Moreover, V is congruence distributive; so in $F_V(\{x, y, z\})$ we have

$$\theta(x, z) \wedge (\theta(y, z) \vee \theta(x, y)) = (\theta(x, z) \wedge \theta(y, z)) \vee (\theta(x, z) \wedge \theta(x, y)) .$$

Since (x, z) is in the relation on the left side of this equation, it is also in the relation on the right side. By congruence permutability this relation is equal to $(\theta(x, z) \wedge \theta(y, z)) \circ (\theta(x, z) \wedge \theta(x, y))$. But then there is a term m with $(x, m(x, y, z)) \in \theta(x, z) \wedge \theta(x, y)$ and $(m(x, y, z), z) \in \theta(x, z) \wedge \theta(y, z)$, and this term m satisfies the identities needed to make it a majority term.

(ii) $\Rightarrow$ (iii): Given terms p and m as in (ii), we take $q(x, y, z)$ to be the term $p(x, m(x, y, z), z)$. Then q has the desired properties, since we have

$$
\begin{aligned}
q(x, x, y) &\approx p(x, m(x, y, y), y) \approx p(x, y, y) \approx x, \\
q(x, y, x) &\approx p(x, m(x, y, x), x) \approx p(x, x, x) \approx x, \quad \text{and} \\
q(y, y, x) &\approx p(y, m(y, y, x), x) \approx p(y, y, x) \approx x.
\end{aligned}
$$

(iii) $\Rightarrow$ (ii): Given a term q as in (iii), we define terms $p(x, y, z) := q(x, y, z)$ and $m(x, y, z) := q(x, q(x, y, z), z)$. Then p is a Mal'cev term, making V congruence permutable by Theorem 9.1.1; and m is a majority term, making V congruence distributive by Theorem 9.2.3, since we have

$$
\begin{aligned}
m(x, x, y) &\approx q(x, q(x, x, y), y) \approx q(x, y, y) \approx x, \\
m(x, y, x) &\approx q(x, q(x, y, x), x) \approx q(x, x, x) \approx x, \quad \text{and} \\
m(y, x, x) &\approx q(y, q(y, x, x), x) \approx q(y, y, x) \approx x. \quad \blacksquare
\end{aligned}
$$

Example 9.3.3 In this example we will show that the variety of all Boolean algebras is arithmetical. It can be shown that the variety of all Boolean algebras is generated by the two-element Boolean algebra $2_B = (\{0, 1\}; \wedge, \vee, \neg, \Rightarrow, 0, 1)$, where the operation symbols here denote the usual Boolean operations on the set $\{0, 1\}$. This means that every identity which is satisfied in the two-element Boolean algebra 2_B is satisfied in any Boolean algebra. To prove arithmeticity by Theorem 9.3.2 (iii), we look for a ternary term $q(x, y, z)$ on the set $\{0, 1\}$ which satisfies the identities $q(x, y, y) \approx q(x, y, x) \approx q(y, y, x) \approx x$ in 2_B. Considering the truth table of such an operation, as shown below, we see that there is exactly one such ternary operation q. This operation q is indeed a term operation of 2_B, since it can be expressed as $q(x, y, z) = (z \wedge (xy \vee (\neg x)(\neg y))) \vee (x \wedge \neg y)$. Thus the variety is arithmetical.

y	0	0	0	0	1	1	1	1
y	0	0	1	1	0	0	1	1
z	0	1	0	1	0	1	0	1
$q(x,y,z)$	0	1	0	0	1	1	0	1

9.4 n-Modularity and n-Permutability

In this section we examine the property of modularity for congruence lattices, as well as variations of modularity and permutability.

Definition 9.4.1 An algebra $\mathcal{A}$ is called *congruence modular* if $Con\,\mathcal{A}$ is a modular lattice, so that any three congruences θ_1, θ_2 and θ_3 on $\mathcal{A}$ satisfy the modular law:

$$\theta_0 \supseteq \theta_2 \quad \Rightarrow \quad \theta_0 \cap (\theta_1 \cup \theta_2) \subseteq (\theta_0 \cap \theta_1) \cup \theta_2.$$

There are several Mal'cev-type conditions for congruence modularity, which we will state here without proof. The first result, due to A. Day, uses quaternary terms, while the second one, due to H. P. Gumm, uses ternary terms. When the identities (M_n) or (G_n) of these Theorems are satisfied by terms of a variety V, the variety is said to be n-modular.

Theorem 9.4.2 *([16]) A variety V of algebras is congruence modular iff there are a natural number $n \geq 2$ and quaternary terms $q_0,\dots,q_n \in W_\tau(X)/Id\,V$ such that the following list (M_n) of identities is satisfied in V:*

$$
\begin{aligned}
q_0(x,y,z,u) &\approx x\,, \\
q_i(x,y,y,x) &\approx x & \text{for all } 0 \leq i \leq n\,, \\
q_i(x,y,y,u) &\approx q_{i+1}(x,y,y,u) & \text{for all } 0 < i < n,\ i \text{ odd}, \\
q_i(x,x,z,z) &\approx q_{i+1}(x,x,z,z) & \text{for all } 0 \leq i < n\,, i \text{ even}, \\
q_n(x,y,z,u) &\approx u.
\end{aligned}
$$

■

Theorem 9.4.3 *([52]) A variety V of algebras is congruence modular iff there are a natural number n and ternary terms $g_0, \ldots, g_n$, $d \in W_\tau(X)/Id\,V$ such that the following list (G_n) of identities is satisfied in V:*

$$
\begin{aligned}
g_0(x,y,z) &\approx x \\
g_i(x,y,x) &\approx x && \text{for all } 0 \le i \le n, \\
g_i(x,x,y) &\approx g_{i+1}(x,x,y) && \text{if } i \text{ is even,} \\
g_i(x,y,y) &\approx g_{i+1}(x,y,y) && \text{if } i \text{ is odd,} \\
g_i(x,y,y) &\approx d(x,y,y) \\
d(x,x,y) &\approx y.
\end{aligned}
$$

∎

Lemma 9.4.4 *If an algebra $\mathcal{A}$ is congruence-permutable, then it is congruence-modular.*

Proof: Let θ, ψ and $\Phi \in Con\,\mathcal{A}$ with $\theta \supseteq \Phi$. To show that $\theta \wedge (\psi \vee \Phi) \subseteq (\theta \wedge \psi) \vee \Phi$, let $(a,b) \in \theta \wedge (\psi \vee \Phi)$. Then $(a,b) \in \theta$, and by congruence permutability there is an element $c \in A$ with $(a,c) \in \Phi$ and $(c,b) \in \psi$. Since $(a,c) \in \Phi$ and $\Phi \subseteq \theta$, we have $(a,c) \in \theta$. Now by symmetry and transitivity we also have $(c,b) \in \theta$. This gives $(c,b) \in \theta \wedge \psi$. Combining this with $(a,c) \in \Phi$ gives us $(a,b) \in (\theta \wedge \psi) \vee \Phi$.   ∎

We noted in Section 9.1 that if two congruences θ and ψ are permutable, then $\theta \cup \psi = \theta \circ \psi$. In this case we say that the congruences are 2-permutable. We can generalize this, to say that two congruences θ and ψ are n-permutable, for $n \ge 2$, if

$$
\theta \cup \psi = \theta_1 \circ \ldots \circ \theta_n \quad \text{where} \quad \theta_i = \begin{cases} \theta & \text{if } i \text{ is even} \\ \psi & \text{if } i \text{ is odd} \end{cases}
$$

If any two congruences of $Con\,\mathcal{A}$ have this property we call $\mathcal{A}$ n-permutable. A variety V is said to be n-permutable if each algebra from V has this property. It can be shown that a variety which is n-permutable is also k-permutable for any $k > n$. A Mal'cev-type condition for n-permutability was given by J. A. Hagemann and A. Mitschke.

Theorem 9.4.5 *([53]) Let $n \ge 2$. A variety V of algebras is n-permutable iff there are ternary terms $p_0, \ldots, p_n \in W_\tau(X)/Id\,V$ such that in V the following identities are satisfied:*

$$\left.\begin{array}{rcl} p_0(x,y,z) & \approx & x, \\ p_i(x,x,y) & \approx & p_{i+1}(x,y,y) \text{ for all } 0 \le i < n, \\ p_n(x,y,z) & \approx & z. \end{array}\right\} (\mathrm{P_n})$$

∎

9.5 Congruence Regular Varieties

In some algebras, any congruence is uniquely determined by one of its congruence classes. For instance, each congruence of a group is uniquely determined by a normal subgroup, which is the congruence class of the identity element of the group. Similarly, each congruence of a ring is uniquely determined by an ideal. This motivates the following definition:

Definition 9.5.1 A congruence θ of an algebra $\mathcal{A}$ is called *regular* if it is completely determined by one of its congruence classes. An algebra $\mathcal{A}$ is called *congruence regular* if every congruence on $\mathcal{A}$ is regular, and a variety V is called *congruence regular* if every $\mathcal{A} \in V$ is congruence regular.

Congruence regularity of a variety can be characterized by a Mal'cev-type-condition which was given by R. Wille in [121].

Theorem 9.5.2 *([121]) A variety V is congruence-regular iff there are natural numbers n and m, and ternary terms $p_0, \ldots, p_n$ and quaternary terms $q_1, \ldots, q_m$ in V, such that for all $1 \le k \le m$ and suitable numbers $0 \le i_k, j_k \le n$, all of*

$$p_0(x,y,z) \approx z, \; p_i(x,x,z) \approx z \text{ for all } 1 \le i \le n,$$
$$q_1(p_{i_1}(x,y,z),x,y,z) \approx x,$$
$$q_{k-1}(p_{j_{k-1}}(x,y,z),x,y,z) \approx q_k(p_{i_k}(x,y,z),x,y,z) \; (2 \le k \le m),$$
$$q_m(p_{i_m}(x,y,z),x,y,z) \approx y$$

are identities in V. ∎

Note that congruence regularity of a variety V implies n-permutability for some number n (see [53]). A simple consequence of Theorem 9.5.2 is the following.

Corollary 9.5.3 *Let V be a variety. If there exists a ternary term l in V with*

$$l(x, y, z) = \begin{cases} z & \text{if } x = y \\ y & \text{if } x = z, \\ x & \text{otherwise} \end{cases} \qquad \text{then } V \text{ is congruence regular.}$$

Proof: Given such a term l, we set $n = m = 1$, $p_0(x, y, z) = z$, $p_1(x, y, z) = l(x, y, z)$, and $q_1(u, x, y, z) = l(u, y, z)$. Then we have $q_1(p_0(x, y, z), x, q, z) = q_1(l(x, y, z), x, y, z) = l(l(x, y, z), y, z) = x$ and $p_1(x, x, z) = l(x, x, z) = z$. ∎

In [21] K. Denecke proved the following result for the special case of varieties generated by two-element algebras.

Corollary 9.5.4 *A variety generated by a two-element algebra is congruence regular if and only if it is congruence permutable (2-permutable).* ∎

9.6   Two-Element Algebras

Some important varieties are generated by two element algebras: the varieties of distributive lattices, of semilattices, of implicative algebras and of Boolean algebras are all generated by a two element algebra. In the next Chapter, we shall give a complete survey and classification of all two element algebras. We set the foundations for this later work in this section, where we survey some of the properties of congruences, from the previous sections, for the special case of two element algebras $\mathcal{A}$. Our presentation is based on the work of M. Reschke, O. Lüders and K. Denecke in [100].

The analysis is made easier by the fact that properties such as congruence distributivity and congruence modularity can be characterized by the presence of certain ternary terms. When A has cardinality two, there are a finite number of ternary operations defined on A, and a complete search for terms with desired properties is possible and in fact, as we shall see, not very difficult.

When an algebra $\mathcal{A}$ has a base set A of cardinality two, we usually denote A by the set $\{0, 1\}$, and we refer to operations on A as *Boolean operations*. Throughout this section, we assume $\mathcal{A} = (\{0, 1\}; F)$, where F is some set of Boolean operations. A complete system of representatives of all two-element algebras was given by E. L. Post ([95]) in 1941. We shall study Post's lattice of representatives in the next chapter, making use of the congruence properties we determine here.

It should be emphasized that two-element algebras are studied here only up to equivalence, that is, equality of the sets of term operations, and up to isomorphism. We will use the following standard notation for Boolean operations: $\neg$ for negation, $\Rightarrow$ for implication, $\wedge$ for conjunction, $\vee$ for disjunction, $+$ for addition modulo 2, and $\Leftrightarrow$ for equivalence. We sometimes also denote conjunction by juxtaposition.

We begin our study of the properties of congruence lattices of two-element algebras with the property of congruence distributivity, starting with 2-distributivity. Suppose that the two-element algebra $\mathcal{A}$ is 2-distributive. Then by Theorem 9.2.3 we must have terms t_0, t_1 and t_2 which satisfy the following identities:

$$t_0(x,y,z) \approx x, \; t_1(x,y,x) \approx x, \; x \approx t_0(x,x,y) \approx t_1(x,x,y),$$
$$t_1(x,y,y) \approx t_2(x,y,y) \approx y, \; t_2(x,y,z) \approx z.$$

In particular, this tells us that t_1 satisfies the identities $t_1(x,y,x) \approx t_1(x,x,y) \approx t_1(y,x,x) \approx x$, and hence acts as a majority term m. Thus for 2-distributivity we need a ternary majority term. When we look at a table of values to see what ternary operations are possible, we see that the majority term properties completely restrict the values $t_1(x,y,z)$ on all eight triples (x,y,z) in A^3.

x	0	0	0	0	1	1	1	1
y	0	0	1	1	0	0	1	1
z	0	1	0	1	0	1	0	1
$t_1(x,y,z)$	0	0	0	1	0	1	1	1

Exactly one Boolean operation is determined, namely

$$t_1(x,y,z) = (x \wedge y) \vee (x \wedge z) \vee (y \wedge z) = m(x,y,z).$$

This proves the following result for 2-distributivity.

Lemma 9.6.1 *A two-element algebra $\mathcal{A}$ is 2-distributive iff the ternary operation $t_1(x, y, z) = (x \wedge y) \vee (x \wedge z) \vee (y \wedge z)$ is one of its term operations.*

Next we look for two-element algebras which are 3-distributive. Any such algebra must have terms $t_0, \ldots, t_3$ which satisfy the (Δ_3) identities from Theorem 9.2.5:

$$
\begin{aligned}
t_0(x, y, z) &\approx x, \quad t_1(x, x, y) \approx t_0(x, x, y) \approx x, \quad t_1(x, y, x) \approx x, \\
t_1(x, y, y) &\approx t_2(x, y, y), \quad t_2(x, x, y) \approx t_3(x, x, y) \approx y, \\
t_3(x, y, x) &\approx x, \quad t_3(x, y, z) \approx z.
\end{aligned}
$$

We can take t_0 to be the term x_1, and t_3 to be x_3; so we are looking for Boolean operations t_1 and t_2 satisfying

$$
\begin{aligned}
t_1(x, x, y) &\approx t_1(x, y, x) \approx x \\
t_2(x, x, y) &\approx y, \quad t_2(x, y, x) \approx x.
\end{aligned}
$$

Again we turn to tables of values to see what choices are possible for t_1 and t_2. Note that the identities above determine the value of each of t_1 and t_2 on six of the eight triples in A^3, leaving us to consider possible values on the two remaining triples $(0, 1, 1)$ and $(1, 0, 0)$.

x	0	0	0	0	1	1	1	1
y	0	0	1	1	0	0	1	1
z	0	1	0	1	0	1	0	1
$t_1(x, y, z)$	0	0	0	?	?	1	1	1
$t_2(x, y, z)$	0	1	0	?	?	1	0	1

Moreover, our identity $t_1(x, y, y) \approx t_2(x, y, y)$ forces t_1 and t_2 to have the same value on these last two triples, leaving us with 4 cases to consider:

Case 1: $t_1(0, 1, 1) = 0$, $t_1(1, 0, 0) = 0$ (and the same values for t_2).
Here we have $t_1(x, y, z) = x \wedge (y \vee z)$ and $t_2(x, y, z) = (x \vee \neg y) \wedge z$. Thus if both of these operations are term operations of $\mathcal{A}$, then $\mathcal{A}$ is 3-distributive.

Case 2: $t_1(0, 1, 1) = 0$, $t_1(1, 0, 0) = 1$ (and the same values for t_2).
In this case we have $t_1(x, y, z) = x$ and $t_2(x, y, z) = (z \wedge (xy \vee (\neg x)(\neg y))) \vee$

$(x \wedge \neg y)$. But using t_2 we can define a term $m(x, y, z) = t_2(x, t_2(x, y, z), z)$, which we can easily see acts as a majority term:

$$
\begin{aligned}
m(x, x, y) &= t_2(x, t_2(x, x, y), y) \approx t_2(x, y, y) \approx x, \\
m(x, y, x) &= t_2(x, t_2(x, y, x), x) \approx t_2(x, x, x) \approx x, \quad \text{and} \\
m(y, x, x) &= t_2(y, t_2(y, x, x), x) \approx t_2(y, y, x) \approx x.
\end{aligned}
$$

Thus in this case our algebra is in fact 2-distributive, by the previous Lemma, and so 3-distributive as well.

Case 3: $t_1(0, 1, 1) = 1$, $t_1(1, 0, 0) = 0$ (and the same values for t_2).
In this case, t_1 is the operation encountered in the 2-distributivity case above, with the properties of a ternary majority term $m(x, y, z)$, and $t_2(x, y, z) = z$. Thus when this operation t_1 is a term, we know that the algebra is 2-distributive and hence also 3-distributive.

Case 4: $t_1(0, 1, 1) = 1$, $t_1(1, 0, 0) = 1$ (and the same values for t_2).
Here $t_1(x, y, z) = x \vee (y \wedge z)$, and $t_2(x, y, z) = (x \wedge \neg y) \vee z$. This is the dual of Case 1, with a similar result.

Thus for 3-distributivity, we have two pairs of terms which gives us 3-distributivity, along with two cases which reduce to 2-distributivity. In fact, a stronger result may be shown for the Cases 1 and 4 above. We consider Case 1 only, with Case 4 being dual. We have shown that if both t_1 and t_2 of this case are terms, our algebra is 3-distributive. We will show below that if (only) $t_1 = x \wedge (y \vee z)$ is a term, we can deduce 4-distributivity. Conversely, if $t_2 = (x \vee \neg y) \wedge z$ is a term, we can in fact produce the operation t_1 as a term, giving us 3-distributivity: when $(x \vee \neg y) \wedge z$ is a term, so is $x \wedge (y \vee \neg z)$, and then t_1 can be expressed as the term $x \wedge (y \vee \neg((x \wedge (y \vee \neg z))))$.

For 4-distributivity, we can start with the same argument for t_1 as in the 3-distributive case. If a two-element algebra has terms $t_0, \ldots, t_4$ which satisfy conditions (Δ_4), then in particular the term t_1 has its value constrained on six of the eight triples, and we have 4 possibilities for t_1, as we did in the 3-distributive case.

Case 1: $t_1 = x \wedge (y \vee z)$. If this operation is a term operation, then so are $x \wedge (z \vee y)$ and $z \wedge (y \vee x)$. Taking these for t_2 and t_3 respectively, and taking $t_0 = x$ and $t_4 = z$, we have five terms $t_0, \ldots, t_4$ which satisfy (Δ_4). Thus in this case our algebra is 4-distributive.

Case 2: $t_1 = x$. Here we must consider the values possible for the remaining terms t_2 and t_3 (with $t_4 = z$). Making a table of values possible for these two, and using the 3-distributive identities forcing values of t_3, shows that their values are also determined on six triples, and must be equal on the remaining two triples. This leads to four subcases:

(i) $t_2 = x \wedge (\neg y \vee z)$ and $t_3 = (x \vee y) \wedge z$. Thus the presence of these two terms gives 4-distributivity. But it is easy to see that the presence of these two as terms is equivalent to Case 4 of the 3-distributive case (with variables x and z interchanged).

(ii) $t_2 = x$ but $t_3 = m$. In this case we have 2-distributivity.

(iii) $t_2 = (z \wedge (xy \vee (\neg x)(\neg y))) \vee (x \wedge \neg y)$. Then as in Case 2 for 3-distributivity, we can use t_2 to define a majority term and obtain 2-distributivity again.

(iv) $t_2 = x \vee (\neg y \wedge z)$ and $t_3 = (x \wedge y) \vee z$. This is dual to Case 2(i), and again gives 3-distributivity.

Case 3: $t_1 = m$, the majority term. In this case our algebra is 2-distributive.

Case 4: $t_1 = x \vee (y \wedge z)$. This case is dual to Case 1.

All remaining instances of n-distributivity, for $n \geq 5$, are dealt with by the following Lemma.

Lemma 9.6.2 *If a two-element algebra $\mathcal{A}$ is n-distributive for $n \geq 2$, then it is also 4-distributive.*

Proof: Let $\mathcal{A}$ be n-distributive for $n \geq 2$. By Theorem 9.2.5 there is a term operation t_1 of $\mathcal{A}$ satisfying $t_1(x, x, z) \approx x$ and $t_1(x, y, x) \approx x$. As in the previous proofs, the value of t_1 is defined on all triples except $(0, 1, 1)$ and $(1, 0, 0)$; so there are only four possibilities for the operation t_1. Three such possibilities lead to n-distributivity for n equal to one of 2, 3 or 4, exactly as in Cases 1, 3 and 4 for 3-distributivity above. The only remaining case is the one in which $t_1 = x$. Using tables of values as before, we see that there are four possibilities for t_2 in this case. Again three of these lead to either 2- or 3-distributivity. The remaining case is when both t_1 and t_2 equal x. But in this case, we see that the n-distributive case reduces to $(n - 2)$-distributive, since t_3 now acts like t_1, t_4 like t_2, and so on. ∎

Combining our analysis of all the cases for n-distributivity, we have the following conclusion.

Theorem 9.6.3 *A two-element algebra $\mathcal{A}$ is congruence-distributive iff one of the following Boolean operations is a term operation of $\mathcal{A}$:*
$$m(x, y, z) = (x \wedge y) \vee (x \wedge z) \vee (y \wedge z),$$
$$x \wedge (y \vee z), \qquad (x \vee \neg y) \wedge z,$$
$$x \vee (y \wedge z), \qquad or \qquad (x \wedge \neg y) \vee z.$$
∎

We now turn to the question of n-permutability of two-element algebras, beginning with 2-permutability. We recall that in a two-element algebra we use $+$ to denote addition modulo two, and abbreviate $x \wedge y$ by xy.

Lemma 9.6.4 *A two-element algebra $\mathcal{A}$ is 2-permutable iff the operation p with $p(x, y, z) = x + y + z$ is a term operation of $\mathcal{A}$.*

Proof: The operation p fits the conditions from Mal'cev's characterization in Theorem 9.1.1 of congruence permutability; so if p is a term operation then $\mathcal{A}$ is congruence permutable and 2-permutable. Conversely, suppose that $\mathcal{A}$ is 2-permutable. Then the identities (P_2) from Theorem 9.4.5 are satisfied for some ternary terms p_0, p_1 and p_2. The conditions for the term p_1 constrain its value on all triples in A^3 other than $(0,1,0)$ and $(1,0,1)$, which means that there are exactly four possible term operations p_1 which fit the conditions (P_2). One of these, the operation with $p_1(0, 1, 0) = 1$ and $p_1(1, 0, 1) = 0$, is the Mal'cev term p, and so guarantees 2-permutability.

We will show that in the three remaining cases for p_1, we can use the existence of the term p_1 to produce a Mal'cev term p.

Case 1: If $p_1(0, 1, 0) = 0 = p_1(1, 0, 1)$, then we can write p_1 in the form $p_1(x, y, z) = x + z + xy + yz + xyz$. From such a term we can produce a Mal'cev term p, by $p(x, y, z) = p_1(x, p_1(x, y, x), p_1(y, p_1(x, y, x), z))$, and hence p is a term operation of $\mathcal{A}$.

Case 2: If $p_1(0, 1, 0) = 0$ and $p_1(1, 0, 1) = 1$, then p_1 can be written in the form $p_1(x, y, z) = x + z + xy + xz + yz$. From this we again can obtain p as a term, by $p(x, y, z) = p_1(p_1(x, z, y), x, p_1(x, y, z))$.

Case 3: If $p_1(0, 1, 0) = 1 = p_1(1, 0, 1)$, then p_1 can be written as $p_1(x, y, z) = x + y + z + xz + xyz$, and then we obtain p by $p(x, y, z) = p_1(p_1(y, x, z), y, p_1(x, z, y))$. ∎

Our next lemma reduces the investigation of n-permutability to the cases $n = 2$ and $n = 3$.

Lemma 9.6.5 *If a two-element algebra* $\mathcal{A}$ *is n-permutable for* $n > 2$, *then it is either 2-permutable or 3-permutable.*

Proof: Let $\mathcal{A}$ be n-permutable, for some $n \geq 3$. Then there is a term operation p_1 of $\mathcal{A}$ which satisfies the identity $p_1(x, z, z) \approx x$. This identity uniquely determines the value of p_1 on four of the eight triples in A^3, leaving the value undefined on (0,0,1), (0,1,0), (1,0,1), and (1,1,0). This gives us 16 possible choices for the values of p_1. But we can eliminate a number of these cases immediately. First, p_1 cannot equal the first projection function x_1, since in that case n-permutability just reduces to (n-1)-permutability. In one case p_1 equals the Mal'cev term p, and the algebra is 2-permutable by the previous Lemma. Moreover, a ternary operation $p_1(x, y, z)$ is a term operation of the algebra $\mathcal{A}$ iff every operation arising from p_1 by permuting y and z is a term operation of $\mathcal{A}$ too; and identities satisfied by p_1 are also satisfied by the operation dual to p_1. With this information we are left with exactly five possibilities to consider for p_1.

Case 1: $p_1(0, 0, 1) = p_1(0, 1, 0) = p_1(1, 1, 0) = 0$ and $p_1(1, 0, 1) = 1$.
Then $p_1(x, y, z) = x + xy + xyz$, and by superposition we obtain a term p_2 by $p_2(x, y, z) = z + zy + xyz$. Now taking $p_0 = x$, $p_1 = x + xy + xyz$, $p_2 = z + zy + xyz$ and $p_3 = z$, we have terms which satisfy condition (P_3) from Theorem 9.4.5, and our algebra is 3-permutable.

Case 2: $p_1(0, 0, 1) = p_1(0, 1, 0) = p_1(1, 0, 1) = p_1(1, 1, 0) = 0$.
Then $p_1(x, y, z) = x + xy + xz$. By superposition we can produce a term q with $q(x, y, z) = p_1(x, p_1(p_1(x, z, y), y, x), y)$. But then q is exactly the term obtained for p_1 in Case 1, and as before the algebra is 3-permutable.

Case 3: $p_1(0, 0, 1) = p_1(1, 1, 0) = 0$, $p_1(0, 1, 0) = p_1(1, 0, 1) = 1$.
Here we have $p_1(x, y, z) = x + y + yz$. Again we can make the term used in

Case 1, by $p_1(x, y, p_1(y, x, z))$, so that 3-permutability is satisfied.

Case 4: $p_1(0, 0, 1) = p_1(1, 0, 1) = 0$, $p_1(0, 1, 0) = p_1(1, 1, 0) = 1$.
Here we can write $p_1(x, y, z) = x + y + xy + yz + xz$. As in the proof of the previous Lemma, we can obtain by superposition a Mal'cev term p, and $\mathcal{A}$ is 2-permutable.

Case 5: $p_1(0, 0, 1) = p_1(1, 0, 1) = p_1(1, 1, 0) = 0$ and $p_1(0, 1, 0) = 1$.
As in Case 4, we can use $p_1(x, y, z) = x + y + xz + yz + xyz$ to produce a Mal'cev term p, again giving us 2-permutability. ∎

Since 2-permutability also implies n-permutability for any $n > 2$, we have shown that any n-permutable two-element algebra is 3-permutable. In fact we have exactly two possibilities:

(i) $\mathcal{A}$ is 2-permutable,   or

(ii) $\mathcal{A}$ is 3-permutable but not 2-permutable.

9.7 Exercises

9.7.1. Prove Corollary 9.5.4.

9.7.2. Prove the assertion following the statement of Theorem 9.2.4, that when V has a majority term m, we may take $n = 2$, and terms $t_0 = x_1$, $t_1 = m$ and $t_2 = x_3$, to obtain the identities (Δ_2) of Jonsson's Condition.

9.7.3. Prove that for any two-element algebra $\mathcal{A}$, the variety $V(\mathcal{A})$ generated by $\mathcal{A}$ is congruence modular iff it is congruence distributive or congruence permutable.

9.7.4. Prove that a two-element algebra is 3-distributive but not 2-distributive iff $(x \vee \neg y) \wedge z$ or $(x \wedge \neg y) \vee z$ is a term operation, but m is not a term operation.

9.7.5. Prove that a two-element algebra is 4-distributive but not 3-distributive if either

(i) $x \vee (y \wedge z)$, but not $(x \wedge \neg y) \vee z$,   or

(ii) $x \wedge (y \vee z)$, but not $(x \vee \neg y) \wedge z$

are term operations, but m is not a term operation.

9.7.6. Prove that a two-element algebra $\mathcal{A}$ is 3-permutable but not 2-permutable iff either r or r', with

$$r(x, y, z) = x + xz + xyz, \quad r'(x, y, z) = x + y + xy + yz + xyz,$$

are term operations of $\mathcal{A}$, but p is not a term operation of $\mathcal{A}$.

9.7.7. Prove that in a congruence permutable variety V there is a term p such that $p(x, x, y) \approx y$ is an identity of V.

Chapter 10

Clones and Completeness

In Example 1.2.14, and again in Chapters 2 and 5, we studied the concept of a clone as a set of operations defined on a base set A which is closed under composition and contains all the projection operations. In Section 10.1 we describe several ways in which we can regard a clone as an algebraic structure. In Section 10.2 we describe clones using relations, since clones occur as sets of operations which preserve relations on a set A. This description is based on a Galois-connection, with the clones corresponding to closed sets, and thus the set of all clones of operations defined on a fixed finite set forms a lattice. In Section 10.3 we shall use our results from Chapter 9 on the congruence lattice properties of two-element algebras, to give a complete description of the lattice of all clones of operations defined on a two-element set. An important question is to decide when a set of operations defined on a fixed set generates the set of all operations defined on this set, the so-called *functional completeness problem*. We will also describe the algebraic properties of certain classes of finite algebras connected with the functional completeness problem.

10.1 Clones as Algebraic Structures

There are several ways to regard a clone of operations as an algebraic structure, that is, as a pair consisting of one (or more) sets and of sets of operations defined on these sets. The problem is how to handle the different arities of the operations. A. I. Mal'cev ([74]) defined clones using what is called the *full iterative algebra* $\mathcal{O}(A) = (O(A); *, \zeta, \tau, \Delta, e_1^2)$, as in Example 1.2.14. Here we have a (homogeneous) algebra with one set of objects, but with five different operations on our set, one binary, three unary, and one nullary. Subalgebras of this algebra are then called *iterative algebras*. The binary operation $*$ is clearly associative. This makes any iterative algebra a semigroup, with three additional unary operations and a nullary operation. It is not hard to prove that universes of subalgebras of $\mathcal{O}(\mathcal{A})$ are clones in our sense: they are closed under composition and contain all the projections. Conversely, any clone of operations can be shown to be closed under the iterative algebra operations $*$, ζ, τ, Δ and e_1^2. Thus any clone occurs as the base set of a subalgebra of a full iterative algebra. Universes of clones are also called *closed classes* or *superposition-closed classes*, and we refer to superposition of functions.

Another approach to defining clones algebraically is to use heterogeneous algebras, also known as many-sorted or multi-based algebras (see Higgins, [55] and G. Birkhoff and J. D. Lipson, [9]). In such algebras, we have more than one base set of objects, and operations between different sets of objects. This approach is useful for clones of operations, where we can separate the operations of different arities into different base sets. Thus we have a base set $O^n(A)$ for each $n \in \mathbb{N}^+$; with composition operations S_m^n for each pair $n, m \in \mathbb{N}^+$ and nullary operations e_i^n for each $n \in \mathbb{N}^+$ and $1 \leq i \leq n$, picking out the projection operations. We write

$$O(A) = ((O^n(A))_{n \in \mathbb{N}^+}; (S_m^n)_{m,n \in \mathbb{N}^+}, (e_i^n)_{n \in \mathbb{N}^+, 1 \leq i \leq n}),$$

for the (heterogeneous) full clone of all operations defined on the set A. A clone on A is then any subalgebra of this algebra. Each such clone belongs to a variety K_0 of heterogeneous algebras defined by the following identities $(C1)$, $(C2)$, $(C3)$:

(C1) $S_m^p(z, S_m^n(y_1, x_1, \ldots, x_n), \ldots, S_m^n(y_p, x_1, \ldots, x_n)) \approx$
      $S_m^n(S_n^p(z, y_1, \ldots, y_p), x_1, \ldots, x_n),$ $(m, n, p \in \mathbb{N}^+),$

(C2) $S_m^n(e_i^n, x_1, \ldots, x_n) \approx x_i,$ $(m \in \mathbb{N}^+,\ 1 \leq i \leq n),$

(C3) $S_n^n(y, e_1^n, \ldots, e_n^n) \approx y,$ $(n \in \mathbb{N}^+).$

Up to isomorphism, the elements of the variety K_0 are exactly the clones regarded as heterogeneous algebras (see for instance W. Taylor, [112]). Our heterogeneous clones correspond to algebraic theories, or particular categories, in the sense of F. W. Lawvere ([71]).

10.2 Operations and Relations

In this section we recall and expand on the Galois-connection introduced in Section 2.2 between sets of operations and sets of relations on a base set A. For any positive integer h, an h-ary relation on the set A is a subset ρ of A^h (a set of h-tuples consisting of elements of A). We say that an n-ary operation $f \in O^n(A)$ *preserves* ρ if for every $h \times n$ matrix Y whose column vectors $\underline{y}_1, \ldots, \underline{y}_n$ all belong to ρ, the vector $(f(\underline{y}^1), \ldots, f(\underline{y}^n))$ belongs to ρ. In this case we also say that ρ is *invariant* with respect to f, or f is a *polymorphism* of ρ, or f is *compatible* with ρ.

The set of all operations defined on A preserving a given relation $\rho \subseteq A^h$ is denoted by $Pol_A\rho$. It is very easy to show that sets of operations of the form $Pol_A\varrho$ are clones. It is also wellknown that all clones occur in this way, using the following notation. For a set A let $R^h(A)$ be the set of all h-ary relations on A and let $R(A) = \bigcup\limits_{h=1}^{\infty} R^h(A)$, the set of all finitary relations on A.

For any set $F \subseteq O(A)$ of operations and any set $Q \subseteq R(A)$ of relations we consider

$$
\begin{aligned}
Pol_A Q &:= \{f \mid f \in O(A) \text{ and } \forall \varrho \in Q \ (f \text{ preserves } \varrho)\} \quad \text{and} \\
Inv_A F &:= \{\varrho \mid \varrho \in R(A) \text{ and } \forall f \in F \ (f \text{ preserves } \varrho)\}.
\end{aligned}
$$

(For convenience we usually write $Pol_A\varrho$ and $Inv_A f$ for $Pol_A\{\varrho\}$ and $Inv_A\{f\}$, and if the base set is clear from the context we usually omit the subscript A.)

In Section 2.2 we considered the operators

$$
\begin{aligned}
Inv &: \ O(A) \to R(A) \ \vdash F Inv_A F \quad \text{and} \\
Pol &: \ R(A) \to O(A) \ \vdash Q Pol_A Q
\end{aligned}
$$

$$
R := \{(f, \varrho) \mid f \in O(A), \ \varrho \in R(A) \text{ and } f \text{ preserves } \varrho\}.
$$

Clones can then be characterized as Galois-closed sets of operations, that is, sets F having the property that $PolInvF = F$ (see R. Pöschel and L. A. Kalužnin, [96]). Dually, closed sets R of relations satisfying $InvPolR = R$ are called *relational clones*. There is also an algebraic characterization of relational clones using certain operations defined on sets of relations ([96]). As we saw in Section 6.1, the two classes of closed sets of a Galois-connection form complete lattices which are dually isomorphic to each other. In the specific case of clones of operations, we see that the set of all clones of operations defined on a base set A forms a complete lattice $\mathcal{L}_A$; moreover this lattice is dually isomorphic to the lattice of all relational clones on A.

In the next section we will give a complete description of this lattice $\mathcal{L}_A$, in the special case that A is a two-element set. First, we give some examples of relations and the corresponding clones in this special case. We fix $A = \{0, 1\}$, and refer to operations on A as Boolean operations. For the first example, we take ρ to be the binary relation $\{(0,1),(1,0)\}$. We will denote subtraction modulo 2 by $-$, and, as before, negation by $\neg$. Then a $2 \times n$ matrix $Y = (y_{ij})$ has all columns in ρ iff $y_{2j} = 1 - y_{1j} = \neg y_{1j}$ holds for all $j = 1, \ldots, n$. Thus the condition is that

$$(f(y_{11}, \ldots, y_{1n}), f(\neg y_{11}, \ldots, \neg y_{1n})) \in \rho$$

holds for all $y_{11}, \ldots, y_{1n} \in \{0, 1\}$. This means that

$$\neg f(y_{11}, \ldots, y_{1n}) = f(\neg y_{11}, \ldots, \neg y_{1n})$$

or

$$f(y_{11}, \ldots, y_{1n}) = \neg(f(\neg y_{11}, \ldots, \neg y_{1n}))$$

for all $y_{11}, \ldots, y_{1n} \in \{0, 1\}$. A Boolean function f^* is called *dual* to the Boolean function f if $f^*(y_{11}, \ldots, y_{1n}) = \neg(f(\neg y_{11}, \ldots, \neg y_{1n}))$; Boolean functions with $f^* = f$ are called *self-dual*. We can express this more algebraically by saying that f is self-dual iff the permutation which interchanges 0 and 1 is an automorphism of the algebra $(\{0,1\}; f)$. Using Post's original notation ([95]), we will denote by D_3 the set of all self-dual Boolean functions. Thus we see that for the given relation ρ on $A = \{0, 1\}$, we have $Pol_A\rho = D_3$.

Next we consider the relation $\rho := \{(0,0),(0,1),(1,1)\}$, which can be defined by $(x, y) \in \rho$ iff $x \leq y$. A $2 \times n$ matrix has all columns in ρ whenever $y_{1j} \leq y_{2j}$ for all $j = 1, \ldots, n$ and thus f preserves ρ if

$$f(y_{11}, \ldots, y_{1n}) \leq f(y_{21}, \ldots, y_{2n}) \text{ provided } y_{11} \leq y_{21}, \ldots, y_{1n} \leq y_{2n}.$$

This is the standard definition of a monotone Boolean function. The set of all monotone Boolean functions is denoted by A_1, and we have $Pol_A\rho = A_1$ in this example.

A unary relation on A is just a subset of A. For ρ equal to the singleton set $\{0\}$, there is a single $1 \times n$ matrix with all columns in $\{0\}$, namely $Y = (0, \ldots, 0)$. The Boolean function f preserves $\{0\}$ iff $f(0, \ldots, 0) = 0$. Similarly, f preserves $\{1\}$ iff $f(1, \ldots, 1) = 1$. We use Post's names C_2 for the set of all 1-preserving Boolean functions and C_3 for the set of all 0-preserving Boolean functions.

As a last example consider

$$\rho := \{(y_1, y_2, y_3, y_4) \in \{0,1\}^4 \mid y_1 + y_2 = y_3 + y_4\},$$

where $+$ as usual is addition modulo 2. A Boolean function is linear if there are elements $c_0, c_1, \ldots, c_n \in \{0, 1\}$ such that $f(x_1, \ldots, x_n) = c_0 + c_1 x_1 + \cdots + c_n x_n$. It can be shown that a Boolean function is linear iff it preserves the relation ρ. The clone of all linear Boolean functions is denoted by L_1.

10.3   The Lattice of All Boolean Clones

As we saw in the previous section, the class of all clones on a fixed set A forms a complete lattice $\mathcal{L}_A$. In the case that A is the two-element set $\{0, 1\}$, this lattice is called the lattice of *Boolean clones*. It was first described by E. L. Post in 1941, and is sometimes also called *Post's lattice*. The lattice is countably infinite, complete, algebraic, atomic and dually atomic. It is also known that every clone in the lattice is finitely generated. Post's original proof of the structure of the lattice requires several combinatorial considerations, and simpler proofs have been given since then (see for instance J. Berman, [7] or D. Lau, [69]).

In this section we will give a proof using our results from Section 9.6 on the properties of the congruence lattices of two-element algebras. Clearly, if all two-element algebras are known, then all closed classes of Boolean operations are known. For the set $A = \{0, 1\}$, we will give a complete classification of the varieties $V(\mathcal{A})$ generated by algebras with base set A. Our analysis will be broken into a number of cases, beginning with whether $V(\mathcal{A})$ is congruence modular or not. In the case that $V(\mathcal{A})$ is congruence modular, we will consider two further cases, depending on whether $V(\mathcal{A})$ is congruence distributive or not. The case that $V(\mathcal{A})$ is congruence distributive will again be broken into cases, based on whether $V(\mathcal{A})$ is 2-distributive, 3-distributive but not 2-distributive or finally 4-distributive but not 3-distributive. Note that as in Section 9.6 we do not distinguish between isomorphic or equivalent (having the same clone of term operations) algebras. We will use the notation *clone*$\mathcal{A}$ for the term clone, or clone of all term operations, of $\mathcal{A}$. We also use the standard notation for Boolean operations, as introduced in Section 9.6. In addition, we will denote by c_0^2 and c_1^2 the binary constant operations with values 0 and 1, respectively.

Lemma 10.3.1 *Let $\mathcal{A}$ be a two-element algebra, with term clone clone$\mathcal{A}$. If $V(\mathcal{A})$ is not congruence modular, then clone$\mathcal{A}$ is a subset of the clone generated by one of the following sets of Boolean operations:*

$$\begin{aligned}
M_1 &= \{\wedge, c_0^2, c_1^2, e_1^2, e_2^2\}, \\
M_2 &= \{\vee, c_0^2, c_1^2, e_1^2, e_2^2\}, \\
M_3 &= \{c_0^2, c_1^2, e_1^2, e_2^2, \neg e_1^2, \neg e_2^2\}.
\end{aligned}$$

Proof: We note first that none of the Boolean operations g_1 to g_7 in the following list can be contained in *clone*$\mathcal{A}$:

$$\begin{aligned}
g_1(x, y) &= \neg x \wedge y, & g_2(x, y) &= x \wedge \neg y, & g_3(x, y) &= \neg x \wedge \neg y, \\
g_4(x, y) &= x + y + 1, & g_5(x, y) &= x \vee \neg y, & g_6(x, y) &= \neg x \vee y, \\
g_7(x, y) &= \neg x \vee \neg y.
\end{aligned}$$

This is because otherwise we could construct by superposition one of the operations $x + y + z$, $(x \wedge y) \vee (x \wedge z) \vee (y \wedge z)$, $x \wedge (y \vee z)$, $x \vee (y \wedge z)$, $(x \vee \neg y) \wedge z$, $(x \wedge \neg y) \vee z$, any of which we know from Theorem 9.6.3 guarantees congruence modularity. Similarly, $x \wedge y$ and $x \vee y$ cannot both be term operations of $\mathcal{A}$, since from them we could produce as a term the operation $(x \wedge y) \vee (x \wedge z) \vee (y \wedge z)$. Also, if either of $\wedge$ or $\vee$ is an element of *clone*$\mathcal{A}$ then $\neg$ cannot be a term operation, since otherwise we could produce one of the operations g_1 to g_7 above.

For any set F of Boolean operations, we denote by F^d the set of all operations dual to those from F. It is easy to prove that the algebras $\mathcal{A} = (\{0, 1\}; (f_i^A)_{i \in I})$ and $\mathcal{A}^d = (\{0, 1\}; ((f_i^A)^d)_{i \in I})$ are always isomorphic. Obviously $M_2^d = M_1$, so we can restrict ourselves to M_1 and M_3. We will show that the set of all binary term operations of $\mathcal{A}$ is contained in either M_1 or M_3.

Next we claim that any at least essentially ternary operation g from *clone* $\mathcal{A}$ must be monotone. Monotonicity means that if we define $(a_1, a_2, \ldots, a_n) \preceq (b_1, b_2, \ldots, b_n)$ iff $a_i \leq b_i$ for all $i \in \{1, \ldots, n\}$, for $a_i, b_i \in \{0, 1\}$, then for arbitrary n-tuples $(a_1, a_2, \ldots, a_n)$ and $(b_1, b_2, \ldots, b_n)$, if $(a_1, \ldots, a_n) \preceq (b_1, \ldots, b_n)$ then $g(a_1, \ldots, a_n) \leq g(b_1, \ldots, b_n)$. Let g be an essentially at least ternary operation. If g was not monotone, there would be a pair of n-tuples

$$\underline{a} = (a_1, a_2, \ldots, a_{i-1}, 0, a_{i+1}, \ldots, a_n)$$
$$\underline{b} = (a_1, a_2, \ldots, a_{i-1}, 1, a_{i+1}, \ldots, a_n)$$

for which $g(\underline{a}) = 1 > 0 = g(\underline{b})$. We can assume that g is not constant, and therefore that $g(0, \ldots, 0) = 0$ and $g(1, \ldots, 1) = 1$. But then by identification of variables we can make a ternary operation f with $f(0, 0, 0) = 0$, $f(1, 0, 1) = 0$, $f(0, 0, 1) = 1$ and $f(1, 1, 1) = 1$. Consideration of all possible values for f on the remaining four triples in A^3 shows that we would get one of the operations $x + y + z$, $(x \vee \neg y) \wedge z$ or $(x \wedge \neg y) \vee z$. But these are impossible, as we saw above. Hence g must be monotone.

Now we will show that the n-ary term operations of *clone* $\mathcal{A}$ are n-ary projections, negations of n-ary projections, n-ary constant operations, or can be written in the form $f(x_1, \ldots, x_n) = x_{i_1} \wedge \ldots \wedge x_{i_n}$ with $\{i_1, \ldots, i_n\} \subseteq \{1, \ldots, n\}$. For the binary term operations this is clear. For $n > 2$, we have seen that any essentially n-ary term operation h is monotone. Suppose that h is not constant and not an n-ary projection. Then $h(\underline{0}) = 0$ and $h(\underline{1}) = 1$. In addition, there must exist an n-tuple $\underline{a} = (a_1, \ldots, a_i, \ldots, a_j, \ldots, a_n)$ with $\underline{a} \neq \underline{0}$ and $\underline{a} \neq \underline{1}$, but $h(\underline{a}) = 1$; that is, there is at least one i and a number j, with $1 \leq i, \ j \leq n$, such that either $a_i = 0$ and $a_j = 1$ or $a_j = 0$ and $a_i = 1$. Since h is not an n-ary projection, there is an n-tuple $\underline{b} = (b_1, \ldots, b_i, \ldots, b_j, \ldots, b_n)$ with $h(\underline{b}) \neq b_j$. By identification and permutation of variables we can make a binary term operation h' with $h'(0, 0) = 0$, $h'(0, 1) = 1$, $h'(b_i, b_j) \neq b_j$ and $h'(1, 1) = 1$. Since h' is monotone we must have $b_j = 0$ and thus $b_i = 1$, and so $h'(x, y) = x \vee y$.

But this means we can produce the join operation $\vee$ as a term operation of $\mathcal{A}$. If the binary term operations of $\mathcal{A}$ are contained in M_1, we have both $\vee$ and $\wedge$, and from these we obtain $(x \wedge y) \vee (x \wedge z) \vee (y \wedge z)$. Similarly, if the set of all binary term operations of $\mathcal{A}$ is contained in the set M_3, we have both $\neg$ and $\vee$ as term operations, and again can produce $(x \wedge y) \vee (x \wedge z) \vee (y \wedge z)$. But this makes $\mathcal{A}$ congruence modular.

This shows that $clone\mathcal{A}$ is contained in the clone generated by M_1. If the binary term operations of $\mathcal{A}$ are contained in M_3 then the n-ary term operations of $\mathcal{A}$ are n-ary constants, n-ary projections, or negations of n-ary projections. Operations of the form $f(x_1, \ldots, x_n) = x_{i_1} \wedge \ldots \wedge x_{i_n}$, $\{i_1, \ldots, i_n\} \subseteq \{1, \ldots, n\}$, cannot be term operations since otherwise $\wedge$ is a term operation. Thus $clone\mathcal{A}$ is included in the clone generated by M_3 in this case. ∎

Analysis of the sets M_1 and M_3 yields the following description of all two-element algebras which generate varieties which are not congruence modular. As usual, we use Post's names for the algebras.

Theorem 10.3.2 *Let $\mathcal{A}$ be a two-element algebra generating a variety which is not congruence modular. Then (up to isomorphism and equivalence) $\mathcal{A}$ is one of the following algebras:*

$$\mathcal{P}_6 = (\{0,1\}; \wedge, c_0^2, c_1^2), \quad \mathcal{P}_3 = (\{0,1\}; \wedge, c_0^2), \quad \mathcal{P}_5 = (\{0,1\}; \wedge, c_1^2),$$

$$\mathcal{P}_1 = (\{0,1\}; \wedge), \quad\quad\quad \mathcal{O}_9 = (\{0,1\}; \neg, c_0^2),$$

$$\mathcal{O}_8 = (\{0,1\}; e_1^2, c_0^2, c_1^2), \quad \mathcal{O}_6 = (\{0,1\}; e_1^2, c_0^2),$$

$$\mathcal{O}_4 = (\{0,1\}; \neg), \quad\quad\quad \mathcal{O}_1 = (\{0,1\}; e_1^2).$$

∎

Having found all two-element algebras which generate varieties which are not congruence modular, we turn now to the other case, of two-element algebras which generate congruence modular varieties. This case will also be treated in two subcases, depending on whether the variety generated is congruence distributive or not. First, suppose that $V(\mathcal{A})$ is congruence modular but not congruence distributive. In this case, we know that $V(\mathcal{A})$ is congruence permutable. By Theorems 9.6.3 and 9.6.4, the operation $x+y+z$ is a term operation of $\mathcal{A}$, but none of $(x \wedge y) \vee (x \wedge z) \vee (y \wedge z)$, $x \wedge (y \vee z)$, $x \vee (y \wedge z)$, $(x \vee \neg y) \wedge z$ and $(x \wedge \neg y) \vee z$ are term operations of $\mathcal{A}$. Clearly, the clone $\langle \{x + y + z\} \rangle$ is the least clone of Boolean operations to contain the operation $x + y + z$. From a well-known result of R. McKenzie in [78] it follows that the greatest clone of Boolean operations to have this property, that is, to contain $x + y + z$ but none of $(x \wedge y) \vee (x \wedge z) \vee (y \wedge z)$, $x \wedge (y \vee z)$, $x \vee (y \wedge z)$, $(x \vee \neg y) \wedge z$ and $(x \wedge \neg y) \vee z$, is the clone consisting of all linear Boolean operations. Linear Boolean operations are those of the form $f(x_1, \ldots, x_n) = c_0 + c_1 x_1 + \ldots + c_n x_n$, for $c_i \in \{0,1\}$. It is easy to show that the interval $[\langle\{x+y+z\}\rangle, \langle\{+, c_1^2\}\rangle)]$ contains exactly the following clones of Boolean operations:

$$\langle\{+, c_1^2\}\rangle, \quad \langle\{x+y+1\}\rangle, \quad \langle\{\neg, x+y+z\}\rangle, \quad \langle\{+\}\rangle, \quad \langle\{x+y+z\}\rangle.$$

This tells us all the two-element algebras which generate a congruence modular but not congruence distributive variety, namely

$$\mathcal{L}_1 = (\{0,1\}; +, c_1^2), \quad \mathcal{L}_2 = (\{0,1\}; x+y+1), \quad \mathcal{L}_3 = (\{0,1\}; +),$$
$$\mathcal{L}_5 = (\{0,1\}; x+y+z, \neg), \quad \mathcal{L}_4 = (\{0,1\}; x+y+z).$$

At this point in our analysis, any remaining two-element algebras generate congruence distributive varieties. As we know from Section 9.6, there are exactly the following three cases for congruence distributivity: $V(\mathcal{A})$ may be 2-distributive, or 3-distributive but not 2-distributive, or 4-distributive but not 3-distributive.

Consider first the case that $V(\mathcal{A})$ is 2-distributive. Then by Lemma 9.6.1, the majority operation $(x \wedge y) \vee (x \wedge z) \vee (y \wedge z)$ is a term operation of $\mathcal{A}$. Using the Baker-Pixley Theorem, Theorem 9.2.6, we can describe $clone\mathcal{A}$ as the clone consisting of exactly all operations defined on A which preserve all subalgebras of the direct square $\mathcal{A}^2$. By examining all sublattices of the lattice of all subsets of A^2 which can be subalgebra lattices of $\mathcal{A}^2$ we can determine all two-element algebras with a majority term operation $(x \wedge y) \vee (x \wedge z) \vee (y \wedge z)$. To describe them all, we need the following five relations on A:

$$\leq := \{(00), (01), (11)\}, \quad s_2 := \{(01), (10)\},$$
$$R_2 := \{(00), (01), (10)\}, \quad R_2' := \{(11), (01), (10)\},$$
$$\Delta := \{(00), (10)\}.$$

Then we get exactly the following clones:

$Pol\ \Delta, \quad Pol\{0\}, \quad Pol\{1\}, \quad Pol(\{0\}, \{1\}), \quad Pol \leq, \quad PolR_2, \quad PolR_2',$
$Pols_2, \quad Pol(\{0\}, \leq), \quad Pol(\{0\}, \{1\}, \leq), \quad Pol(\{1\}, \leq^{-1}), \quad Pol(\{0\}, \{1\}, \leq^{-1}),$
$Pol(\{1\}, R_2), \quad Pol(\{0\}, R_2'), \quad Pol(\{0\}, \{1\}, s_2), \quad Pol(\leq, R_2, R_2').$

It is not hard to determine a generating system for each clone in this list, and from this we produce our list of all two-element algebras which generate a 2-distributive variety:

$\mathcal{A}_1 = (\{0,1\}; \wedge, \vee, c_0^2, c_1^2),$
$\mathcal{A}_2 = (\{0,1\}, \wedge, \vee, c_1^2),$
$\mathcal{A}_3 = (\{0,1\}; \wedge, \vee, c_0^2),$
$\mathcal{A}_4 = (\{0,1\}; \wedge, \vee),$
$\mathcal{C}_1 = (\{0,1\}; \wedge, \neg),$
$\mathcal{C}_3 = (\{0,1\}); \vee, g_2),$ where $g_2 = x \wedge \neg y,$
$\mathcal{C}_2 = (\{0,1\}; \vee, x + y + 1),$
$\mathcal{C}_4 = (\{0,1\}; \vee, t),$ with $t(x,y,z) = x \wedge (y + z + 1),$
$\mathcal{D}_2 = (\{0,1\}; (x \wedge y) \vee (x \wedge z) \vee (y \wedge z)),$
$\mathcal{D}_1 = (\{0,1\}; (x \wedge y) \vee (x \wedge z) \vee (y \wedge z), x + y + z),$
$\mathcal{D}_3 = (\{0,1\}; (x \wedge y) \vee (x \wedge z) \vee (y \wedge z), x + y + z, \neg),$

$\mathcal{F}_5^2 = (\{0,1\}; (x \wedge y) \vee (x \wedge z) \vee (y \wedge z), (x \wedge \neg y) \vee z),$
$\mathcal{F}_6^2 = (\{0,1\}; (x \wedge y) \vee (x \wedge z) \vee (y \wedge z), x \wedge (y \vee z)),$
$\mathcal{F}_7^2 = (\{0,1\}; (x \wedge y) \vee (x \wedge z) \vee (y \wedge z), c_0^2),$
$\mathcal{F}_9^2 = (\{0,1\}; (x \wedge y) \vee (x \wedge z) \vee (y \wedge z), g_2),$
$\mathcal{F}_1^2 = (\{0,1\}; (x \wedge y) \vee (x \wedge z) \vee (y \wedge z), (x \vee \neg y) \wedge z),$
$\mathcal{F}_2^2 = (\{0,1\}; (x \wedge y) \vee (x \wedge z) \vee (y \wedge z), x \vee (y \wedge z)),$
$\mathcal{F}_3^2 = (\{0,1\}; (x \wedge y) \vee (y \wedge z) \vee (x \wedge z), c_1^2),$
$\mathcal{F}_4^2 = (\{0,1\}; (x \wedge y) \vee (y \wedge z) \vee (x \wedge z), x \vee \neg y).$

For the next case, we assume now that $V(\mathcal{A})$ is congruence distributive, but not 2-distributive. Then $x \wedge (y \vee z)$ or $(x \vee (y \wedge z))$ is a term operation, but $(x \wedge y) \vee (x \wedge z) \vee (y \wedge z)$ is not a term operation. We consider two further subcases here, depending on whether the term clone $clone\mathcal{A}$ can be represented as $Pol\ \rho$ for some u-ary relation ρ on $\{0,1\}$. The next lemma, due to M. Reschke and K. Denecke ([99]), handles the case where $clone\mathcal{A} = Pol\rho$ for some at least ternary relation. It uses a function h_μ with properties corresponding to the generalized Baker-Pixley Theorem (see Remark 9.2.7). Such functions are called near-unanimity functions. The generalized Baker-Pixley Theorem shows that a function f is a term function of $\mathcal{A}$ if and only if f preserves all subalgebras of $\mathcal{A}^\mu$.

Lemma 10.3.3 *([99]) Let $clone\mathcal{A} = Pol\ \rho$ for a μ-ary relation $\rho \subseteq \{0,1\}^\mu$, where $\mu \geq 3$. Then $V(\mathcal{A})$ is congruence distributive but not 2-distributive if and only if the following operation h_μ or its dual h_μ' is a term operation in* *clone*$\mathcal{A}$:

$$h_\mu(x_1,\ldots,x_{\mu+1}) = \bigvee_{i=1}^{\mu+1} (x_1 \wedge \ldots \wedge x_{i-1} \wedge x_{i+1} \wedge \ldots \wedge x_{\mu+1}).$$

Proof: We will prove that $x \wedge (y \vee z) \in Pol\ \rho$ if and only if $h_\mu \in Pol\ \rho$. The dual case, that $x \vee (y \wedge z) \in Pol\ \rho$ if and only if $h'_\mu \in clone\mathcal{A}$, is similar. In one direction, if $h_\mu \in Pol\ \rho$ for some natural number $\mu \geq 3$, then $h_\mu(x_1,x_2,x_3,x_1,\ldots,x_1) = x_1 \wedge (x_2 \vee x_3) \in Pol\ \rho$. Conversely, assume that $x_1 \wedge (x_2 \vee x_3) \in Pol\ \rho$ and that

$$\underline{a_1} = \begin{pmatrix} a_{11} \\ \vdots \\ a_{1\,\mu} \end{pmatrix} \in \rho,\ldots, \underline{a_{\mu+1}} = \begin{pmatrix} a_{\mu+1\,1} \\ \vdots \\ a_{\mu+1\,\mu} \end{pmatrix} \in \rho.$$

Since $clone\mathcal{A} = Pol\rho$, it is enough to show that

$$h_\mu(\underline{a_1},\ldots,\underline{a_{\mu+1}}) = \begin{pmatrix} b_1 \\ \vdots \\ b_\mu \end{pmatrix} = \underline{b} \in \rho.$$

Let $i_1, i_2, \ldots, i_l$ be the indices k from $\{1,\ldots,\mu\}$ for which $b_k = 1$. Since h_μ is a join of meets, this means that there must be an index j, with $1 \leq j \leq \mu+1$, for which the corresponding entries $a_{ji_1}, \ldots, a_{ji_l}$ are also equal to 1. (If $\underline{b}$ is the zero vector, we can pick any index j.) If $\underline{a_j}$ agrees with $\underline{b}$ in all other entries as well, then $\underline{b} = \underline{a_j} \in \rho$. Otherwise, let α be the least index from $1,\ldots,\mu$ such that $b_\alpha \neq a_{j\alpha}$. This means that $b_\alpha = 0$, while $a_{j\alpha} = 1$. From the structure of h_μ again, it is clear that there are two elements $\underline{a_r}$ and $\underline{a_s}$, with $r \neq s \neq j$ and $r, s \in \{1,\ldots,\mu+1\}$, such that $a_{r\alpha} = a_{s\alpha} = 0$. Now since the term $x_1 \wedge (x_2 \vee x_3)$ is in $Pol\rho$, we must have $\underline{a_j} \wedge (\underline{a_r} \vee \underline{a_s}) = \underline{b_1} \in \rho$. This b_1 is a μ-tuple which agrees with $\underline{b}$ at the positions $i_1, i_2, \ldots, i_l$ and α. If $\underline{b} = \underline{b_1} \in \rho$, we are done. Otherwise, we repeat this process: we let $\beta \in \{1,\ldots,\mu\}$ be the least index for which $b_\beta \neq b_{1\beta}$, so that $b_\beta = 0$ and $b_{1\beta} = 1$. Continuing in this way, after finitely many steps, we get a μ-tuple $\underline{b_k} \in \rho$ which agrees with $\underline{b}$.   ∎

The least clone of the form $clone\mathcal{A} = Pol\ \rho$ such that $V(\mathcal{A})$ is 4-distributive but not 2-distributive is the clone generated by h_μ. It can be shown that this clone is $Pol\{R_\mu, \leq, \{1\}\}$, where R_μ is the relation $\{0,1\}^\mu \setminus \{(1,\ldots,1)\}$

for $\mu \geq 3$. Dually, we have $\langle \{h'_\mu\} \rangle = Pol\{R'_\mu, \leq, \{0\}\}$ for the relation $R'_\mu = \{0,1\}^\mu \setminus \{(0,\dots,)\}$. There are four clones of this kind, for $\mu \geq 3$:

1. $clone\mathcal{A} = Pol\{R_\mu, \leq\}$ (dually $Pol\{R'\mu, \leq\}$),
2. $clone\mathcal{A} = Pol\{R_\mu, \leq, \{1\}\}$ (dually $Pol\{R'_\mu, \leq, \{0\}\}$),
3. $clone\mathcal{A} = Pol\{R_\mu, \{1\}\}$ (dually $Pol\{R'_\mu, \{0\}\}$),
4. $clone\mathcal{A} = Pol\ R_\mu$ (dually $Pol\ R'_\mu$).

In cases 3 and 4 the terms $(x \wedge \neg y) \vee z$ or dually $(x \vee \neg y) \wedge z$ are elements of the clone, and thus $V(\mathcal{A})$ is 3-distributive, while in cases 1 and 2 $V(\mathcal{A})$ is 4-distributive but not 3-distributive. These are exactly all cases for which $V(\mathcal{A})$ is congruence distributive but not 2-distributive and $clone\mathcal{A} = Pol\ \rho$ for some finitary ρ.

We turn now to our last case, where we have $V(\mathcal{A})$ congruence distributive but not 2-distributive, but there is no finitary relation ρ on $\{0,1\}$ such that $clone\mathcal{A} = Pol\ \rho$. By Lemma 10.3.3 we know that $h_\mu \notin clone\mathcal{A}$. The least clone $clone\mathcal{A}$ such that $V(\mathcal{A})$ is 4-distributive and not 3-distributive is the clone generated by $x \wedge (y \vee z)$ or $x \vee (y \wedge z)$, and the least clone such that $V(\mathcal{A})$ is 3-distributive and not 2-distributive is the clone generated by $(x \wedge \neg y) \vee z$ or $(x \vee \neg y) \wedge z$. It is easy to show moreover that the only other clones with the first property are $\langle \{x \wedge (y \vee z), c_0^1\} \rangle$ and $\langle \{x \vee (y \wedge z), c_1^1\} \rangle$, and the only other clones of the second type are $\langle \{(x \wedge \neg y) \vee z, c_0^1\} \rangle$ and $\langle \{(x \vee \neg y) \wedge z, c_1^1\} \rangle$.

Theorem 10.3.4   *(i) Let $\mathcal{A}$ be a two-element algebra generating a 4-distributive but not 3-distributive variety $V(\mathcal{A})$, and assume that $clone\mathcal{A} = Pol\ \rho$ for some μ-ary relation ρ (with $\mu \geq 3$ minimal). Then $\mathcal{A}$ is (up to equivalence and dual isomorphism) one of the algebras $\mathcal{F}_6^\mu = (\{0,1\}; h_\mu)$ or $\mathcal{F}_7^\mu = (\{0,1\}; c_0^2)$.*

(ii) Let $\mathcal{A}$ be a two-element algebra generating a 4-distributive but not 3-distributive variety $V(\mathcal{A})$, and assume that there is no finitary relation ρ for which $clone\mathcal{A} = Pol\ \rho$. Then $\mathcal{A}$ is (up to equivalence and dual isomorphism) one of the algebras $\mathcal{F}_6^\infty = (\{0,1\}; x \wedge (y \vee z))$ or $\mathcal{F}_7^\infty = (\{0,1\} : x \wedge (y \vee z), c_0^1)$.

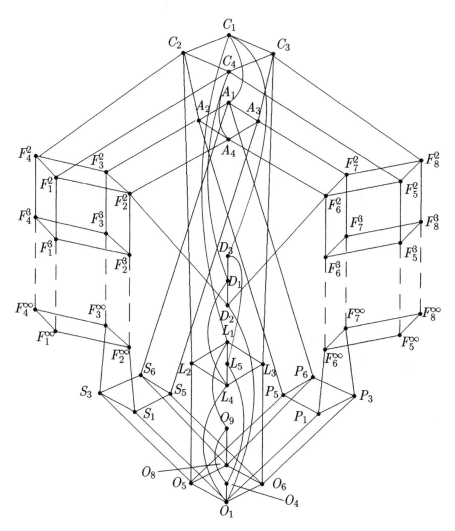

(iii) Let $\mathcal{A}$ be a two-element algebra generating a 3-distributive but not 2-distributive variety, and assume that clone$\mathcal{A} = Pol\ \rho$ for some μ-ary relation ρ (with $\mu \geq 3$ minimal). Then $\mathcal{A}$ is (up to equivalence and dual isomorphism) one of the algebras $\mathcal{F}_5^\mu = (\{0,1\}; (x \wedge \neg y) \vee z, h_\mu)$ or $\mathcal{F}_8^\mu = (\{0,1\}; (x \wedge \neg y) \vee z, h_\mu, c_0^2) = (\{0,1\}; h_\mu, g_2)$.

(iv) Let $\mathcal{A}$ be a two-element algebra generating a 3-distributive but not 2-

distributive variety, and assume that there is no finitary relation ρ for which clone$\mathcal{A} = Pol\ \rho$. Then $\mathcal{A}$ is (up to equivalence and dual isomorphism) one of the algebras $\mathcal{F}_5^\infty = (\{0,1\}; (x \wedge \neg y) \vee z)$ or $\mathcal{F}_8^\infty = (\{0,1\}; (x \wedge \neg y) \vee z, c_0^2) = (\{0,1\}; g_2).$ ∎

This completes our survey of all two-element algebras, since all our cases have now been considered. The diagram shows the lattice of all clones of Boolean operations.

10.4   The Functional Completeness Problem

The functional completeness problem is that of deciding, for a given subset C of the set $O(A)$ of all operations on a base set A, whether C generates all of $O(A)$. That is, we want to know when the clone generated by a subset C is the whole clone $O(A)$. Such a set C is said to be *functionally complete*. To solve this problem, we apply the General Completeness Criterion for finitely generated algebras, Theorem 1.3.16. Throughout this section, we assume that the set A is finite, with $\mid A \mid \geq 2$.

In the case that A is a two-element set, we consider the clone of all Boolean functions. We know already that in this case $O(A)$ is generated by $\{\vee, \neg\}$ and thus is finitely generated. Post's results show that there are exactly 5 maximal subclones of $O(A)$: these are C_2, C_3, A_1, D_3 and L_1. Hence a set C of Boolean functions is functionally complete iff for each of these five maximal subclones there exists a function $f \in C$ which is not an element of this maximal subclone.

To prove a similar functional completeness criterion for $O(A)$ where A is a finite set with $|A| > 2$, we have to prove that $O(A)$ is finitely generated, and then to determine all maximal subclones of $O(A)$. For the first step we have the following theorem, first proved by E. L. Post ([95]), which shows that on a finite set A, the clone $O(A)$ is indeed finitely generated.

Theorem 10.4.1 *Let A be a finite set of cardinality ≥ 2, and let 0 and 1 be two different elements in A. Let $+$ and $\cdot$ be two binary operations on A, with the properties that $0 + x \approx x \approx x + 0$ and $x \cdot 1 \approx x$, $x \cdot 0 \approx 0$. For each*

$a \in A$, let $\chi_a : A \to A$ be a *unary operation with*

$$\chi_a(x) = \begin{cases} 1 & if \ x = a \\ 0 & otherwise. \end{cases}$$

Then $\langle \{+, \cdot, (\chi_a)_{a \in A}\} \rangle_{O(A)} = O(A)$, *and any operation* $f : A^n \to A$ *may be expressed as*

$$f(x_1, \ldots, x_n) = \sum_{(a_1, \ldots, a_n) \in A^n} f(a_1, \ldots, a_n) \cdot \prod_{i \leq n} \chi_{a_i}(x_i),$$

where $\sum$ *means iteration of* $+$ *in a canonical way and* $\prod_{i \geq 1} y_i$ *is defined similarly by* $\prod_{i \leq 1} y_i := y_1$ *and* $\prod_{i \leq n+1} y_i := (\prod_{i \leq n} y_i) \cdot y_{n+1}$.

Proof: If $f : A^n \to A$ is an arbitrary n-ary operation and $(a_{1j}, \ldots, a_{nj})$ is an arbitrary n-tuple, then $f(a_{1j}, \ldots, a_{nj}) = f(a_{1j}, \ldots, a_{nj}) \, \chi_{a_{nj}}(a_{nj})$. This shows that any f can be expressed as $f(x_1, \ldots, x_n) = \sum_{(a_1, \ldots, a_n) \in A^n} f(a_1, \ldots, a_n) \cdot \prod_{i \leq 1} \chi_{a_i}(x_i)$, as claimed. Moreover, this representation requires only the operations $+$ and $\cdot$ and the functions χ_A for each a in A. Since A is finite, we now have a finite generating system for $O(A)$. ∎

Note that in the Boolean case $A = \{0, 1\}$, we have $\chi_0(x) = \neg x$ and $\chi_1(x) = x$. If we take $+$ to represent disjunction and $\cdot$ to represent conjunction, the representation given in Theorem 10.4.1 is just the usual disjunctive normal form.

Now that we know that $O(A)$ is finitely generated when A is finite, we must look for the maximal subclones of $O(A)$. These are in fact known for every finite A. The deep theorem which describes explicitly all maximal clones on $O(A)$ is due to I. G. Rosenberg ([103], [102]). The special cases where $|A|$ is 2, 3 or 4 were solved earlier by E. L. Post ([95]), S.V. Jablonskij ([59]), and A.I. Mal'cev (unpublished). New proofs of Rosenberg's Theorem were also given by R. W. Quackenbush ([97]) and D. Lau ([70]).

This description of the maximal subclones of $O(A)$ gives us the following completeness criterion.

Corollary 10.4.2 *Let A be a finite set and let F be a subset of $O(A)$. Then $\langle F \rangle_{O(A)} = O(A)$ if and only if F is not contained in one of the maximal subclones of $O(A)$.*   ∎

Rosenberg's Theorem classifying all maximal clones is too complex to be proven here. Instead we merely describe his classification. Every maximal clone on a finite set A is of the form $Pol\rho$ for some h-ary relation ρ. So the description of the maximal clones amounts to the determination of the corresponding h-ary relations $\rho \subseteq A^h$. We have the following six classes of relations.

(i) Let $\mathcal{S}_A$ be the full symmetric group of all permutations on A. Let $s \in \mathcal{S}_A$ be a fixed-point free permutation with $r = n/p$ cycles of equal prime length p. We set $\rho_s = \{(a,b) \in A^2 \mid s(a) = b\}$. Then $Pol\rho_s$ is a maximal clone.

(ii) Let $\rho \subseteq A^2$ be a partial order with least and greatest elements 0 and 1, respectively. Then $Pol\rho$ is a maximal clone.

(iii) Let $\mathcal{G} = (A; +, -, 0)$ be an abelian group. A function $f \in O^k(A)$ is called quasilinear with respect to $\mathcal{G}$ if for all $x_1, \ldots, x_k$ and $y_1, \ldots, y_k \in A$ we have $f(x_1, \ldots, x_k) + f(y_1, \ldots, y_k) = f(x_1 + y_1, \ldots, x_k + y_k) + f(0, \ldots, 0)$. The set of all quasilinear functions is a clone which can be characterized as $Pol\chi_G$ for the relation $\chi_G := \{(x, y, z, u) \in A^4 : x + y = z + u\}$. This clone is maximal iff $\mathcal{G}$ is p-elementary (meaning that $px = 0$ for all $x \in A$) or, equivalently, if $\mathcal{G}$ is the additive group of an m-dimensional vector space over the p-element field $GF(p)$. Thus $|A| = p^m$ for some prime p and $m \in \mathbb{N}$.

(iv) Let $|A| \geq 3$. Let ϑ be a non-trivial equivalence relation on A, distinct from A^2 and the diagonal relation Δ_A. Then $Pol_A\vartheta$ is a maximal clone. For B a proper subset of A with at least two elements, let ϑ_B denote the (non-trivial) equivalence relation having B as unique non-singleton block (that is, the blocks of ϑ_B are B and all $\{c\}$ with $c \in A \setminus B$). Later we will distinguish the two cases $\vartheta = \vartheta_B$ and $\vartheta \neq \vartheta_B$.

(v) A relation $\varrho \subseteq A^h$ is called *totally reflexive* if ϱ contains each h-tuple $(a_1, \ldots, a_h) \in A^h$ with a repetition of coordinates (that is, with $a_i = a_j$ for some $1 \leq i < j \leq h$). A relation ρ is called *totally symmetric* if $(a_1, \ldots, a_h) \in \varrho \Leftrightarrow (a_{\pi(1)}, \ldots, a_{\pi(h)}) \in \varrho$ for every permutation π on the set $\{1, 2, \ldots, h\}$. The center $C(\varrho)$ of ϱ is the set of all elements $c \in A$ with the property that $(c, a_2, \ldots, a_h) \in \varrho$ for all $a_2, \ldots, a_h \in A$. A totally reflexive and totally symmetric relation ϱ with a non-trivial (not all of A) center is called *central*. $Pol_A\varrho$ is a maximal clone for every central relation ϱ. Later we distinguish two cases depending on whether ϱ is unary or not.

(vi) Let $|A| = n$. For each $3 \le t \le n$, let
$E_t := \{1, 2, \ldots, t\}$,
$\iota_t := \{(c_1, \ldots, c_t) \in E_t^t \mid c_i = c_j$ for some $1 \le i < j \le t\}$,   and
$\iota_t^{\otimes m} := \iota_t \otimes \ldots \otimes \iota_t$
$= \{((c_{11}, \ldots, c_{1m}), \ldots, (c_{t1}, \ldots, c_{tm})) \in (E_t^m)^t \mid (c_{1i}, \ldots, c_{ti}) \in \iota_t$ for $i = 1, \ldots, m\}$.
For $t \ge 3$, a t-ary relation $\varrho \subset A^t$ is called *t-universal* if there are an $m \ge 1$ and a surjective mapping $\mu : A \longrightarrow E_t^m$ such that

$$\varrho = \varrho(\mu) := \{(a_1, \ldots, a_t) \in A^t \mid (\mu(a_1), \ldots, \mu(a_t)) \in \iota_t^{\otimes m}\}.$$

There is only one non-trivial n-universal relation ($t = |A| = n, m = 1$), namely

$$\iota_n(A) = \{(a_1, \ldots, a_n) \in A^n \mid a_i = a_j \text{ for some } 1 \le i < j \le n\}.$$

The well-known Słupecki criterion says that $f \in Pol\iota_n$ iff f is not surjective or f depends essentially on at most one variable ([109]). $Pol\varrho$ is a maximal clone for every t-universal relation.

10.5   Primal Algebras

In this section we describe algebraic properties and characterizations of primal and functionally complete algebras. We recall from Section 10.3 the notation *clone*$\mathcal{A}$ for the clone of term operations of an algebra $\mathcal{A}$. From Theorem 5.2.3, this clone is generated by the set $\{f_i^A \mid i \in I\}$ of fundamental operations of the algebra $\mathcal{A}$. Throughout this chapter we will denote this set by F^A, so that *clone*$\mathcal{A} = \langle F^A \rangle$. We also have the clone of all polynomial operations on $\mathcal{A}$, denoted by $P(\mathcal{A})$ or *Pclone*$\mathcal{A}$ and generated by the set of fundamental operations of $\mathcal{A}$ plus the constant functions c_a for each $a \in A$.

Definition 10.5.1 A finite algebra $\mathcal{A}$ is called *primal* if every operation $f \in O(A)$ is a term operation of $\mathcal{A}$, so that *clone*$\mathcal{A} = O(A)$. A finite algebra is called *functionally complete* if *Pclone*$\mathcal{A} = O(A)$.

Primality of $\mathcal{A}$ thus means that for every $n \geq 1$ and every operation $f :$ $A^n \to A$, there exists a term operation $t^{\mathcal{A}}$ of $\mathcal{A}$ such that $f = t^{\mathcal{A}}$; so for all n-tuples $(a_1, \ldots, a_n) \in A^n$ the equation

$$f(a_1, \ldots, a_n) = t^{\mathcal{A}}(a_1, \ldots, a_n)$$

is satisfied. Functional completeness of an algebra $\mathcal{A}$ means that for every $f \in O(A)$ there exists a polynomial operation $p^{\mathcal{A}}$ on $\mathcal{A}$ with $f = p^{\mathcal{A}}$. This property is a generalization of the *interpolation* property in ring theory, where an arbitrary unary operation is interpolable by a polynomial over a ring.

It follows from the definition that any primal algebra is functionally complete. Another connection between these two properties is based on the following construction. For any finite algebra $\mathcal{A}$, we can form a new algebra $\mathcal{A}^+$ from $\mathcal{A}$ by adding as fundamental operations every constant operation c_a, for $a \in A$. (The finiteness of A means that we still have only finitely many fundamental operations in this algebra.) That is, $\mathcal{A}^+ = (A; F^A \cup \{c_a \mid a \in A\})$. Then the following result is easily verified, and its proof is left as an exercise for the reader.

Lemma 10.5.2 *A finite algebra $\mathcal{A}$ is functionally complete if and only if $\mathcal{A}^+$ is primal.*   ∎

The functional completeness criterion from Section 10.4 can be used to obtain examples of primal algebras. It is very easy to see in this way that the two-element Boolean algebra is primal. We list here some more examples of primal and functionally complete algebras; we leave the verification, using Theorem 10.4.1 and Corollary 10.4.2, to the reader.

Example 10.5.3 1. For $k \geq 2$, let $\mathcal{A}_k = (\{0, \ldots, k-1\}; min, g)$ be the type $(2, 1)$ algebra with min the minimum with respect to the usual order of the natural numbers $0, 1, \ldots, k-1$ and g defined by

$$g(x) = \begin{cases} x+1 & \text{if } x \neq k-1 \\ 0 & \text{if } x = k-1. \end{cases}$$

E. L. Post showed in [94] that $\mathcal{A}_k$ is primal. (Note that for $k = 2$, this is the algebra $(\{0, 1\}; \wedge, \neg)$.)

2. For $k \geq 2$, let $\mathcal{B}_k = (\{0, \ldots, k-1\}; /)$ be the type (2) algebra with

$$x/y = \begin{cases} 0 & \text{if } x = y = k - 1 \\ min(x, y) + 1 & \text{otherwise.} \end{cases}$$

The algebras $\mathcal{B}_k$ were shown to be primal by D. L. Webb in [117].

3. Let $\mathcal{A} = (A; \cdot, g)$ be an algebra of type $(2, 1)$ in which there is an element $0 \in A$ such that

(i) $(A \setminus \{0\}; \cdot)$ is a group,

(ii) $a \cdot 0 = 0 \cdot a = 0$ for all $a \in A$, and

(iii) g is a cyclic permutation on A.

This algebra was shown to be primal by A. L. Foster in [45].

Using the General Completeness Criterion of Theorem 1.3.16, Corollary 10.4.2 and the list from Section 10.4 of all relations determining maximal clones, we get the following result.

Theorem 10.5.4 *A finite non-trivial algebra* $\mathcal{A} = (A; F^A)$ *is primal if and only if* F^A *is not a subset of* $Pol\varrho$ *for any of the relations* ϱ *from the list of all relations determining maximal clones.* ∎

Another important example of primal algebras is the k-element *Post algebra of order* k. These are defined, for $k \geq 2$, by

$$\mathcal{A}_k = (\{0, \ldots, k-1\}; \cup, \cap, C, D_1, \ldots, D_{k-1}, 0, \ldots, k-1),$$

with $i \cup j := max(i, j)$ and $i \cap j := min(i, j)$ for $i, j \in \{0, \ldots, k-1\}$, and

$$D_i(j) := \begin{cases} k - 1 & \text{if } i \leq j \\ 0 & \text{if } i > j, \end{cases}$$

$$C(i) := \begin{cases} k - 1 & \text{if } i = 0 \\ 0 & \text{if } i > 0. \end{cases}$$

These algebras play the same role for k-valued propositional calculi with $k > 2$ as the two-element Boolean algebra plays for the two-element propositional calculus: the term operations of the Post algebras of order k are precisely the truth-value functions of the k-valued propositional calculus. Traczyk ([116]) investigated the variety generated by the Post algebra of order k, and gave a description of this variety by axioms.

Another useful example of primal algebras is given in the next theorem.

Theorem 10.5.5 *For every prime number p, the prime field modulo p is primal.*

Proof: The prime field modulo p is up to isomorphism the algebra $\mathcal{Z}_p = (Z_p; +, -, \cdot, 0, e)$ of type $\tau = (2, 1, 2, 0, 0)$ where Z_p is the set of the residue classes modulo p. We can use the fundamental operations of $\mathcal{Z}_p$ to construct, for each $k \in Z_p$, the unary operation χ_k defined by $\chi_k(x) := e - (x - k)^{p-1}$. These operations satisfy the following property:

$$\chi_k(x) = \begin{cases} e & \text{if } x = k \\ 0 & \text{otherwise.} \end{cases}$$

This is because $\chi_k(k) = e - (k - k)^{p-1} = e$, while for all $x \neq k$, we have $(x - k)^{p-1} = e$ (the multiplicative group of $\mathcal{Z}_p$ has order $p - 1$), and so $\chi_k(x) = e - e = 0$ for $x \neq k$. Now by Theorem 10.4.1 the fundamental operations together with these terms χ_k for each $k \in Z_p$ generate the whole term clone of the algebra, making $\mathcal{Z}_p$ primal. ∎

Note that the Baker-Pixley Theorem, Theorem 9.2.6, also gives a characterization of primal algebras. It can be formulated as follows:

Theorem 10.5.6 *A finite algebra $\mathcal{A}$ is primal iff there is a majority term which induces a term operation on $\mathcal{A}$ and $\mathcal{A}^2$ has only itself and the diagonal Δ_A as subalgebras.* ∎

Primal and functionally complete algebras have the following properties:

Proposition 10.5.7
(i) Let $\mathcal{A}$ be a primal algebra. Then
    1. $\mathcal{A}$ has no proper subalgebras,
    2. $\mathcal{A}$ has no non-identical automorphisms,
    3. $\mathcal{A}$ is simple,   and
    4. $\mathcal{A}$ generates an arithmetical variety.
(ii) Every functionally complete algebra is simple.

Proof: (i) Since every operation defined on the set A is a term operation of the algebra $\mathcal{A}$, for each proper subset $B \subset A$, each non-identical permutation φ on A and each non-trivial equivalence relation $\theta \subseteq A^2$ there exist term operations t_1^A, t_2^A, t_3^A of $\mathcal{A}$ such that

$t_1^A(b_1, \ldots, b_n) \notin B$ for some elements $b_1, \ldots, b_n \in B$,

$\varphi(t_2^A(a_1, \ldots, a_n)) \neq t_2^A(\varphi(a_1), \ldots, \varphi(a_n))$ for some $a_1, \ldots, a_n \in A$, and

$(t_3^A(a_1, \ldots, a_n), t_3^A(b_1, \ldots, b_n)) \notin \theta$ for some pairs $(a_i, b_i) \in \theta, i = 1, \ldots, n$.

The variety $V(\mathcal{A})$ is then arithmetical since there is a term q satisfying the identities of Theorem 9.3.2 (iii) which induces a term operation on $\mathcal{A}$.

(ii) Let $\mathcal{A}$ be functionally complete. Then by Lemma 10.5.2 $\mathcal{A}^+$ is primal and therefore (by part (i)) simple. But then $\mathcal{A}$ is also simple, since the constant operations $c_a, a \in A$ preserve all equivalence relations on A. ∎

It turns out that the four properties of primal algebras given in Proposition 10.5.7 are sufficient to characterize primal algebras. An additional characterization was given by H. Werner in [118] (see also A. F. Pixley, [87]), using the so-called ternary discriminator term:

$$t(x, y, z) = \begin{cases} z & \text{if } x = y \\ x & \text{otherwise.} \end{cases}$$

Theorem 10.5.8 *For a finite algebra $\mathcal{A}$ the following propositions are equivalent:*

(i) $\mathcal{A}$ is primal.

(ii) $\mathcal{A}$ generates an arithmetical variety, has no non-identical automorphisms, has no proper subalgebras and is simple.

(iii) There is a ternary discriminator term which induces a term operation on $\mathcal{A}$, and the algebra $\mathcal{A}$ has no proper subalgebras and no non-identical automorphisms.

Proof: (i) $\Rightarrow$ (iii): This is clear since when $\mathcal{A}$ is primal every operation, including the ternary discriminator, induces a term operation on $\mathcal{A}$. The other two properties were proved in the previous theorem.

(iii) $\Rightarrow$ (ii). The ternary discriminator term satisfies the identities

$$t(x, x, y) \approx t(x, y, x) \approx t(y, y, x) \approx x,$$

making the variety $V(\mathcal{A})$ arithmetical by Theorem 9.3.2. To show that $\mathcal{A}$ is simple, let $\theta \neq \Delta_A$ be a congruence relation of $\mathcal{A}$. Then there are elements $a, b \in A, a \neq b$ with $(a, b) \in \theta$. But for every $c \in A$ we have $(t(a, b, c), t(a, a, c)) = (a, c) \in \theta$, and therefore $\theta = A^2$.

(ii) $\Rightarrow$ (i): By Theorem 9.3.2 there is a majority term operation in $\mathcal{A}$; so by Theorem 10.5.6 it will suffice to show that $\mathcal{A}^2$ has only itself and Δ_A as subalgebras. Suppose that $\mathcal{B}$ is a proper subalgebra of $\mathcal{A}^2$; we will show in several steps that $\mathcal{B}$ must equal Δ_A.

Step 1. We will show that there are subalgebras $\mathcal{B}_1$ and $\mathcal{B}_2$ of $\mathcal{A}$, congruence relations $\theta_1 \in Con\mathcal{B}_1$ and $\theta_2 \in Con\mathcal{B}_2$ and an isomorphism $\varphi : \mathcal{B}_1/\theta_1 \to \mathcal{B}_2/\theta_2$ such that the universe of $\mathcal{B}$ can be written as

$$B = \bigcup \{X \times \varphi(X) \mid X \in B_1/\theta_1\}.$$

To show this, we also need several steps.

Step 1a). We show first that B has the following property, which is called rectangularity:

$$\text{If } (a, d), (b, d), (b, c) \in B \text{ then } (a, c) \in B.$$

Let $p_j : \mathcal{B} \to \mathcal{A}$, for $j = 1, 2$ be the projection homomorphisms. Then we have

$$((a, d), (b, d)) \in kerp_2 \quad \text{and} \quad ((b, d), (b, c)) \in kerp_1,$$

and thus $((a, d), (b, c)) \in kerp_1 \circ kerp_2$. Since $V(\mathcal{A})$ is an arithmetical and therefore congruence permutable variety, this composition is equal to $kerp_2 \circ kerp_1$. But this means there exists a pair $(y, z) \in B$ with $((a, d), (y, z)) \in kerp_1$ and $((y, z), (b, c)) \in kerp_2$. From this we have $a = y$ and $z = c$, giving $(a, e) \in B$.

Step 1b). Using the projection homomorphisms we define our algebras and congruences:

$$\mathcal{B}_1 := p_1(\mathcal{B}) \text{ and } \mathcal{B}_2 := p_2(\mathcal{B}),$$
$$\theta_1 := \{(a, b) \in A^2 \mid \exists d \in A((a, d), (b, d)) \in B\},$$
$$\theta_2 := \{(d, e) \in A^2 \mid \exists a \in A((a, d), (a, e)) \in B\}.$$

Now we must show that θ_1 and θ_2 are congruences on $\mathcal{B}_1$ and $\mathcal{B}_2$ respectively. By definition $\theta_1 \subseteq B_1^2$ and θ_1 is both symmetric and reflexive. For transitivity, suppose that $(a, b), (b, c) \in \theta_1$. Then there are elements d and e in A with (a, d), (b, d), (b, e) and (c, e) all in B. From the rectangularity of Step 1a) it follows that $(a, e) \in B$, and from this we get $(a, c) \in \theta_1$. This shows that θ_1 is an equivalence relation on B_1. For the congruence property, let f be an n-ary operation symbol of the language of $\mathcal{A}$ and let (a_1, b_1), ..., (a_n, b_n) be in θ_1. Then there exist elements $d_1, \ldots, d_n \in A$ such that (a_1, d_1), (b_1, d_1), ..., (a_n, d_n) and (b_n, d_n) are all in B. Since $\mathcal{B}$ is a subalgebra of $\mathcal{A}^2$, we have
$$f^{\mathcal{A}^2}((a_1, d_1), \ldots, (a_n, d_n)) = (f^{\mathcal{B}_1}(a_1, \ldots, a_n), f^{\mathcal{B}_2}(d_1, \ldots, d_n)) \in B \quad \text{and}$$
$$f^{\mathcal{A}^2}((b_1, d_1), \ldots, (b_n, d_n)) = (f^{\mathcal{B}_1}(b_1, \ldots, b_n), f^{\mathcal{B}_2}(d_1, \ldots, d_n)) \in B,$$
and therefore $(f^{\mathcal{B}_1}(a_1, \ldots, a_n), f^{\mathcal{B}_2}(b_1, \ldots, b_n)) \in \theta_1$. Therefore θ_1 is a congruence relation on $\mathcal{B}_1$. In the same way, θ_2 is a congruence relation on $\mathcal{B}_2$.

Step 1c). Now we define a mapping $\varphi : \mathcal{B}_1/\theta_1 \to \mathcal{B}_2/\theta_2$, by $[a]_{\theta_1} \mapsto [d]_{\theta_2}$ for all $(a, d) \in B$. We will show in this step that φ is an isomorphism.

First we check that φ is well defined. If $[a]_{\theta_1} = [a']_{\theta_1}$, and (a, d) is in B, then $(a, a') \in \theta_1$ and so (a, e) and $(a', e) \in B$ for some $e \in A$. Then by definition of θ_2, having both (a, d) and (a, e) in B means that $(d, e) \in \theta_2$. From this we have $\varphi([a]_{\theta_1}) = [d]_{\theta_2} = [e]_{\theta_2} = \varphi([a']_{\theta_1})$.
To see that the mapping φ is injective, suppose that $[d]_{\theta_2} = \varphi([a]_{\theta_1}) = \varphi([a']_{\theta_1}) = [e]_{\theta_2}$. Then (a, d) and (a', e) are in B, and $(d, e) \in \theta_2$ means that (c, d) and (c, e) are in B for some $c \in A$. Then (a, c) and (a', c) are both in θ_1. By symmetry and transitivity of θ_1 we get $(a, a') \in \theta_1$, and so $[a]_{\theta_1} = [a']_{\theta_1}$.

The mapping φ is also surjective, since for every $d \in B_2$ there exists an $a \in A$ with $(a, d) \in B$. Finally, we show that φ is a homomorphism. Let f be an n-ary operation symbol and assume that $\varphi([a_i]_{\theta_1}) = [d_i]_{\theta_2}$, so $(a_i, d_i) \in B$, for $i = 1, \ldots, n$. Since $\mathcal{B}$ is a subalgebra of $\mathcal{A}^2$,

$$\varphi([f^{\mathcal{B}_1}(a_1, \ldots, a_n)]_{\theta_1}) = [f^{\mathcal{B}_2}(d_1, \ldots, d_n)]_{\theta_2}$$

and

$$f^{\mathcal{B}_2}(\varphi([a_1]_{\theta_1}), \ldots, \varphi([a_n]_{\theta_1})) = f^{\mathcal{B}_2}([d_1]_{\theta_2}, \ldots, [d_n]_{\theta_2}).$$

Step 1d). Now we show that $B = \bigcup\{X \times \varphi(X) \mid X \in B_1/\theta_1\}$. First, let $(a, d) \in B$, so that $\varphi([a]_{\theta_1}) = [d]_{\theta_2}$. Taking $X := [a]_{\theta_1}$, we have $(a, d) \in X \times \varphi(X)$. This shows $B \subseteq \bigcup\{X \times \varphi(X) \mid X \in B_1/\theta_1\}$. For the opposite inclusion, assume now that $(a, d) \in X \times \varphi(X)$ for $X \in B_1/\theta_1$, say $X = [b]_{\theta_1}$ and $\varphi(X) = [e]_{\theta_2}$ with $(b, e) \in B$. Then $(a, d) \in [b]_{\theta_1} \times [e]_{\theta_2}$, so that $(a, b) \in \theta_1$ and $(d, e) \in \theta_2$. The definitions of θ_1 and of θ_2 respectively give us elements d' and $a' \in A$ for which (a, d'), (b, d'), (a', d) and (a', e) are in B, and by the rectangularity property from Step 1a) we have $(b, d) \in B$ and $(a, d) \in B$. This means that $X \times \varphi(X) \in B$ and $\bigcup \{X \times \varphi(X) \mid X \in B_1/\theta_1\} \subseteq B$. Altogether we have equality.

Step 2. Next we show that every non-empty proper subalgebra B of $\mathcal{A}^2$ has a universe of the form $\{(a, \varphi(a)) \mid a \in A\}$, where φ is an automorphism of $\mathcal{A}$.

Combining the result of Step 1 with our assumption that $\mathcal{A}$ has no proper subalgebras, we see that $B_1 = B_2 = A$. This means that θ_1 and θ_2 are actually congruences on $\mathcal{A}$, and since $\mathcal{A}$ is simple we see that θ_1 and θ_2 must each be one of Δ_A or A^2. If $\theta_1 = A^2$, then the isomorphism $\varphi : \mathcal{A}/\theta_1 \to \mathcal{A}/\theta_2$ shows that $\theta_2 = A^2$ also, and in this case we have $B = A^2$. We also have $B = \{(a, \varphi(a)) \mid a \in A\}$. If instead $\theta_1 = \Delta_A$, then the isomorphism φ shows that θ_2 must also equal Δ_A. Again B has the form $\bigcup \{\{a\} \times \varphi(\{a\}) \mid a \in A\}$, so $B = \{(a, \varphi(a)) \mid a \in A\}$.

Step 3. Now we apply the fact that $\mathcal{A}$ has no non-identical automorphism to the result of Step 2, to conclude that the only subalgebra of $\mathcal{A}^2$ is Δ_A. As we remarked above, this is enough to complete our proof. ∎

We point out that for the proof of the claim of Step 1 we used only congruence permutability. We will make use of this fact later, in the proof of Theorem 10.6.5.

For functionally complete algebras we have the following simpler characterization, based on the equivalent conditions of Theorem 10.5.8.

Corollary 10.5.9 *A finite algebra $\mathcal{A}$ is functionally complete iff the ternary discriminator t, with*

$$t(x, y, z) = \begin{cases} z & \text{if } x = y \\ x & \text{otherwise,} \end{cases}$$

induces a polynomial operation on $\mathcal{A}$.

Proof: First, we note that if $\mathcal{A}$ is functionally complete then every operation defined on the set A, including the ternary discriminator, is a polynomial operation of $\mathcal{A}$. For the converse we use Lemma 10.5.2, which tells us that it suffices to prove that the related algebra $\mathcal{A}^+$ is primal. We show the primality of this algebra by verifying that the conditions of Theorem 10.5.8 hold for it. The algebra $\mathcal{A}^+$ has no proper subalgebras, since all elements of A are nullary term operations and must be included in any subalgebra. For the same reason $\mathcal{A}^+$ admits only the identical automorphism. Since the ternary discriminator term t is a polynomial operation of $\mathcal{A}$ it is a term operation of $\mathcal{A}^+$. The algebra $\mathcal{A}$ is then simple: if θ is a congruence on $\mathcal{A}$ which contains a pair (a, b) with $a \neq b$, using (a, b), (a, a) and (c, c) gives $(c, b) = (t^{\mathcal{A}}(a, a, c), t^{\mathcal{A}}(b, a, c)) \in \theta$ for any $c \in A$, so that $\theta = A^2$. But then the algebra $\mathcal{A}^+$ is also simple. By Theorem 10.5.8, $\mathcal{A}^+$ is primal and so $\mathcal{A}$ is functionally complete. ∎

Remark 10.5.10 We leave it as an exercise to show that another proof of Corollary 10.5.9 may be obtained by using Theorem 10.4.1, with the following polynomials:

$$x + y := t(x, 0, y),$$
$$x \cdot y := t(y, 1, x),$$
$$\chi_0(x) := t(y, 1, x),$$
$$\chi_a(x) := t(t(0, a, x), 0, 1), \text{ for } a \in A \setminus \{0\}.$$

R. McKenzie proved in [78] a criterion for functional completeness of an algebra $\mathcal{A}$ such that $V(\mathcal{A})$ is congruence permutable, using the following concept:

Definition 10.5.11 An algebra $\mathcal{A} = (A; F^{\mathcal{A}})$ is said to be *affine with respect to an abelian group* if there is an abelian group $(A; +, 0)$ such that for every $n \geq 1$ and every n-ary term operation $t^{\mathcal{A}}$ of $\mathcal{A}$, and for every pair of n-tuples $((a_1, \ldots, a_n), (b_1, \ldots, b_n))$ of elements from A, we have

$$t^{\mathcal{A}}(a_1, \ldots, a_n) + t^{\mathcal{A}}(b_1, \ldots, b_n) - t^{\mathcal{A}}(0, \ldots, 0) = t^{\mathcal{A}}(a_1 + b_1, \ldots, a_n + b_n).$$

Theorem 10.5.12 *Let $\mathcal{A}$ be a finite non-trivial algebra which generates a congruence permutable variety. Then $\mathcal{A}$ is functionally complete iff $\mathcal{A}$ is simple but is not affine with respect to any elementary abelian p-group.*

This theorem can be proved very easily using Rosenberg's description of all maximal classes of operations defined on a finite set by relations, and we leave it as an exercise for the reader.

There is an especially simple characterization of primal algebras with only one at least binary fundamental operation given by G. Rousseau in [104].

Theorem 10.5.13 *([104]) A finite non-trivial algebra $\mathcal{A} = (A; f^A)$ with one single fundamental operation f^A which is at least binary is primal if and only if $\mathcal{A}$ is simple, has no proper subalgebra and has no non-identical automorphisms.* ∎

We can look for properties of a variety $V(\mathcal{A})$ generated by a primal algebra $\mathcal{A}$. In particular, we want to know what the subdirectly irreducible algebras in $\mathcal{A}$ are, and how $V(\mathcal{A})$ is located in the lattice of all varieties of the type of $\mathcal{A}$. Using our results on primal algebras, we have the following answers.

Corollary 10.5.14 *If $\mathcal{A}$ is primal, then $V(\mathcal{A})$ has no non-trivial subvarieties and $\mathcal{A}$ is the only subdirectly irreducible algebra in $V(\mathcal{A})$.* ∎

10.6 Different Generalizations of Primality

A finite algebra $\mathcal{A}$ is primal when every operation which is definable on its universe set A is a term operation of the algebra $\mathcal{A}$. In this section we consider several variations and generalizations of primality. All have the same basic definition: instead of requiring that all operations are term operations, we require only that certain operations, having some common property, are all term operations. The concept of functional completeness can be similarly weakened.

Definition 10.6.1 A finite non-trivial algebra $\mathcal{A} = (A; F^A)$ is called *semiprimal* if every operation on A which preserves all subalgebras of $\mathcal{A}$ is a term operation of $\mathcal{A}$.

We recall that an n-ary operation $f : A^n \to A$ preserves a subalgebra $\mathcal{B} \subseteq \mathcal{A}$ of $\mathcal{A}$ if $f(b_1, \ldots, b_n) \in B$ for all $b_1, \ldots, b_n \in B$. If $\mathcal{A}$ is semiprimal and has precisely one proper subalgebra it is called *subprimal*. A further distinction is made based on the cardinality of the subalgebra of $\mathcal{A}$: the algebra $\mathcal{A}$ is called *regular subprimal* if the cardinality of this subalgebra is greater than 1 and *singular subprimal* otherwise.

Operations on an algebra $\mathcal{A}$ can also preserve, or be compatible with, homomorphisms and congruence relations of $\mathcal{A}$. We say that $f : A^n \to A$ is compatible with an endomorphism (or isomorphism) $\varphi : \mathcal{A} \to \mathcal{B}$ if $\varphi(f(a_1, \ldots, a_n)) = f(\varphi(a_1), \ldots, \varphi(a_n))$ for all $a_1, \ldots, a_n \in A$. As a consequence of the General Homomorphism Theorem, we know that operations f which preserve all homomorphisms of $\mathcal{A}$ have the property that for all $\theta \in Con\mathcal{A}$ and for all $(a_1, b_1), \ldots, (a_n, b_n) \in \theta$, the pair $(f(a_1, \ldots, a_n), f(b_1, \ldots, b_n))$ is in θ.

Definition 10.6.2 A finite non-trivial algebra $\mathcal{A} = (A; F^A)$ is called *demiprimal* if $\mathcal{A}$ has no proper subalgebras and every operation from $O(A)$ which preserves all automorphisms of $\mathcal{A}$ is a term operation of $\mathcal{A}$.

Definition 10.6.3 A finite non-trivial algebra $\mathcal{A} = (A; F^A)$ is called *quasiprimal* if every operation on A which preserves all subalgebras and all isomorphisms between non-trivial subalgebras of $\mathcal{A}$ is a term operation of $\mathcal{A}$.

Definition 10.6.4 A finite non-trivial algebra $\mathcal{A} = (A; F^A)$ is called *hemiprimal* if every operation on A which preserves all congruence relations on $\mathcal{A}$ is a term operation of $\mathcal{A}$. The algebra $\mathcal{A}$ is called *affine complete* if every such operation is a polynomial operation of $\mathcal{A}$.

Clearly, demiprimal and semiprimal algebras are examples of quasiprimal algebras. All of these types of algebras can be determined using relations, in the following sense. In each case there is a set of relations $R \subseteq A^2$ such that $clone\mathcal{A}$ is precisely the clone $PolR$.

There are two other kinds of algebras which are defined in a different way. A finite non-trivial algebra is called *cryptoprimal* if $\mathcal{A}$ is simple, has no proper subalgebra and generates a congruence distributive variety. Also, $\mathcal{A}$ is called *paraprimal* if each subalgebra of $\mathcal{A}$ is simple and $\mathcal{A}$ generates a congruence permutable variety.

We now present some characterizations of these various kinds of algebras, beginning with quasiprimal algebras.

Theorem 10.6.5 *For a finite algebra $\mathcal{A} = (A; F^A)$ the following conditions are equivalent:*

(i) *$\mathcal{A}$ is quasiprimal.*

(ii) *The ternary discriminator operation induces a term operation of $\mathcal{A}$.*

(iii) *Each subalgebra of $\mathcal{A}$ is simple and there is a term q in the language of $\mathcal{A}$ such that $q(y, y, x) \approx q(x, y, y) \approx q(x, y, x) \approx x$ are identities in $\mathcal{A}$.*

Proof. (i) $\Rightarrow$ (ii): Since the ternary discriminator t of A is an operation which preserves all subalgebras of $\mathcal{A}$ and all isomorphisms between non-trivial subalgebras of $\mathcal{A}$, it is a term operation of $\mathcal{A}$ when $\mathcal{A}$ is quasiprimal.

(ii) $\Rightarrow$ (iii): Let $\theta \neq \Delta_B$ be a non-trivial congruence relation of a non-trivial subalgebra $\mathcal{B}$ of $\mathcal{A}$. Then there is a pair (a, b) in θ with $a \neq b$. For any element $c \in B$, the ternary discriminator properties give $(t(a, b, c), t(a, a, c)) \in \theta$. Hence $(a, c) \in \theta$ for every c in B, making $\theta = B^2$. The ternary discriminator term also satisfies the identities $t(x, x, z) \approx t(z, x, x) \approx t(z, x, z) \approx z$.

(iii) $\Rightarrow$ (i): By Theorem 9.3.2 the variety $V(\mathcal{A})$ is arithmetical and there is a majority term operation in $\mathcal{A}$. Therefore we can apply Theorem 9.2.6. The variety $V(\mathcal{A})$ is congruence permutable since it is arithmetical, and the claim from Step 1 in the proof of Theorem 10.5.8 is satisfied. Thus there are subalgebras $\mathcal{B}_1$ and $\mathcal{B}_2$ of $\mathcal{A}$, congruence relations $\theta_1 \in Con\mathcal{B}_1$ and $\theta_2 \in Con\mathcal{B}_2$ and an isomorphism $\varphi : B_1/\theta_1 \to B_2/\theta_2$ such that the universe of $\mathcal{B}$ can be written as $B = \bigcup\{X \times \varphi(X) | X \in B_1/\theta_1\}$. Since by assumption each subalgebra of $\mathcal{A}$ is simple, we have $\theta_i \in \{\Delta_B, B_i^2\}$, for $i = 1, 2$. But this means we have either isomorphisms between the subalgebras $B_i = B_i/\Delta_B$, or isomorphisms between trivial (one-element) algebras B_i/B_i^2. Theorem 9.2.6 then shows that every operation on A which preserves all subalgebras and all isomorphisms between non-trivial subalgebras of $\mathcal{A}$ is a term operation of $\mathcal{A}$, and thus $\mathcal{A}$ is quasiprimal. ∎

Semiprimal algebras can be characterized as follows:

Theorem 10.6.6 *A finite non-trivial algebra $\mathcal{A} = (A; F^A)$ is semiprimal iff the following conditions are all satisfied:*

(i) *$V(\mathcal{A})$ is arithmetical;*

(ii) Every non-trivial subalgebra of $\mathcal{A}$ is simple;

(iii) Every subalgebra of $\mathcal{A}$ has no non-identical automorphisms; and

(iv) No two distinct subalgebras with more than one element are isomorphic.

Proof. First suppose that $\mathcal{A}$ is semiprimal. Then $\mathcal{A}$ is also quasiprimal and Theorem 10.6.5 can be applied. Consequently, by Theorem 9.3.2 the variety $V(\mathcal{A})$ is arithmetical and every non-trivial subalgebra of $\mathcal{A}$ is simple. From the definition of semiprimal algebras it follows that a non-trivial subalgebra of a semiprimal algebra is semiprimal as well. Suppose that φ is a non-identical automorphism of $\mathcal{A}$, so that there are distinct elements a and b of A with $\varphi(a) = b \neq a$. Since φ is injective, this also means that $\varphi(b) = b' \neq b$. Consider the following operation f defined on A: we set $f(x,y) = x$ if $x \neq a$ or $y \neq b$ and $f(a,b) = b$. This operation preserves all subalgebras of $\mathcal{A}$ and therefore it is a term operation of $\mathcal{A}$. Then $f(\varphi(a), \varphi(b)) = f(b, b') = b$ while $\varphi(f(a,b)) = \varphi(b) = b'$. This shows that $\mathcal{A}$ cannot have non-identical automorphisms. In the same way we can show that every non-trivial subalgebra of $\mathcal{A}$ has no non-identical automorphisms.

To show (iv), let $\mathcal{S}_1$ and $\mathcal{S}_2$ be two distinct isomorphic subalgebras of $\mathcal{A}$. Since $\mathcal{A}$ is finite, no proper subalgebra of $\mathcal{S}_2$ can be isomorphic to $\mathcal{S}_2$, and we may suppose that the set $S_1 \setminus S_2$ is non-empty. Let $a \in S_1 \setminus S_2$, and let $b \in S_2$ be the image of a under the isomorphism φ. Furthermore, let $c \neq a$ be any other element of S_1 and let $\varphi(c) = d \in S_2$. By injectivity, $d \neq b$. Obviously, the operation f defined by $f(x,y) = x$ if $x = a$ and $f(x,y) = y$ otherwise is a term operation of $\mathcal{A}$; but we have $f(\varphi(a), \varphi(c)) = f(b,d) = d$ yet $\varphi(f(a,c)) = \varphi(a) = b$, which is a contradiction. This completes the first direction of the proof.

Now we will prove that from conditions (i) - (iv) the semiprimality of $\mathcal{A}$ can be deduced. By Theorem 10.6.5 the algebra $\mathcal{A}$ is quasiprimal. There are no isomorphisms between non-trivial subalgebras of $\mathcal{A}$, and therefore each operation preserving all subalgebras of $\mathcal{A}$ is a term operation of $\mathcal{A}$. Consequently, $\mathcal{A}$ is indeed semiprimal. ∎

Clearly, a semiprimal algebra with no proper subalgebra is primal. A semiprimal algebra $\mathcal{A}$ which is not primal has at least one minimal subalgebra, that is, a subalgebra which itself has no proper subalgebras. It can also be shown that each minimal subalgebra of a semiprimal algebra is either primal or is a one-element subalgebra. If $\mathcal{A}$ has more than one minimal subalgebra, any two of them are disjoint.

The next theorem characterizes demiprimal algebras.

Theorem 10.6.7 *A finite non-trivial algebra is demiprimal iff the following conditions are satisfied:*

(i) *$V(\mathcal{A})$ is arithmetical;*

(ii) *$\mathcal{A}$ has no proper subalgebra;*

(iii) *$\mathcal{A}$ is simple;   and*

(iv) *Every non-identical automorphism of $\mathcal{A}$ has no fixed points.*

Proof. If $\mathcal{A}$ is demiprimal then it is also quasiprimal. Therefore, by Theorem 10.6.5 $\mathcal{A}$ is simple and $V(\mathcal{A})$ is arithmetical. Clearly, the set of all fixed points of an automorphism forms a subalgebra of $\mathcal{A}$. But a demiprimal algebra has no proper subalgebra, so we have (iv).

Conversely, from conditions (i) - (iv) and Theorem 10.6.5 we see that $\mathcal{A}$ is quasiprimal. Since $\mathcal{A}$ has no proper subalgebras, each isomorphism between non-trivial subalgebras of $\mathcal{A}$ is an automorphism of $\mathcal{A}$, and $\mathcal{A}$ is demiprimal. ∎

The following characterization of hemiprimal algebras is due to A. F. Pixley, from [88].

Theorem 10.6.8 *Let $\mathcal{A} = (A; F^A)$ be a finite non-trivial algebra, for which $Con\mathcal{A}$ is arithmetical. Then $\mathcal{A}$ is hemiprimal iff the following conditions are satisfied:*

(i) *$V(\mathcal{A})$ is arithmetical;*

(ii) *$\mathcal{A}$ has no proper subalgebra;   and*

(iii) *If θ_1, θ_2 are congruences of $\mathcal{A}$ then any isomorphism $\varphi : \mathcal{A}/\theta_1 \to \mathcal{A}/\theta_2$ is the identity mapping, and in particular $\theta_1 = \theta_2$.* ∎

For more results on affine complete and functionally complete algebras we refer the reader to the work of A. F. Pixley in [89]. Using the properties of semi-, demi-, and hemiprimal algebras and Theorem 6.5.7 we obtain the following structure theorem:

Theorem 10.6.9 *Let $\mathcal{A}$ be a finite non-trivial algebra and let $V(\mathcal{A})$ be the variety generated by $\mathcal{A}$. Then:*

(i) *If $\mathcal{A}$ is semiprimal then $V(\mathcal{A}) = \mathbf{IP}_s\mathbf{S}(\mathcal{A})$. (If $\mathcal{A}$ is regular subprimal then $V(\mathcal{A}) = \mathbf{IP}_s\{\mathcal{A}, \mathcal{B}\}$ where $\mathcal{B}$ is the only non-trivial subalgebra of $\mathcal{A}$, and if $\mathcal{A}$ is singular subprimal then $V(\mathcal{A}) = \mathbf{IP}_s\mathbf{S}(\mathcal{A})$.)*

(ii) *If $\mathcal{A}$ is hemiprimal then $V(\mathcal{A}) = \mathbf{IP}_s(\mathcal{A})$.*

(iii) *If $\mathcal{A}$ is demiprimal then $V(\mathcal{A}) = \mathbf{IP}_s(\mathcal{A})$. When $\mathcal{A}$ is singular subprimal or demiprimal, the algebra $\mathcal{A}$ is the only subdirectly irreducible algebra in $V(\mathcal{A})$ and $V(\mathcal{A})$ has no non-trivial subvariety (so $V(\mathcal{A})$ is a minimal variety).* ■

10.7 Preprimal Algebras

Definition 10.7.1 A finite non-trivial algebra $\mathcal{A}$ is called *preprimal* if $clone\mathcal{A}$ is one of the maximal clones described in Section 6.4.

By Corollary 1.3.14, the clone M is a maximal subclone of $O(A)$ if for every $f \in O(A) \setminus M$, the clone $\langle M \cup \{f\}\rangle_{O(A)}$ generated by $M \cup \{f\}$ equals the whole clone $O(A)$. Let $\mathcal{A}$ be a preprimal algebra. If not all the nullary operations defined on the set A are term operations of $\mathcal{A}$, then $\mathcal{A}$ is functionally complete. If however all the nullary operations are term operations of the algebra, then $\mathcal{A}$ admits no non-identical automorphisms and has no proper subalgebras. Functionally complete preprimal algebras are simple. By Post's classification of all two-element algebras ([95]), there exist exactly five preprimal two-element algebras. These are :

$$C_3 = (\{0,1\}; \vee, g_2), \text{ for } g_2 := x \wedge \neg y,$$
$$\mathcal{D}_3 = (\{0,1\}; (x \wedge y) \vee (x \wedge z) \vee (y \wedge z), x + y + z, \neg),$$
$$C_2 = (\{0,1\}; \vee, x + y + 1),$$
$$\mathcal{A}_1 = (\{0,1\}; \wedge, \vee, c_0^2, c_1^2),$$

$$\mathcal{L}_1 = (\{0,1\}; +, c_1^2).$$

Here $\mathcal{C}_3$ and $\mathcal{C}_2$ are dually isomorphic, and $\mathcal{C}_3$, $\mathcal{C}_2$ and $\mathcal{D}_3$ are functionally complete.

We will use the following concept defined by A. L. Foster ([46]). An element $a \in A$ is called a *centroid element* of $\mathcal{A}$ if there exists a unary term operation c_a of $\mathcal{A}$ with $c_a(x) = a$ for all $x \in A$. The set $C(\mathcal{A})$ of all centroid elements of $\mathcal{A}$ is called the *centroid* of $\mathcal{A}$. An algebra is called *constantive* if $C(\mathcal{A}) = \mathcal{A}$. Then we have

Theorem 10.7.2 *A preprimal algebra $\mathcal{A}$ is either quasiprimal or constantive.*

Proof. Let $\mathcal{A}$ be preprimal. Clearly, every constant operation preserves the diagonal relation Δ_A. By definition $clone \mathcal{A}$ is a maximal subclone of $O(A)$, and is determined by one of the relations described in Rosenberg's list from Section 10.4. Moreover, it is known that constantive preprimal algebras correspond to relations which are reflexive or totally reflexive. Then bounded partial order relations (type (ii) in the list from Section 10.4), non-trivial equivalence relations (type (iv)) and central relations for $h = 2$ (type (v)) are reflexive; and central relations for $h \geq 3$ (type (v)), t-universal relations (type (vi)) and the relations of type (iii) are totally reflexive. Preprimal algebras corresponding to relations of type (i) are demiprimal; preprimal algebras corresponding to unary central relations are semiprimal (subprimal). Therefore, these algebras are quasiprimal. There are no other kinds of preprimal algebras, showing that every preprimal algebra is either constantive or quasiprimal.   ∎

As a consequence, we see that preprimal algebras are functionally complete iff they are quasiprimal.

Preprimal algebras corresponding to type (iv) are hemiprimal with exactly one non-trivial congruence relation. It was shown by K. Denecke in [20] that such algebras generate arithmetical varieties. It is also easy to see that every preprimal algebra which does not correspond to a relation of type (iv) is simple. Every preprimal algebra except those corresponding to type (v) for $h = 1$ has no non-trivial subalgebra, and every preprimal algebra which does not correspond to type (i) has no non-trivial automorphisms. For algebras generating arithmetical varieties we have the following result.

Theorem 10.7.3 *A non-trivial algebra $\mathcal{A}$ which generates an arithmetical variety is preprimal if and only if one of the following mutually exclusive cases is satisfied:*

Case 1.   *(i′) $\mathcal{A}$ has exactly one proper subalgebra $\mathcal{B}$, which is either primal or trivial,*
      (ii) $\mathcal{A}$ has no non-identical automorphisms,   and
      (iii) $\mathcal{A}$ is simple.

Case 2.   *(i) $\mathcal{A}$ has no proper subalgebras,*
      (ii′) The automorphism group of $\mathcal{A}$ is cyclic and of prime order,   and
      (iii) $\mathcal{A}$ is simple.

Case 3.   *(i) $\mathcal{A}$ has no proper subalgebras,*
      (ii) $\mathcal{A}$ has no non-trivial automorphisms,   and
      (iii′) $\mathcal{A}$ has one non-trivial homomorphic image, and this homomorphic image is a primal algebra.

Proof. "⇒": Assume that $\mathcal{A}$ generates an arithmetical variety and the three conditions of Case 1 are satisfied. Since $\mathcal{A}$ is simple and the only other subalgebra $\mathcal{B}$ of $\mathcal{A}$ is also simple, since it is either primal or trivial, we conclude that all subalgebras of $\mathcal{A}$ are simple. The algebra $\mathcal{A}$ has no non-identical automorphisms and the subalgebra $\mathcal{B}$, as a trivial or primal algebra, also has no non-identical automorphisms; hence all subalgebras of $\mathcal{A}$ have no proper automorphisms. This means that all conditions of Theorem 10.6.6 are satisfied, and $\mathcal{A}$ is semiprimal. Since $\mathcal{A}$ has only one proper subalgebra, by the definition of semiprimality every operation defined on A which preserves the subset $B \subset A$ is a term operation of $\mathcal{A}$, and *clone*$\mathcal{A}$ is a clone of operations preserving a central relation with $h = 1$ (class (v)), making $\mathcal{A}$ preprimal.

Next assume that $\mathcal{A}$ generates an arithmetical variety and that the three conditions of Case 2 are satisfied. Let the automorphism group of $\mathcal{A}$ be the cyclic group of prime order p. We want to apply Theorem 10.6.7. Clearly, Aut$\mathcal{A}$ is generated by an automorphism s of prime order $p \neq 1$. The automorphism s has no fixed points, since the fixed points of an automorphism form a subalgebra but $\mathcal{A}$ has no proper subalgebras, forcing $p = 1$, a contradiction. Therefore $\mathcal{A}$ is demiprimal.

Now assume that $\mathcal{A}$ generates an arithmetical variety and that the conditions of Case 3 are satisfied. Then all conditions of Theorem 10.6.8 are fulfilled; so $\mathcal{A}$ is hemiprimal and $clone\mathcal{A} = Pol\theta$ for a non-trivial congruence relation θ of $\mathcal{A}$. Then $\mathcal{A}$ is preprimal of type (iv).

"$\Leftarrow$" By Theorem 10.7.2 a preprimal algebra is either quasiprimal or constantive. Quasiprimal algebras generate arithmetical varieties, and correspond to class (i) or to class (vi) with $h = 1$. In the first case Case 2 is satisfied, while in the second case Case 1 is satisfied. All preprimal algebras other than those corresponding to class (iv) are simple, since if $\mathcal{A}$ has a non-trivial congruence relation, then $clone\mathcal{A} \subset Pol\theta$ and $\mathcal{A}$ is not preprimal. Preprimal algebras $\mathcal{A}$ with $clone\mathcal{A} = Pol\theta$, for θ non-trivial, generate arithmetical varieties by Theorem 10.6.8. These algebras fit Case 3. No other preprimal algebras generate arithmetical varieties since constantive algebras have no non-identical automorphisms and no proper subalgebras, and since an algebra having no proper subalgebras, no non-identical automorphisms and no non-trivial congruence relations which generates an arithmetical variety is primal by Theorem 10.5.8.   ■

We remark that it was shown by K. Denecke in [20] that preprimal algebras corresponding to class (iii) generate varieties which are congruence permutable but not congruence distributive. Preprimal algebras $\mathcal{A}$ for which $clone\mathcal{A}$ is in class (v) generate 3-distributive but not 2-distributive varieties if $h > 2$, and 2-distributive varieties if $h = 2$. Preprimal algebras for which $clone\mathcal{A}$ is in class (ii) and the relation ϱ is a lattice order generate 2-distributive varieties. We note also that in the last case there are bounded partial orders ϱ such that $Pol\varrho$ is not finitely generated (see G. Tardos, [110]). In this case we do not get a finite algebra in the usual sense of having both a finite universe and finitely many fundamental operations.

Preprimal algebras $\mathcal{A}$ for which $clone\mathcal{A} = Pol\theta$ for some non-trivial equivalence relation θ are the only non-simple preprimal algebras. But such algebras have only one non-trivial congruence relation; so they are subdirectly irreducible. This shows that all preprimal algebras are subdirectly irreducible. The question of describing all other subdirectly irreducible algebras in the variety $V(\mathcal{A})$ generated by a preprimal algebra $\mathcal{A}$ was solved for most classes of preprimal algebras, independently by K. Denecke in [20] and A. Knoebel in [65]. This is closely connected to the problem of describing the subvariety lattice of varieties $V(\mathcal{A})$ generated by preprimal algebras $\mathcal{A}$. The results are summarized in the following table; the proofs may be found in [20] and [65].

Class	Relation	Number of Subvarieties of $V(\mathcal{A})$	Number of Subdirectly Irreducibles
(i)	fixed point free permutation	2	1
(ii)	bounded partial order	2	?
(iii)	elementary abelian p-group	3	2
(iv)	equivalence relation	3	2
(v)	$h = 1$, subset $|B| = 1$	2	1
(v)	$h = 1$, subset $|B| > 1$	3	2
(v)	$h > 2$	2	1
(vi)		5 or 6	finite

10.8 Exercises

10.8.1. Prove directly that the pair (Pol, Inv) of operators introduced in Section 10.2 forms a Galois connection.

10.8.2. Prove Lemma 10.5.2.

10.8.3. Verify that the algebra 3 in Example 10.5.3, is primal.

10.8.4. Prove Remark 10.5.10.

10.8.5. Prove Theorem 10.5.12.

10.8.6. Determine all two-element preprimal algebras which generate an arithmetical variety.

10.8.7. Prove that for preprimal algebras $\mathcal{A}$ corresponding to class (i), the variety $V(\mathcal{A})$ has no non-trivial subvarieties and $\mathcal{A}$ is the only subdirectly irreducible algebra in $V(\mathcal{A})$.

10.8.8. Prove that for preprimal algebras $\mathcal{A}$ corresponding to class (iv), the variety $V(\mathcal{A})$ has only one non-trivial subvariety and two subdirectly irreducible algebras.

10.8.9. Prove that for preprimal algebras $\mathcal{A}$ corresponding to class (v), when $|B| = 1$ the variety $V(\mathcal{A})$ has no non-trivial subvarieties.

10.8.10. Let A be the two-element set $\{0, 1\}$. Prove that an operation $f : A^n \to A$ is a linear Boolean function iff it is in $Pol\rho$, where

$$\rho = \{(y_1, y_2, y_3, y_4) \in A^4 \mid y_1 + y_2 = y_3 + y_4\}.$$

Chapter 11

Tame Congruence Theory

An algebra $\mathcal{A}$ is called finite when its universe A is finite and every fundamental operation is finitary. Finite algebras are important in many areas where finiteness plays a crucial role, for instance in Computer Science. A major area of research activity has been to try to classify all finite algebras of a given type. For instance, the classification of all finite groups has been a longstanding mathematical problem.

In the early 1980's, R. McKenzie and D. Hobby developed a new theory called "Tame Congruence Theory," which offers a structure theory for finite algebras ([57]). In this chapter we present the main ideas and results of this important theory. A central concept in tame congruence theory is the notion of a *minimal algebra*. We will give a classification of all minimal algebras, using the properties of the congruence lattices of the algebras in a variety generated by a minimal algebra.

11.1 Minimal Algebras

Let A and B be sets, let $f : A \to B$ be a mapping and let U be a subset of A. Then the *restriction* of f to the set U is the mapping $f|_U$ from U to B defined by $f|_U(x) := f(x)$ for all $x \in U$. We will be particularly interested in the restrictions of polynomial operations on an algebra $\mathcal{A}$ to certain special subsets of the base set A. We will denote by $P^n(\mathcal{A})$ and $P(\mathcal{A})$ the sets of all n-ary polynomial operations and of all finitary polynomial operations, respectively, of the algebra $\mathcal{A}$. We also use $Eq\, A$ for the lattice of all equivalence relations on a set A.

251

Definition 11.1.1 Let A be a set with U a subset of A, and let $\mathcal{A}$ be an algebra with A as its universe. We define the following restrictions to set U:

(i) For any equivalence relation θ in $Eq\,A$, we set $\theta|_U := \theta \cap U^2$.

(ii) For any n-ary operation g on A, we set $g|_U := g|_{U^n}$.

(iii) We set $P(\mathcal{A})|_U := \bigcup\limits_{n=0}^{\infty} P^n(\mathcal{A})|_U$, and
$$P^n(\mathcal{A})|_U = \{g|_U \mid g \in P^n(\mathcal{A}) \text{ and } g(U^n) \subseteq U\}.$$

(iv) We define $\mathcal{A}|_U$ to be the algebra $(U; P(\mathcal{A})|_U)$.

It is easy to check that when θ is an equivalence relation on A, the relation $\theta|_U$ is also an equivalence relation on the set U, called the *restriction* of θ to U. An operation $g|_U$ is also called the *restriction* of g to U. The algebra $\mathcal{A}|_U$ is called the algebra *induced* on U by $\mathcal{A}$, or the restriction of the algebra $\mathcal{A}$ to the set U.

To obtain our minimal algebras, we want to consider restrictions of algebras to certain special subsets, those which occur as images of mappings, particularly of idempotent polynomial mappings on the algebra.

A mapping $e : A \to A$ is called *idempotent* if $e^2 := e \circ e = e$. For $\mathcal{A}$ an algebra, we shall denote by $E(\mathcal{A})$ the set of all idempotent polynomial operations of $\mathcal{A}$.

Our first theorem of this chapter shows that under certain conditions, the restriction of a congruence on $\mathcal{A}$ to a subset U of A is also a congruence relation, on the induced algebra $\mathcal{A}|_U$.

Theorem 11.1.2 *Let $\mathcal{A}$ be an algebra and let $e \in E(\mathcal{A})$. Let U be the image set $U := e(A)$. Then the mapping $\varphi_U : Con\,\mathcal{A} \to Con(\mathcal{A}|_U)$, defined by $\theta \mapsto \theta|_U$, is a surjective lattice homomorphism.*

Proof: We must show first that if $\theta \in Con\,\mathcal{A}$, then $\theta|_U$ is in $Con(\mathcal{A}|_U)$, so that our mapping φ_U is well defined. Let $(a_1, b_1), \ldots, (a_n, b_n)$ be in $\theta|_U$, so that $(a_1, b_1), \ldots, (a_n, b_n)$ are all in $\theta \cap U^2$. The fundamental operations of the induced algebra $\mathcal{A}|_U$ have the form $f^{\mathcal{A}}|_U$, where $f^{\mathcal{A}}$ is a polynomial operation on $\mathcal{A}$. Then we have

$$(f^{\mathcal{A}|_U}(a_1,\ldots,a_n),\ f^{\mathcal{A}|_U}(b_1,\ldots,b_n))$$
$$= (f^{\mathcal{A}}|_U(a_1,\ldots,a_n),\ f^{\mathcal{A}}|_U(b_1,\ldots,b_n))$$
$$= (f^{\mathcal{A}}(a_1,\ldots,a_n),\ f^{\mathcal{A}}(b_1,\ldots,b_n)),$$

which we claim is in $\theta \cap U$. This is because $P(\mathcal{A})$ is the clone generated by the set $clone\mathcal{A} \cup \{c_a \mid a \in A\}$, and our claim is true for every f in $clone\mathcal{A}$ by Theorem 5.2.4 and obviously true for each constant mapping c_a.

Next we must show that the mapping φ_U is compatible with the lattice operations $\wedge$ (which is just $\cap$) and $\vee$ on the congruence lattices. For $\wedge$, we let θ and ψ be in $Con\mathcal{A}$, and let $a, b \in A$. Then $(a,b) \in \theta|_U \cap \psi|_U \Leftrightarrow (a,b) \in (\theta \cap U^2) \cap (\psi \cap U^2) \Leftrightarrow (a,b) \in \theta \cap \psi \cap U^2 \Leftrightarrow (a,b) \in (\theta \cap \psi)|_U$. It follows that $\varphi_U(\theta) \cap \varphi_U(\psi) = \varphi_U(\theta \cap \psi)$.

For $\vee$, we again take θ and ψ to be in $Con\mathcal{A}$. We have $\theta|_U = \theta \cap U^2 \subseteq (\theta \vee \psi) \cap U^2$ and $\psi|_U = \psi \cap U^2 \subseteq (\theta \vee \psi) \cap U^2$, and thus we have the containment $\theta|_U \vee \psi|_U \subseteq (\theta \vee \psi)|_U$. For the opposite inclusion, let $(a,b) \in (\theta \vee \psi) \cap U^2$. Then by our characterization of joins of congruences in Lemma 9.2.2 we know that there are elements $a_0, a_1, \ldots, a_n \in A$ such that $a = a_0$, $b = a_n$, and $(a_{2i}, a_{2i+1}) \in \theta$ and $(a_{2i+1}, a_{2i+2}) \in \psi$, for $i \geq 0$. Since e is a unary polynomial on $\mathcal{A}$, it preserves the congruences θ and ψ (see Exercise 5.4.7), so we have $(e(a_{2i}), e(a_{2i+1})) \in \theta$ and $(e(a_{2i+1}), e(a_{2i+2})) \in \psi$, for $i \geq 0$.

Now we use the idempotence of our polynomial mapping e, which up till now we have not made use of. This idempotence, and our choice of U as $e(A)$, means precisely that $e|_U$ is the identity mapping on U: for any $x \in U$, we have $x = e(y)$ for some y in A, and then $e(x) = e(e(y)) = e(y) = x$. The elements $e(a_j)$ are all in U, and moreover a and b are in U so $e(a) = a$ and $e(b) = b$. From this, applying our join criterion in the opposite direction now, we conclude that (a,b) is in $\theta|_U \vee \psi|_U$. This gives us the equality $\varphi_U(\theta) \vee \varphi_U(\psi) = \varphi_U(\theta \vee \psi)$ we need for $\vee$.

Finally we show that the mapping φ_U is surjective. Let Φ be any congruence in $Con(\mathcal{A}|_U)$. We define the relation

$$\overline{\Phi} := \{(x,y) \in A^2 \mid \forall f \in P^1(\mathcal{A}),\ ((e(f(x)), e(f(y))) \in \Phi)\},$$

which we shall show is a congruence on $\mathcal{A}$ whose image under φ_U is Φ. The relation $\overline{\Phi}$ is clearly an equivalence relation on A. For the congruence property, we show that $\overline{\Phi}$ is compatible with any unary polynomial operation

g of $\mathcal{A}$, which is sufficient to guarantee a congruence by Theorem 1.4.5. If $(x, y) \in \overline{\Phi}$, then for any unary polynomial operation f on $\mathcal{A}$, the composition operation $f \circ g$ is also a unary polynomial, and so by definition of $\overline{\Phi}$ we have $(e(f \circ g)(x), e(f \circ g)(y)) \in \Phi$. But then $(e(f(g(x))), e(f(g(y)))) \in \Phi$ for all $f \in P^1(\mathcal{A})$, and thus $(g(x), g(y)) \in \overline{\Phi}$.

To finish our proof of surjectivity, we show that $\overline{\Phi}|_U = \Phi$. Let (x, y) be in $\overline{\Phi}|_U$, so that $(x, y) \in \overline{\Phi} \cap U^2$. Taking f to be the unary identity polynomial on $\mathcal{A}$ we get $(e(x), e(y)) \in \Phi$, and the fact that e is the identity mapping on U shows that (x, y) is then in Φ. Conversely, suppose that $(x, y) \in \Phi$. For all $f \in P^1(\mathcal{A})$ we have $e f(U) \subseteq U$, so that $(ef)|_U$ is a fundamental operation of the algebra $\mathcal{A}|_U$. But then $(e(f(x)), e(f(y))) \in \Phi$, and by definition of $\overline{\Phi}$ the pair (x, y) belongs to $\overline{\Phi}$. This gives the containment and hence the equality we need.   ■

Thus we have proved that if e is an idempotent unary polynomial operation on an algebra $\mathcal{A}$, and our subset U is the image of A under e, then every congruence on the induced algebra $\mathcal{A}|_U$ is the restriction of a congruence on $\mathcal{A}$, with a lattice homomorphism between the two congruence lattices. Our next example, while it illustrates the restriction of congruences, also shows that it is not necessary to use the image of an idempotent polynomial.

Example 11.1.3 Consider the algebra $\mathcal{A} = (\{a, b, c, d\}; f)$, where f is defined by the table

x	a	b	c	d
$f(x)$	b	c	c	d

The mapping f is not idempotent, but nevertheless we can form the set $U := f(A) = \{b, c, d\}$, and compare the lattices of congruences on the original algebra $\mathcal{A}$ and the induced algebra $\mathcal{A}|_U$. It is straightforward to work out that lattice $Con\ \mathcal{A}$ has the Hasse diagram shown below, with θ_0 and θ_1 denoting the trivial congruences Δ_A and $A \times A$, respectively.

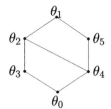

Our induced algebra $\mathcal{A}|_U$ has universe U, and fundamental operation set $P(A)|_U$. There is only one unary term operation on $\mathcal{A}$ which preserves the subset $U = \{b, c, d\}$, namely the term $g := f|_U = (f \circ f)|_U$, given by the table below.

x	b	c	d
$g(x)$	c	c	d

Thus $P(\mathcal{A})|_U$ is generated by g and the three constant mappings with values in U. Using this, we can work out that this algebra $\mathcal{A}|_U$ has four congruences:

$\theta_0' = \Delta_U$ and $\theta_1' = U \times U$,
$\theta_2' = \{(b,b), (c,c), (d,d), (b,c), (c,b)\}$, and
$\theta_3' = \{(b,b), (c,c), (d,d), (c,d), (d,c)\}$.

Thus $Con\,\mathcal{A}|_U$ has the Hasse diagram:

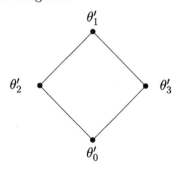

As in Theorem 11.1.2, we can consider the mapping $\varphi_U \colon Con\,\mathcal{A} \to Con(\mathcal{A}|_U)$, taking each θ to $\theta|_U$. Then we have

$$\theta_1 \mapsto \theta_1|_U = \theta_1', \quad \theta_0 \mapsto \theta_0|_U = \theta_0',$$
$$\theta_2 \mapsto \theta_2|_U = \theta_1', \quad \theta_3 \mapsto \theta_3|_U = \theta_3'$$
$$\theta_4 \mapsto \theta_4|_U = \theta_2', \quad \theta_5 \mapsto \theta_5|_U = \theta_2'.$$

It is easy to check that this is a lattice homomorphism, in spite of the fact that our subset U was the image of a non-idempotent polynomial operation.

We are now ready to define minimal sets of an algebra $\mathcal{A}$. These will be sets which are images of A of the form $f(A)$, where f is a unary polynomial operation. As we have seen, f need not be idempotent, but we do want the image set $f(A)$ to have certain useful properties.

Definition 11.1.4 Let $\mathcal{A}$ be an algebra and let $f \in P^1(\mathcal{A})$ be a unary polynomial operation of $\mathcal{A}$. The image set $f(A)$ is called a *minimal set* of $\mathcal{A}$ if the following conditions are satisfied:

(i) $|f(A)| > 1,$ and

(ii) for all $g \in P^1(\mathcal{A})$ with $g(A) \subseteq f(A)$ and $|g(A)| > 1$, we have $g(A) = f(A)$.

The collection of all minimal sets of an algebra $\mathcal{A}$ will be denoted by $Min(\mathcal{A})$. If A is a finite set of cardinality at least two, then $\mathcal{A}$ has some minimal sets, so that $Min(\mathcal{A})$ is non-empty. An induced algebra $\mathcal{A}|_U$ induced by a minimal set $U \in Min(\mathcal{A})$ is called a *minimal algebra* of $\mathcal{A}$.

Example 11.1.5 Consider the algebra

$$\mathcal{A} = (\{0, 1, 2\}; \max(x, y), \min(x, y), 0, 1, 2, m_1, m_2),$$

where m_1 and m_2 are unary operations given by the tables

x	$m_1(x)$	$m_2(x)$
0	0	0
1	2	0
2	2	2

Then $clone\mathcal{A} = P(\mathcal{A})$ is the set of all operations which are monotone with respect to the usual order $0 \le 1 \le 2$. For each of the following monotone operations $f_1, \ldots, f_6$,

x	f_1	f_2	f_3	f_4	f_5	f_6
0	0	0	0	0	1	1
1	0	1	0	2	1	2
2	1	1	2	2	2	2

the corresponding image set $f_i(\{0,1,2,\})$ is a minimal set. Notice that these are all idempotent operations.

As usual, a bijective unary operation $f : A \to A$ is called a *permutation* of the set A. Let $\mathcal{S}_A = (S_A; \circ)$ be the *full permutation group* on the set A. Any subgroup $\mathcal{G}$ of this group is called a *permutation group* on A.

Theorem 11.1.6 *Let A be a finite algebra and let $U \in Min(A)$. Then every unary polynomial operation of the induced algebra $A|_U$ is either a permutation or a constant operation. Moreover, the non-constant unary polynomial operations of $A|_U$ form a permutation group on U.*

Proof: By definition of $A|_U$, all its polynomial operations are fundamental operations. To show this, suppose that $f \in P(A|_U)$, so that $f \in \langle P(A)|_U \cup \{c_a \mid a \in U\}\rangle = \langle P(A)|_U \rangle$. But the set $P(A)|_U = \{g|_U \mid n \in \mathbb{N}$ and $g \in P^n(A), g(U^n) \subseteq U\}$ is a clone. Therefore, $\langle P(A)|_U \rangle = P(A)|_U$ and thus $P(A|_U) \subseteq P(A)|_U$. The converse inclusion is also satisfied since the elements of $P(A)|_U$ are the fundamental operations of $A|_U$.

This means that the unary polynomial operations of $A|_U$ are exactly the mappings of the form $g|_U$ for some $g \in P^1(A)$ such that $g(U) \subseteq U$. Now suppose that such a mapping is not a permutation of the set U. Since U is minimal, it has the form $U = f(A)$ for some $f \in P^1(A)$. Then $g(f(A)) = g(U) \subset U = f(A)$. But again since U is minimal we must have $|g(U)| = |g(f(A))| = 1$. Therefore $g|_U$ is constant on U. ∎

Algebras with this property, that every non-constant unary polynomial operation is a permutation, are called *permutation algebras*. Our theorem thus says that any minimal algebra of a finite algebra A is a permutation algebra. Later we will show that there are exactly five types of permutation algebras, and thus five types of minimal algebras.

Minimal algebras can also be defined using congruence relations.

Definition 11.1.7 Let $\mathcal{A}$ be an algebra and let $\beta \in Con\,\mathcal{A}$. A set of the form $f(A)$ with $f \in P^1(\mathcal{A})$ is called a β-*minimal set* of $\mathcal{A}$, or a *minimal set with respect to* β, if the following conditions are satisfied:

(i) $f(\beta) \not\subseteq \Delta_A$: there is a pair $(x, y) \in \beta$ with $f(x) \neq f(y)$.

(ii) for any $g \in P^1(\mathcal{A})$ with $g(A) \subseteq f(A)$ and $g(\beta) \not\subseteq \Delta_A$, we have $g(A) = f(A)$.

The set of all β-minimal sets of an algebra $\mathcal{A}$ is denoted by $Min_\mathcal{A}(\beta)$, or when the algebra is clear from the context, just $Min(\beta)$. The corresponding induced algebras $\mathcal{A}|_U$, for $U \in Min(\beta)$, are called β-*minimal algebras* of $\mathcal{A}$, or *minimal algebras with respect to* β.

Example 11.1.8 We consider again the algebra $\mathcal{A} = (\{a, b, c, d\}; f)$ from Example 11.1.3, with f defined by the table

x	a	b	c	d
$f(x)$	b	c	c	d

and take as our congruence

$$\beta = \{(a, a), (b, b), (c, c), (d, d), (b, c), (c, b), (b, d), (d, b), (c, d), (d, c)\}.$$

For the operation g given by the table

x	a	b	c	d
$g(x)$	c	c	c	d

we have an image set $U = g(A) = \{c, d\}$. The image $g(\beta) = \{(c, c), (d, d), (c, d), (d, c)\}$ contains a non-diagonal pair, and moreover there is no other unary polynomial with this property which gives a smaller image. Thus our set $g(A)$ meets both conditions regarding our congruence, and is a β-minimal set of $\mathcal{A}$.

Definition 11.1.7 is in fact a generalization of Definition 11.1.4. A minimal set for an algebra $\mathcal{A}$ is always a β-minimal set for β equal to the largest congruence $A \times A$ on $\mathcal{A}$. The first condition for β-minimal sets, in this special case, merely says that the cardinality of the minimal set is greater than one, while the second condition translates directly. We thus have $Min(\mathcal{A}) = Min_\mathcal{A}(\mathcal{A} \times \mathcal{A})$.

We now generalize the characterization from Theorem 11.1.6 of minimal algebras of a finite algebra.

Theorem 11.1.9 *Let $\mathcal{A}$ be a finite algebra, and let $\beta \in \text{Con}\mathcal{A}$. Let $U \in Min_{\mathcal{A}}(\beta)$ be a β-minimal set such that $U = e(A)$ for some idempotent $e \in E(\mathcal{A})$. Then every unary polynomial operation g of the induced minimal algebra $\mathcal{A}|_U$ is either a permutation or satisfies $g(\beta|_U) \subseteq \Delta_U$. For every $\beta|_U$-congruence class N of $\mathcal{A}|_U$, the algebra $(\mathcal{A}|_U)|_N$ induced by $\mathcal{A}|_U$ on N is a permutation algebra, that is, all its non-constant unary polynomial operations are permutations.*

Proof: By definition, every unary polynomial operation g of $\mathcal{A}|_U$ has the form $g = h|_U$ for some unary polynomial h on $\mathcal{A}$ such that $h(e(A)) = h(U) = g(U) \subseteq U$. If g is not a permutation of U, then by the finiteness of A we have $h(e(A)) = h(U) = g(U) \subset U$; and since U is minimal, this means that $h(e(\beta)) \subseteq \Delta_A$. Since e is idempotent, we have $e(\beta) = \beta|_U$ and then $g(\beta|_U) = h(\beta|_U) \subseteq \Delta_U$.

Now let N be a $\beta|_U$ congruence class of $\mathcal{A}|_U$. Every unary polynomial operation of $(\mathcal{A}|_U)|_N$ has the form $h|_N$ for some $h \in P^1(\mathcal{A})$ with $h(U) \subseteq U$ and $h(N) \subseteq N$. Then $h|_U$ is either a permutation on U, or satisfies $h(\beta|_U) \subseteq \Delta_U$. In the first case $h|_N$ is a permutation on N, while in the second case $h|_N$ is constant. ∎

Definition 11.1.10 Let $\mathcal{A}$ be an algebra, let $\beta \in \text{Con } \mathcal{A}$ and let $U \in Min_{\mathcal{A}}(\beta)$. Any $\beta|_U$-congruence class N of $\mathcal{A}|_U$ with at least two elements is called a β-*trace* of U. Note that by Definition 11.1.7 every β-minimal set U contains at least one such β-trace. For such a trace N, the induced algebra $(\mathcal{A}|_U)|_N$ is called a *trace algebra* of $\mathcal{A}$ with respect to β. The set U is then divided into two subsets called the body and tail of U, as follows:
body of U := $\bigcup \{N \mid N \text{ is a } \beta\text{-trace of } U\}$
tail of U := $U \setminus (\text{body of } U)$.

Example 11.1.11 As an example we consider the group $(\mathbb{Z}_6; +, -, 0)$ of equivalence classes modulo 6. Each congruence of a commutative group corresponds to a subgroup of the group. In this example we have two proper subgroups of our group, $\{0, 3\}$ and $\{0, 2, 4\}$, and hence two congruences other than Δ and $\mathbb{Z}_6 \times \mathbb{Z}_6$:

$\alpha = \{(0,0)(1,1),(2,2),(3,3),(4,4),(5,5),(0,3),(3,0),(1,4),(4,1),$
$\qquad (2,5),(5,2)\},$
$\beta = \{(0,0),(1,1),(2,2),(3,3),(4,4),(5,5),(0,2),(2,0),(0,4),(4,0)),$

$$(2,4),(4,2),(1,3),(3,1),(1,5),(5,1),(3,5),(5,3)\}.$$

The unary polynomial operations of our algebra have the form $f(x) = ax+b$, for some $a,b \in \mathbb{Z}_6$. From this information we can calculate all the minimal sets of this algebra:

$$
\begin{aligned}
Min(\mathbb{Z}_6 \times \mathbb{Z}_6) &= \{\{0,3\},\{1,4\},\{2,5\},\{0,2,4\},\{1,3,5\}\}, \\
Min(\alpha) &= \{\{0,3\},\{1,4\},\{2,5\}\}, \\
Min(\beta) &= \{\{0,2,4\},\{1,3,5\}\}.
\end{aligned}
$$

Our characterization of minimal algebras obtained from minimal sets U, in Theorem 11.1.6, showed that for any unary polynomial f for which the set $U = f(A)$ is minimal, the induced minimal algebra is a permutation algebra. The generalized version of this theorem for congruences, Theorem 11.1.9, was more limited, in the sense that Theorem 11.1.9 applies only to β minimal sets U of the form $e(A)$ for an idempotent unary polynomial e. This raises the question of whether all the β-minimal sets of an algebra can be obtained from idempotent unary polynomial operations only. The answer to this question depends on the congruence β and a special property of the congruence lattice of the original algebra. To describe the property involved, we need the following definition.

Definition 11.1.12 Let $\mathcal{L} = (L, \wedge, \vee)$ be any lattice.

 (i) A mapping $\mu : L \to L$ is called a *meet-endomorphism* of $\mathcal{L}$ if it satisfies $\mu(x \wedge y) = \mu(x) \wedge \mu(y)$ for all elements x and y in L.

 (ii) A mapping $\mu : L \to L$ is called a *join-endomorphism* of $\mathcal{L}$ if it satisfies $\mu(x \vee y) = \mu(x) \vee \mu(y)$ for all x and y in L.

 (iii) Recall from Chapter 1 that a mapping $\varphi : L \to L$ is *extensive* if $x \leq \varphi(x)$ for all $x \in L$; now we call a mapping *strongly extensive* if $x \leq \varphi(x)$ but $\varphi(x) \neq x$ for all $x \neq 1$ (where 1 is the greatest element of $\mathcal{L}$, if there is one).

A lattice endomorphism is a mapping of a lattice to itself which preserves both the meet and join operations of the lattice. It is possible for a mapping on a lattice to preserve meets but not joins, or joins but not meets, and hence the properties from Definition 11.1.12 are weaker than the lattice homomorphism property.

Theorem 11.1.13 *Let $\mathcal{A}$ be a finite algebra and let $\beta \in Con\ \mathcal{A}$. Assume that the interval $[\Delta_A, \beta] \subseteq Con\ \mathcal{A}$ has no non constant strongly extensive meet endomorphisms. Then every β-minimal set U of $\mathcal{A}$ has the form $U = e(A)$ for some idempotent unary polynomial $e \in E(\mathcal{A})$.*

Proof: Let U be any β-minimal set of $\mathcal{A}$. We define a set $K := \{f \in P^1(\mathcal{A}) | f(A) \subseteq U\}$. For every congruence relation $\theta \in [\Delta_A, \beta]$ we define a relation

$$\mu(\theta) := \{(x,y) \in \beta \mid \forall f \in K, (f(x), f(y)) \in \theta\}.$$

By definition $\mu(\theta) \subseteq \beta$, and we will show that $\mu(\theta)$ is also a congruence on $\mathcal{A}$. To do this, it suffices by Lemma 5.3.2 to show that $\mu(\theta)$ is compatible with all unary polynomials on $\mathcal{A}$. Suppose that $(x,y) \in \mu(\theta)$ and g is a unary polynomial operation of $\mathcal{A}$. By definition of $\mu(\theta)$ we have (x,y) in β and, for the particular polynomial $f \circ g$ in K, the pair $(f(g(x)), f(g(y)))$ in θ. Since β is a congruence and compatible with g, we also have $(g(x), g(y))$ in β. Now from $(g(x), g(y)) \in \beta$ and $(f(g(x)), f(g(y))) \in \theta$, we conclude that $(g(x), g(y))$ is in $\mu(\theta)$, as required.

Next we show that the mapping μ is an extensive meet-endomorphism on the interval $[\Delta_A, \beta]$. It follows from the definition that for any θ in this interval, the image $\mu(\theta)$ is also in this interval, and that our mapping preserves meets (intersections) of congruences. Let θ be a congruence in this interval and let (x,y) be in θ. By the compatibility of θ with unary polynomials on $\mathcal{A}$ we have $(f(x), f(y)) \in \theta$ for every unary polynomial f. We also have $(x,y) \in \beta$ since $\theta \subseteq \beta$, so that (x,y) in $\mu(\theta)$. This shows that $\theta \subseteq \mu(\theta)$, and our mapping μ is extensive. Since U is β-minimal, it has the form $U = f(A)$ for some unary polynomial $f \in K$ for which $f(\beta) \not\subseteq \Delta_A$. Therefore, $\mu(\Delta_A) \subset \beta$, for if $\mu(\Delta_A) = \beta$ we would have $\mu(\theta) = \beta$ for all θ. This means that μ is not a constant mapping. But our assumption about the interval $[\Delta_A, \beta]$ means that μ cannot be strongly extensive. This means there must exist a congruence θ_0 in the interval, with $\theta_0 \neq \beta$, for which $\mu(\theta_0) = \theta_0$. It follows by extensivity that $\mu(\mu(\Delta_A)) \subseteq \mu(\mu(\theta_0)) = \theta_0 \subset \beta$. But $\mu(\mu(\Delta_A)) = \{(x,y) \in \beta \mid \forall f, g \in K\ (f(g(x)) = f(g(y)))\}$; so for this to be a proper subset of β there must exist polynomials f and g in K and a pair $(x,y) \in \beta$ such that $f(g(x)) \neq f(g(y))$. Then $f(A) = g(A) = f(g(A)) = U$ and $f(U) = U$. Since A and U are finite there is a natural number k for which $(f|_U)^k$ is the identity mapping on U. Now taking $e := f^k$ gives an idempotent $e^2 = e$ for which $e(A) = U$, and hence our β-minimal set U is obtainable by an idempotent unary polynomial. $\blacksquare$

Lemma 11.1.14 *Let $\mathcal{L}$ be a finite lattice with least element 0, greatest element 1 and the property that the meet of arbitrary coatoms (elements directly below 1) is zero. Then $\mathcal{L}$ has no non–constant strongly extensive meet–endomorphisms.*

Proof: Assume that $c_1, \ldots, c_n$ are the coatoms of L and that φ is a strongly extensive meet–endomorphism. For $i = 1, \ldots, n$, since $1 \wedge c_i = c_i$ and $\varphi(1) \wedge \varphi(c_i) = 1 \wedge \varphi(c_i) = \varphi(c_i)$, we have $\varphi(c_i) \leq \varphi(1) = 1$. Now $c_i < \varphi(c_i) \leq 1$, and the fact that c_i is a coatom means that $\varphi(c_i) = 1$. Thus $\varphi(0) = \varphi(c_1 \wedge \cdots \wedge c_n) = \varphi(c_1) \wedge \cdots \wedge \varphi(c_n) = 1$. Since φ is order–preserving we have $\varphi(x) = 1$ for all $x \in L$.   ■

This Lemma can be applied to lattices $\mathcal{M}_n$ of the form

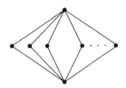

Since $Con\,\mathbb{Z}_6 \cong M_2$, all the minimal algebras of the algebra $\mathcal{Z}_6$ of Example 11.1.11 have the form $e(\mathbb{Z}_6)$.

11.2   Tame Congruence Relations

Definition 11.2.1 A finite algebra $\mathcal{A}$ is called *tame* if it has a minimal set $U \in Min(\mathcal{A})$ satisfying the following three conditions, for all $\theta \in Con\mathcal{A}$:

(i) There is an $e \in E(\mathcal{A})$ with $U = e(\mathcal{A})$,

(ii) $\theta \supset \Delta_A \Rightarrow \theta|_U \supset \Delta_U$,

(iii) $\theta \subset A \times A \Rightarrow \theta|_U \subset U \times U$.

Theorem 11.2.2 *Let $\mathcal{A}$ be a finite algebra. A minimal set $U \in Min(\mathcal{A})$ fulfills conditions (i), (ii) and (iii) from Definition 11.2.1 iff the following two conditions are both satisfied:*

(T1) *For all $x, y \in A$ with $x \neq y$, there exists a unary polynomial $f \in P^1(\mathcal{A})$ such that $f(A) = U$ and $f(x) \neq f(y)$.*

(T2) *The congruence generated by U is equal to $A \times A$; that is, for all $x, y \in A$ there are elements $a_1, \ldots, a_k, b_1, \ldots, b_k \in U$ and $f_1, \ldots, f_k \in P^1(\mathcal{A})$ with $x = f_1(a_1), f_i(b_i) = f_{i+1}(a_{i+1})$ for $i = 1, \ldots k - 1$ and $f_k(b_k) = y$.* ∎

As we saw in the previous section, minimal sets of an algebra $\mathcal{A}$ can be seen as a special case of β-minimal sets of $\mathcal{A}$; for congruences β, a minimal set is in fact a β-minimal set for the largest congruence $A \times A$ on $\mathcal{A}$. Similarly, our definition of tame algebras by means of minimal sets can be generalized to tame congruences, for any congruence on the algebra. In particular, Theorem 11.2.2 above is a particular case of a more general theorem for this more general situation, and thus will be proved later.

Definition 11.2.3 Let $\mathcal{A}$ be a finite algebra. A congruence relation $\beta \neq \Delta_A$ on $\mathcal{A}$ is called *tame* if there is a β-minimal set $U \in Min_\mathcal{A}(\beta)$ such that the following conditions are satisfied, for all $\theta \in [\Delta_A, \beta]$:

  (i) There is an element $e \in E(\mathcal{A})$ with $U = e(A)$,

  (ii) $\theta \supset \Delta_A \Rightarrow \theta|_U \supset \Delta_U$,

  (iii) $\theta \subset \beta \Rightarrow \theta|_U \subset \beta|_U$.

Let us remark that the concept of a tame congruence can be generalized even further, in the following way. Let $\mathcal{A}$ be finite and α and β be congruences on $\mathcal{A}$ with $\alpha \subset \beta$. Then the interval $[\alpha, \beta] \subseteq Con\mathcal{A}$ is called *tame* if the congruence relation β/α of the quotient algebra $\mathcal{A}/\alpha$ is tame. The β/α-minimal sets, and the corresponding induced algebras, of the quotient algebra $\mathcal{A}/\alpha$ are called $[\alpha, \beta]$-minimal. (Note that by the Second Isomorphism Theorem, Theorem 3.2.2, the intervals $[\alpha, \beta] \subseteq Con\mathcal{A}$ and $[\Delta_{\mathcal{A}/\alpha}, \beta/\alpha] \subseteq Con\mathcal{A}/\alpha$ are isomorphic.)

We now present the generalization of Theorem 11.2.2, giving an equivalent characterization of a tame congruence.

Theorem 11.2.4 *Let $\mathcal{A}$ be a finite algebra and let $\beta \in Con\mathcal{A}$ with $\beta \neq \Delta_A$. A β-minimal set $U \in Min_\mathcal{A}(\beta)$ satisfies conditions (i), (ii) and (iii) from Definition 11.2.3 (so that β is a tame congruence) iff the following two conditions are both satisfied:*

*(Z1) For all $(x, y) \in \beta$ with $x \neq y$, there is a unary polynomial $f \in P^1(\mathcal{A})$
       such that $f(A) = U$ and $f(x) \neq f(y)$.*

*(Z2) $\theta(\beta|_U)$ $=$ β; that is, for all (x, y) $\in$ β there are pairs
       $(a_1, b_1), \ldots, (a_k, b_k) \in \beta|_U$ and polynomials $f_1, \ldots, f_k \in P^1(\mathcal{A})$ with
       $x = f_1(a_1)$, $f_i(b_i) = f_{i+1}(a_{i+1})$ for $i = 1, \ldots, k-1$, and $f_k(b_k) = y$.*

Proof: (Z1) $\Rightarrow$ (i): Since U is β-minimal there is a unary polynomial
$g \in P^1(\mathcal{A})$ for which $g(A) = U$, and a pair $(a, b) \in \beta$ with $g(a) \neq g(b)$. Application of (Z1) on the pair $(g(a), g(b))$ gives us another polynomial $f \in P^1(\mathcal{A})$
with $f(A) = U$ and $f(g(a)) \neq f(g(b))$. This means that $f(g(A)) = f(U) \subseteq U$,
and using the second condition from the definition of the β-minimality of U,
we get $f(U) = f(g(A)) = U$. Therefore f is a permutation on U and there is
an $n \in \mathbb{N}$ with $(f^n)|_U = id_U$. Then $e := f^n$ is an idempotent in $E(\mathcal{A})$, and
we have $e(A) = U$, as required for condition (i). To complete our proof, we
shall show that (Z1) $\Leftrightarrow$ (ii) and (Z2) $\Leftrightarrow$ (iii). Moreover, we may now assume
that $U = e(A)$ for an idempotent polynomial $e \in E(\mathcal{A})$.

(ii) $\Rightarrow$ (Z1): Assume that $(x, y) \in \beta$ and $x \neq y$. Then in $\mathcal{A}$ we have
$\theta(x, y) \supset \Delta_A$, where $\theta(x, y)$ is the congruence generated by (x, y). Then
by condition (ii), we have $\theta(x, y)|_U \supset \Delta_U$, and there is a pair $(u, v) \in$
$\theta(x, y)$ with $u, v \in U$ and $u \neq v$. Then by Lemma 5.3.3 there exist
polynomials $g_1, \ldots, g_n \in P^1(\mathcal{A})$ with $u \in \{g_1(x), g_1(y)\}$, $\{g_i(x), g_i(y)\} \cap$
$\{g_{i+1}(x), g_{i+1}(y)\} \neq \emptyset$ for $i = 1, \ldots, n-1$ and $v \in \{g_n(x), g_n(y)\}$. From
$u, v \in U$ and $U = e(A)$ we get $e(u) = u$ and $e(v) = v$. Then we have
$u \in \{e(g_1(x)), e(g_1(y))\}, \{e(g_i(x)), e(g_i(y))\} \cap \{e(g_{i+1}(x)), e(g_{i+1}(y))\} \neq \emptyset$
for $i = 1, \ldots, n-1$ and $v \in \{e(g_n(x)), e(g_n(y))\}$. Since $u \neq v$ we obtain
$e(g_i(x)) \neq e(g_i(y))$ for at least one i. From this and from $U = e(A)$ we get
(Z1).

(Z1) $\Rightarrow$ (ii): Assume now that (Z1) is satisfied and that $\theta \in Con\mathcal{A}$ with
$\Delta_A \subset \theta \subseteq \beta$. Then for any $(x, y) \in \theta$ with $x \neq y$, by (Z1) there is a polyno-
mial $f \in P^1(\mathcal{A})$ with $(f(x), f(y)) \in \theta \cap U^2 = \theta|_U$ and $f(x) \neq f(y)$. Thus (ii)
is satisfied.

(Z2) $\Leftrightarrow$ (iii): The condition (iii), that if $\theta \subset \beta$ then $\theta|_U \subset \beta|_U$, for all
$\theta \in [\Delta_A, \beta]$, is by Theorem 11.1.2 equivalent to the fact that the congruence
relation on $\mathcal{A}$ generated by $\beta \cap U^2$ is equal to β; that is, $\theta(\beta|_U) = \beta$. But
since $\theta(\beta|_U) \subseteq \beta$ and $\beta|_U \subseteq \theta(\beta|_U)$, we do have $\theta(\beta|_U) = \beta$. ∎

Definition 11.2.5 Let $\mathcal{A}$ be an algebra and let B and C be subsets of the universe set A of $\mathcal{A}$. Then the sets B and C are called *polynomially isomorphic* in $\mathcal{A}$ if there exist unary polynomials f and g in $P^1(\mathcal{A})$ such that

$$f(B) = C, \ g(C) = B, \ gf|_B = id_B, \text{ and } fg|_C = id_C.$$

In this case the mapping $f|_B$ is called a *polynomial isomorphism* from B onto C.

Lemma 11.2.6 *Let B and C be polynomially isomorphic subsets in $\mathcal{A}$. Then $\mathcal{A}|_B$ and $\mathcal{A}|_C$ are isomorphic as non-indexed algebras, with $f|_B : B \to C$ as an isomorphism between them. Moreover, for all $\theta \in Con\mathcal{A}$ we have $f(\theta|_B) = \theta|_C$.*

Proof: For convenience, let us denote $\mathcal{A}|_B$ by $\mathcal{B}$ and $\mathcal{A}|_C$ by $\mathcal{C}$, and let $H := \{h \mid h \in P^n(\mathcal{A}), n \in \mathbb{N}, h(B^n) \subseteq B\}$. The fundamental operations of $\mathcal{B}$ are exactly the operations of the form $h|_B$ with $h \in H$. We regard $\mathcal{B}$ as an (unindexed) algebra of type H. For every $h \in H$, we define a mapping $h' \in P^n(\mathcal{A})$ by

$$h'(x_1, \ldots, x_n) := f(h(g(x_1), \ldots, g(x_n))).$$

Then $h'(C^n) \subseteq C$ for any such h', and the operation $h'|_C$ is one of the fundamental operations of $\mathcal{C}$. We now show that any fundamental operation of $\mathcal{C}$ can be represented in this way. For $p \in P^n(\mathcal{A})$ with $p(C^n) \subseteq C$, we define a mapping $h \in P^n(\mathcal{A})$ by

$$h(x_1, \ldots, x_n) := g(p(f(x_1), \ldots, f(x_n))).$$

Then $h(B^n) \subseteq B$, so that $h \in H$, and $p|_C = h'|_C$. Thus $\mathcal{C}$ is an algebra of type H. Also $f|_B$ is an isomorphism of the algebras $\mathcal{B}$ and $\mathcal{C}$, since for all $h \in H \cap P^n(\mathcal{A})$ and for all $b_1, \ldots, b_n \in B$ we have

$$f(h(b_1, \ldots, b_n)) = f(h(gf(b_1), \ldots, gf(b_n))) = h'(f(b_1), \ldots, f(b_n))$$

using the fact that $gf|_B = id_B$.

For all $\theta \in Con\mathcal{A}$, it follows from $f \in P^1(\mathcal{A})$ that $f(\theta) \subseteq \theta$. The equation $f(B) = C$ gives $f(\theta|_B) = f(\theta \cap B^2) \subseteq \theta \cap C^2 = \theta|_C$. Similarly, we obtain $g(\theta|_C) \subseteq \theta_B$ and then $\theta|_C = f(g(\theta|_C)) \subseteq f(\theta|_B)$. Altogether we have equality. ∎

Theorem 11.2.7 *Let β be a tame congruence relation of the finite algebra $\mathcal{A}$. Then the β-minimal sets have the following properties:*

(i) *Any β-minimal sets $U_1, U_2 \in Min_{\mathcal{A}}(\beta)$ are polynomially isomorphic in $\mathcal{A}$.*

(ii) *For all $U \in Min_{\mathcal{A}}(\beta)$, the conditions (Z1), (Z2) from Theorem 11.2.4 and (i),(ii),(iii) from Definition 11.2.3 are satisfied.*

(iii) *Let $U \in Min_{\mathcal{A}}(\beta)$, with $U = f(A)$ for some $f \in P^1(\mathcal{A})$ with $f(\beta|_U) \not\subseteq \Delta_A$. Then $f(U)$ is also in $Min_{\mathcal{A}}(\beta)$, and $f|_U$ is a polynomial isomorphism from U onto $f(U)$.*

Proof: (i) Since β is tame there is a set $W \in Min_{\mathcal{A}}(\beta)$ satisfying conditions (i), (ii), (iii) from Definition 11.2.3 and (Z1), (Z2) from Theorem 11.2.4. Moreover, we may assume that $W = e(A)$ for an idempotent unary polynomial e. We show that for every $U \in Min_{\mathcal{A}}(\beta)$, the sets U and W are polynomially isomorphic. Since U is β-minimal, there is a polynomial operation $s \in P^1(\mathcal{A})$ with $s(A) = U$ and $s(\beta) \not\subseteq \Delta_A$. Then there is a pair $(x, y) \in \beta$ with $s(x) \neq s(y)$. Applying condition (Z2) for W to the pair (x, y), we obtain the existence of a pair $(a, b) \in \beta|_W$ and of a polynomial $h \in P^1(\mathcal{A})$ such that $s(h(a)) \neq s(h(b))$ (since otherwise $s(x) = s(y)$). Now we define s_1 to be the composition mapping she. Then we have $s_1(A) = U$, since $s(h(e(A))) = s(h(W)) \subseteq U$ because of the β-minimality of U. Applying (Z1) to the pair $(s_1(a), s_1(b))$ gives us a polynomial $t \in P^1(\mathcal{A})$ with $t(A) = W$ and $ts_1(a) \neq ts_1(b)$. By the β-minimality of W again, we have $ts_1(W) = W$. Now we use s_1 and t to construct the required polynomial isomorphism between U and W. The mapping $s_1t|_U$ satisfies $s_1(t(U)) \subseteq U$ and because of the minimality of U we have $s_1(t(U)) = U$. This shows that $s_1t|_U$ is a permutation on the finite set U. Therefore there is an element $k \in \mathbb{N}$ with $(s_1t)^k = id_U$. Taking $f := t$ and $g := (s_1t)^{k-1}s_1$, we have $f(U) = t(U) = W$, $g(W) = (s_1t)^{k-1}s_1(W) = (s_1t)^{k-1}(U) = U$ and $gf = (s_1t)^{k-1}s_1t = id_U$. Since every $w \in W$ can be written as $f(u)$ for some $u \in U$, and

$$f(g(w)) = f(g(f(u))) = f(gf(u)) = f(u) = w,$$

we see that $fg|_W = id_W$. This means that $f|_U$ is a polynomial isomorphism of U onto W (and $g|_W$ is the isomorphism which is inverse to $f|_U$).

(ii) Let U be a β-minimal set. With the help of the polynomial isomorphisms $f|_U$ and $g|_W$ from the proof of (i), the properties (Z1) and (Z2) are satisfied for U. For (Z2) we use the additional fact that $f(\beta|_U) = \beta|_W$, which follows from Lemma 11.2.6. The conditions (i), (ii) and (iii) of Definition 11.2.3 then follow by Theorem 11.2.4.

(iii) Since $f(\beta|_U) \not\subseteq \Delta_A$, and since by (i) of this proof all $U \in Min_A(\beta)$ have the same cardinality, we get $f(U) \in Min_A(\beta)$. For the polynomial isomorphism, we have to show the existence of a polynomial $g \in P^1(\mathcal{A})$ with $g(f(U)) = U$, $gf|_U = id_U$, and $fg|_{f(U)} = id_{f(U)}$. Assume that $(a,b) \in \beta|_U$ with $f(a) \neq f(b)$. Then by condition (Z1) there is an element $h \in P^1(\mathcal{A})$ with $h(A) = U$ and $h(f(a)) \neq h(f(b))$. This gives $h(f(U)) = U$. Then as in the proof of part (i) there is a $k \in I\!N$ with $(hf)^k = id_U$ and the mapping $g := (hf)^{k-1}h$ has the desired property.   ∎

Now we look for properties of the congruence lattice of an algebra $\mathcal{A}$ which force its congruences to be tame. The lattice property involved is called *tightness*.

Definition 11.2.8 Let $\mathcal{L}$ be a lattice with least element 0 and greatest element 1. A homomorphism $\varphi : \mathcal{L} \to \mathcal{L}'$ of $\mathcal{L}$ into $\mathcal{L}'$ is called 0-*separating* if

$$\varphi^{-1}(\varphi(0)) = \{0\},$$

and is called 1-*separating* if

$$\varphi^{-1}(\varphi(1)) = \{1\}.$$

A homomorphism which is both 0- and 1-separating is called 0-1-*separating*. A lattice $\mathcal{L}$ is called 0-1-*simple* if $|L| > 1$ and every non-constant homomorphism $f : \mathcal{L} \to \mathcal{L}'$ is 0-1-separating. A lattice $\mathcal{L}$ is called *tight* if L is finite with $|L| > 1$, and is 0-1-simple and has no non-constant strongly extensive meet-endomorphisms.

We know from Chapter 3 that any congruence relation of an algebra (and in particular of a lattice) occurs as a kernel of a homomorphism. The condition for a homomorphism φ on a lattice $\mathcal{L}$ with 0 and 1 to be 0-separating can be expressed as the requirement that the congruence class of the element 0 under the congruence $\ker \varphi$ is a singleton set, $\{0\}$, and similarly for 1-separating. This means that a lattice $\mathcal{L}$ with 0 and 1 is 0-1-simple iff all congruence relations different from $L \times L$ have the singleton sets $\{0\}$ and $\{1\}$ as congruence classes (blocks). In particular, it follows that all simple lattices are 0-1-simple. By Lemma 11.1.14, all finite 0-1-simple lattices whose coatoms have meets equal to zero are tight.

Example 11.2.9 1. The lattices $\mathcal{M}_n$ given by Hasse diagrams

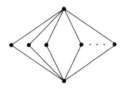

are tight.

2. If A is a finite set then the lattice $Eq(A)$ of all equivalence relations on A is tight.

3. The congruence lattice of any finite vector space is tight.

Theorem 11.2.10 *Let $\mathcal{A}$ be a finite algebra, and let $\beta \in Con\mathcal{A}$ with $\beta \neq \Delta_A$. If the lattice $[\Delta_A, \beta]$ is tight, then β is tame.*

Proof: We consider the lattice $\mathcal{L} := [\Delta_A, \beta]$, with Δ_A as its 0-element and β as its 1-element. By assumption $\mathcal{L}$ has no non-constant strongly extensive meet-endomorphisms and every non-constant homomorphism $f : \mathcal{L} \to \mathcal{L}'$ is 0-1-separating. Then by Theorem 11.1.13, any β-minimal set U of $\mathcal{A}$ has the form $U = e(\mathcal{A})$ for an idempotent $e \in E(\mathcal{A})$. Now we use the mapping $\varphi_U : Con\mathcal{A} \to Con(\mathcal{A}|_U)$, defined by $\theta \mapsto \theta|_U$, from Theorem 11.1.2. The inclusions $\Delta_A \subset \theta$ and $\theta \subset \beta$ imply $\Delta_U = \Delta_A|_U \subseteq \theta|_U$ and $\theta|_U \subseteq \beta|_U$, respectively. Since every non-constant homomorphism of $[\Delta_A, \beta]$ is 0-1-separating, Δ_A has only Δ_U as image and β has only $\beta|U$ as image, so $\Delta_U \neq \theta|_U$ and $\theta|_U \neq \beta|U$. Thus β fits the conditions of Definition 11.2.3, and is tame.   ∎

Example 11.2.11 Consider again the algebra $(\mathbb{Z}_6; +, -, 0)$ from Example 11.1.11. The two congruence relations

$$\alpha = \{(0, 0), (1, 1), (2, 2), (3, 3), (4, 4), (5, 5), (0, 3), (3, 0), (1, 4), (4, 1), (2, 5), (5, 2)\}$$

and
$$\beta = \{((0,0),(1,1),(2,2),(3,3),(4,4),(5,5),(0,2),(2,0),(0,4),(4,0),(2,4),$$
$$(4,2),(1,3),(3,1),(1,5),(5,1),(3,5),(5,3)\}$$

are tame, since both intervals $[\Delta_{\mathbb{Z}_6}, \alpha], [\Delta_{\mathbb{Z}_6}, \beta]$ are simple. But the algebra $\mathbb{Z}_6$ is not tame, since neither congruence has $\{0\}$ or $\{1\}$ as singleton blocks.

All two-element lattices are tight. For any algebra $\mathcal{A}$, we can look at what are called prime intervals $[\alpha, \beta] \in Con\mathcal{A}$, intervals which contain only the two congruences α and β. For minimal sets which correspond to prime intervals, the conditions (Z1) and (Z2) are satisfied. This means that when $\mathcal{A}$ is finite we can consider all prime intervals. In this way tame congruence theory gives a structure theory for all finite algebras.

11.3 Permutation Algebras

Definition 11.3.1 An algebra with the property that all its non constant unary polynomial operations are permutations on its universe set is called a *permutation algebra*.

As we saw in Theorem 11.1.6, for any finite algebra $\mathcal{A}$ the minimal algebras of $\mathcal{A}$ are permutation algebras. Our goal in this section is to classify all finite permutation algebras, which will allow us in the next section to classify all minimal algebras. For two-element permutation algebras, we are able to use our classification results from Chapters 9 and 10. We begin now by considering finite permutation algebras of cardinality at least three.

Definition 11.3.2 Let $f : A^n \to A$ be an n-ary operation on A, and let $1 \leq i \leq n$. Then f depends on the i-th variable if there are elements $a_1, \ldots, a_{i-1}, a_{i+1}, \ldots, a_n$ such that the unary operation defined by $x \mapsto f(a_1, \ldots, a_{i-1}, x, a_{i+1}, \ldots, a_n)$ is not constant. An operation is said to be *essentially binary* if it depends on two of its variables, and to be *essentially at least binary* if it depends on at least two variables.

Theorem 11.3.3 *Let $\mathcal{A}$ be a finite permutation algebra, with $|A| \geq 3$. If $\mathcal{A}$ has a polynomial operation which depends on at least two variables, then $\mathcal{A}$ is polynomially equivalent to a vector space; that is, there is a field $\mathcal{K}$ and a vector space structure $(A; +, -, 0, K)$ on A such that the polynomial operations of $\mathcal{A}$ are exactly the operations of the form*

$$(x_1, \ldots, x_n) \mapsto a + k_1 x_1 + \cdots + k_n x_n$$

for some $n \in \mathbb{N}$, $a \in A$, and $k_1, \ldots, k_n \in K$.

Before we begin the proof of this theorem we need the following fact:

Lemma 11.3.4 *Let $\mathcal{A}$ be an algebra with a polynomial operation which depends on at least two variables. Then $\mathcal{A}$ also has a polynomial operation which is essentially binary.*

Proof: Let f be a polynomial operation of arity $n \geq 3$ which depends on at least two variables. Then there are elements $a_2, \ldots, a_n$, and $b_1, b_3, \ldots, b_n$ in A such that $f(x_1, a_2, \ldots, a_n)$ depends on x_1 and $f(b_1, x_2, b_3, \ldots, b_n)$ depends on x_2. If $f(x_1, x_2, a_3, \ldots, a_n)$ depends on x_2, we can use it as our essentially binary polynomial operation. Similarly, if $f(x_1, a_2, x_3, \ldots, a_n)$ depends on x_3 we can use it.

Otherwise we replace a_3 by b_3, and we have $f(x_1, a_2, b_3, a_4, \ldots, a_n)$ depending on x_1. We continue in this way. If none of the operations $f(x_1, a_2, b_3, \ldots, b_{i-1}, x_i, a_{i+1}, \ldots, a_n)$ depends on x_i, then $f(x_1, a_2, b_3, \ldots, b_n)$ depends on x_1 and because of the choice of $b_3, \ldots, b_n$ we can take $f(x_1, x_2, b_3, \ldots, b_n)$ to be our essentially binary polynomial. ∎

Proof of Theorem 11.3.3: Our construction of a vector space structure on A will proceed via several steps.

Step 1. *Every essentially binary polynomial operation of $\mathcal{A}$ is the operation of a quasigroup.*

Proof: We recall from Example 1.2.7 that a quasigroup $(Q; +)$ is a groupoid with the property that for any $a \in Q$, both left and right addition with a form permutations on Q; that is, the maps $x \mapsto a + x$ and $x \mapsto x + a$ are permutations on Q.

Quasigroups can also be defined as algebras $(Q; +, /, \backslash)$ of type $(2, 2, 2)$ which satisfy the following identities:

(Q1)	$x \backslash (x + y)$	$\approx$	y
(Q2)	$(x + y)/y$	$\approx$	x
(Q3)	$x + (x \backslash y)$	$\approx$	y
(Q4)	$(x/y) + y$	$\approx$	$x.$

We let $x + y$ be any essentially binary polynomial operation of $\mathcal{A}$. We will show that for any $a \in A$, the left addition mapping $L_a(x)$ taking any x to $a + x$ is a permutation. It can be shown similarly that the right addition mapping $R_a(x) := x + a$ is a permutation, and therefore $(A; +)$ is a quasigroup.

Suppose that there is an $a \in A$ for which the mapping L_a is not a permutation. Since L_a is a unary polynomial operation and since by assumption $\mathcal{A}$ is a permutation algebra, the operation L_a must be constant. Hence there is an element $s \in A$ such that $L_a(x) = s$ for all $x \in A$. Since the operation $x + y$ depends on the second variable, there is an element $b \in A$ such that L_b is a permutation. In particular L_b is surjective, so there is an element $c \in A$ with $L_b(c) = s$. Now we have $a + c = s = L_b(c) = b + c$, making $R_c(a) = R_c(b)$. This means that the unary polynomial R_c cannot be a permutation; so by our assumption on $\mathcal{A}$ it must be constant. Since one of its values is s, from $R_c(a) = a + c = s$, all of its values equal s, and $R_c(x) = s$ for all $x \in A$.

Now we claim that this forces all the polynomials $L_{a'}$, for $a' \neq a$, to be non-constant. For suppose that there was some $a' \neq a$ for which $L_{a'}$ was constant. Then $L_{a'}(x) = L_{a'}(c) = a' + c = s$, so that $L_{a'}$ always gives the value s. Thus $L_{a'} = L_a$, and for every $x \in A$ we have $a' + x = a + x$. But then for every $x \in A$ we have $R_x(a') = a' + x = a + x = R_x(a)$. Then R_x is not injective, hence not a permutation, hence must be constant. Now we have all right-additions R_x constant, contradicting the fact that our operation $x + y$ is essentially binary. Therefore we see that every $L_{a'}$, for $a' \neq a$, is non-constant.

Now we choose an element $t \in A \setminus \{s\}$ and define a mapping f on A by $f(x) := L_x^{m!}(t)$ for $m := |A|$. (Here $L_x^{m!}$ means the composition $L_x \circ L_x \circ \cdots \circ L_x$.) Since the mapping $x + x + \cdots + x$ is not constant by assumption, it is a permutation and its order divides the order $m!$ of the permutation group on A. This means that $f(x) = L_x^{m!}(t) = t$ for $x \neq a$, while $f(a) = L_a^{m!}(t) = s$. Thus the unary polynomial f is neither a permutation nor constant, which is a contradiction.

Step 2. *There is a loop operation in $P(\mathcal{A})$.*

Proof: We recall that a loop is an algebra $(L; +, 0)$ of type $(2, 0)$ such that $(L; +)$ is a quasigroup and 0 is a neutral element with respect to the operation

+. We know from Step 1 that the essentially binary polynomial operation $x + y$ of $\mathcal{A}$ is a quasigroup operation. From this we can define two new binary operations, right subtraction and left subtraction: $x/y := R_y^{-1}(x)$ and $x \setminus y := L_x^{-1}(y)$. Taking $m = |A|$ we have $R_y^{-1} = R_y^{m!-1}$ and thus $x/y = R_y^{m!-1}(x)$, so that $/$ is a polynomial operation of $\mathcal{A}$. Similarly, it can be shown that $\setminus$ is also a polynomial operation of $\mathcal{A}$.

Any quasigroup can be made into a loop: we select an element $0 \in A$ and then define a new addition operation $+$ in terms of the old one, by $x + y := (x/(0 \setminus 0)) + (0 \setminus y)$. It is easy to verify that $0 + x = x + 0 = x$, making 0 a neutral element for our operation.

Step 3. *Let G be the set of all non constant unary polynomial operations of $\mathcal{A}$. Then two different polynomials f and g in G agree on at most one element $a \in A$.*

Proof: Suppose that for some f and g in G we have $f(a) = g(a)$ and $f(b) = g(b)$ with $a \neq b$. Let h be the unary polynomial on $\mathcal{A}$ defined by $h(x) := f(x)/g(x)$, where $/$ is the right subtraction belonging to the loop operation $+$, as in Step 2. Then $h(a) = f(a)/g(a) = 0 = h(b) = f(b)/g(b)$, so h must be constant with value 0. But this means $f(x) = g(x)$ for all $x \in A$, and $f = g$.

Step 4. *$(A; +)$ is an abelian group.*

Proof: It suffices to show that $+$ is commutative and associative. For commutativity we start with $R_a(0) = 0 + a = a = a + 0 = L_a(0)$ and $R_a(a) = a + a = L_a(a)$, for every $a \in A$. From Step 1 we know that each L_a is a non-constant polynomial, as is each R_a. For $a \neq 0$, Step 3 then tells us that $R_a = L_a$; that is, $a + x = x + a$ for all $a \neq 0$. For $a = 0$ we have $0 + x = x + 0$ from Step 2.

A similar argument can be used for associativity. For all $a, b \in A$ we have $L_a(L_b(0)) = a+(b+0) = a+b = L_{a+b}(0)$. Because of the commutativity of $+$ we have also $L_a(L_b(a)) = a+(b+a) = (b+a)+a = (a+b)+a = L_{a+b}(a)$. Thus for $a \neq 0$, the polynomials $L_a \circ L_b$ and L_{a+b} agree on at least two elements, 0 and a. By Step 3, they must be equal, so that $a + (b + x) = (a + b) + x$ for all $a, b, x \in A$. For $a = 0$ this equation holds trivially.

Now we have our abelian group $(A; +, -, 0)$. To make a vector space structure, we use the fact that the set of endomorphisms of an abelian group forms a ring. The multiplication operation in this ring is composition of mappings, while addition of operations is defined pointwise, by $(f_1 + f_2)(a) = f_1(a) + f_2(a)$ for all $a \in A$. Let $K := \{k \in G \mid k(0) = 0\} \cup \{\overline{0}\}$, where $\overline{0}$ is the constant mapping with $\overline{0}(x) = 0$ for all $x \in A$.

Step 5. *The set K is the universe of a subring of the ring of all endomorphisms of the abelian group $(A; +, -, 0)$. Moreover since K is finite and all elements of $K \setminus \{\overline{0}\}$ are permutations, $\mathcal{K} = (K, +, -, \overline{0}, \circ, ^{-1}, id_A)$ is a field and $(A; +, -, 0, K)$ is a vector space over $\mathcal{K}$.*

Proof: We show first that every $k \in K \setminus \{\overline{0}\}$ is an endomorphism of $(A; +, -, 0)$. Consider the values of $k(a - x)$ and $k(a) - k(x)$ on the two inputs $x = 0$ and $x = a$; we have $k(a - 0) = k(a) - k(0)$ and $k(a - a) = k(0) = 0 = k(a) - k(a)$. Thus for any $a \in A$, the operations $k(a - x)$ and $k(a) - k(x)$ agree on $x = 0$ and $x = a$. For any $a \neq 0$ these are two different inputs; so by Step 3, $k(a - x) = k(a) - k(x)$. Thus the operations $k(y - b)$ and $k(y) - k(b)$ agree for all $y \neq 0$ and all $b \in A$. Since $|A| \geq 3$ they are equal also for $y = 0$. But this means that k is an endomorphism of $(A; +, -, 0)$. It is clear from the definition of K that for any $k_1, k_2 \in K$ we also have $k_1 + k_2$, $-k_1$ and $k_1 \circ k_2 \in K$, showing that $(K; +, -, \overline{0}, \circ)$ is a subring of the endomorphism ring of $(A; +, -, 0)$. Moreover since K is finite, for any $k \in K \setminus \{\overline{0}\}$ the inverse k^{-1} is also in K. Therefore $\mathcal{K}$ is a field, and $(A; +, -, 0, K)$ is a vector space over $\mathcal{K}$.

Step 6. *The elements of $P(\mathcal{A})$ are exactly the operations of the form*

$$(x_1, \ldots, x_n) \mapsto a + k_1 x_1 + \ldots + k_n x_n,$$

for some $n \in \mathbb{N}$, $a \in A$ and $k_1, \ldots, k_n \in K$.

Proof: By definition every operation of this form belongs to $P(\mathcal{A})$. For the converse, let f be a polynomial on $\mathcal{A}$ of arity $n \geq 0$. For $n = 0$ the claim is clear, so we assume now that $n \geq 1$ and proceed inductively.

Let g be the operation defined by

$$g(x_1, x_2, \ldots, x_n) := f(x_1, x_2, \ldots, x_n) - f(0, x_2, \ldots, x_n).$$

Then for all $x_2, \ldots, x_n \in A$ we have $g(0, x_2, \ldots, x_n) = 0$. For the base case $n = 1$ we take $f(x_1) = f(0) + g(x_1)$, and the claim is proved. Now let $n \geq 2$.

For arbitrary elements $b_3, \ldots, b_n$ we define $x * y := g(x, y, b_3, \ldots, b_n)$. Since $0 * y = g(0, y, b_3, \ldots, b_n) = 0$ for all $y \in A$, the operation $*$ is not the operation of a quasigroup. So by Step 1 the operation $*$ cannot be essentially binary, and must depend on at most one variable. Suppose that $x * y$ depends only on the variable y. Then there are elements a, b_1 and b_2 such that $a * b_1 \neq a * b_2$. This implies that for at least one of $i = 1, 2$, we have $a * b_i \neq 0 = 0 * b_i$. But that means that $x * y$ also depends on x. This contradiction shows that $x * y$ cannot depend on y. The elements $b_3, \ldots, b_n \in A$ were arbitrary; so we have shown that g does not depend on x_2. In the same way we show that g does not depend on $x_3, \ldots, x_n$. Thus g depends at most on x_1. If g does not even depend on x_1 then we take $k_1 = \bar{0}$; otherwise there is an element $k_1 \in K$ such that $g(x_1, \ldots, x_n) = k_1 x_1$ for all $x_1, \ldots, x_n \in A$. Then $f(x_1, x_2, \ldots, x_n) = k_1 x_1 + f(0, x_2, \ldots, x_n)$. By induction on n we get that f has the form claimed. This finishes the proof of Theorem 11.3.3. ∎

Theorem 11.3.3 shows that if $\mathcal{A}$ is a non-trivial minimal algebra with a polynomial operation depending on at least two variables but $\mathcal{A}$ is not polynomially equivalent to a vector space, then $\mathcal{A}$ is a two-element algebra. The two-element case is easy, since any two-element algebra is clearly a permutation algebra: any polynomial which is not constant is a permutation.

In Section 10.2 we determined all two-element algebras, and classified them into the following four cases, depending on the congruence lattice behaviour of the variety $V(\mathcal{A})$ generated by $\mathcal{A}$: $V(\mathcal{A})$ can be neither congruence distributive nor congruence permutable, $V(\mathcal{A})$ can be one of congruence distributive or congruence permutable but not the other, or $V(\mathcal{A})$ can be both congruence distributive and congruence permutable.

In addition, these cases were characterized in Section 9.6 by conditions on the term operations. That is, for a two-element algebra $\mathcal{A}$ the variety $V(\mathcal{A})$ is congruence permutable iff the operation p with $p(x, y, z) = x + y + z$ (where $+$ denotes addition modulo 2) is a term operation of $\mathcal{A}$, while $V(\mathcal{A})$ is congruence distributive iff one of the following operations is a term operation of $\mathcal{A}$: $m(x, y, z) = (x \wedge y) \vee (x \wedge z) \vee (y \wedge z)$, $x \wedge (y \vee z)$, $x \wedge (y \vee \neg z)$ or the dual operations of these last two. From this characterization we obtain the following classification (up to isomorphism) of all two-element algebras (or Boolean clones):

Class 1. $\mathcal{O}_1$, $\mathcal{O}_5$, $\mathcal{O}_8$:
- all two-element algebras with unary term operations;
- polynomially equivalent to $\mathcal{O}_8 = (\{0, 1\}; c_0^1, c_1^1)$.

Class 2. $\mathcal{L}_1$, $\mathcal{L}_3$, $\mathcal{L}_4$, $\mathcal{L}_5$:
- all two-element algebras which generate a congruence permutable, but not congruence distributive variety;
- polynomially equivalent to $\mathcal{L}_1 = (\{0, 1\}; c_0^1, c_1^1, +, \neg)$.

Class 3. $\mathcal{C}_1$, $\mathcal{C}_3$, $\mathcal{C}_4$, $\mathcal{D}_3$, $\mathcal{D}_1$, $\mathcal{F}_5^n$ and $\mathcal{F}_8^n$ for $n \geq 2$, $\mathcal{F}_5^\infty$, $\mathcal{F}_8^\infty$:
- all two-element algebras which generate a congruence distributive and congruence permutable, or a 3-distributive but not 2-distributive and not congruence permutable variety;
- polynomially equivalent to $\mathcal{C}_1 = (\{0, 1\}; c_0^1, c_1^1, \wedge, \neg)$.

Class 4. $\mathcal{A}_1$, $\mathcal{A}_3$, $\mathcal{A}_4$, $\mathcal{D}_2$, $\mathcal{F}_6^n$ and $\mathcal{F}_7^n$ for $n \geq 2$, $\mathcal{F}_6^\infty$, $\mathcal{F}_7^\infty$:
- all two-element algebras which generate a congruence distributive but not congruence permutable variety;
- polynomially equivalent to $\mathcal{A}_1 = (\{0, 1\}; c_0^1, c_1^1, \wedge, \vee)$.

Class 5. $\mathcal{P}_1$, $\mathcal{P}_3$, $\mathcal{P}_6$, $\mathcal{P}_5$:
- polynomially equivalent to $\mathcal{P}_6 = (\{0, 1\}; c_0^1, c_1^1, \wedge)$.

Taking one representative from each class (up to polynomial equivalence), we obtain the following sublattice of the lattice of all Boolean clones.

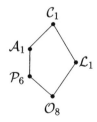

We now have, up to polynomial equivalence, five kinds of permutation algebras. This motivates the following definition.

Definition 11.3.5 Let $\mathcal{M}$ be a minimal algebra. Then we define the *type* of $\mathcal{M}$ as follows:

$\mathcal{M}$ has type 1: $\Leftrightarrow$ $\mathcal{M}$ is polynomially equivalent to an algebra $(M; \pi)$ for some $\pi \subseteq \mathcal{S}_M$ (unary type)

$\mathcal{M}$ has type 2: $\Leftrightarrow$ $\mathcal{M}$ is polynomially equivalent to a vector space (vector space type)

$\mathcal{M}$ has type 3: $\Leftrightarrow$ $\mathcal{M}$ is polynomially equivalent to $\mathcal{C}_1$ (Boolean type)

$\mathcal{M}$ has type 4: $\Leftrightarrow$ $\mathcal{M}$ is polynomially equivalent to $\mathcal{A}_1$ (lattice type)

$\mathcal{M}$ has type 5: $\Leftrightarrow$ $\mathcal{M}$ is polynomially equivalent to a two-element semilattice (semilattice type).

Using Theorem 11.3.3 and our classification of all two-element algebras we have the following result.

Theorem 11.3.6 *A finite algebra $\mathcal{A}$ is minimal iff it is of one of the types 1–5.* ∎

11.4 The Types of Minimal Algebras

In the previous section we characterized all minimal algebras, and assigned each one a type based on which of five representative minimal algebras it is polynomially equivalent to. We now want to extend our classification of types to the finite algebras $\mathcal{A}$ from which we obtain the minimal algebras $\mathcal{A}|U$, for U a minimal or β-minimal set. In Theorem 11.2.7 we showed that when β is a tame congruence relation on a finite algebra $\mathcal{A}$, any two minimal algebras induced on $\mathcal{A}$ are polynomially isomorphic to each other. This means that in this case all the minimal algebras induced on $\mathcal{A}$ have the same type, and we can use that type to assign a type to the algebra $\mathcal{A}$, as follows.

Definition 11.4.1 Let $\mathcal{A}$ be a finite tame algebra. If one (and hence all) of the minimal algebras of $\mathcal{A}$ is of type i, for $i \in \{1, 2, 3, 4, 5\}$, then $\mathcal{A}$ is said to be of *type i*, and we write type $\mathcal{A} = i$.

In the remark following Definition 11.2.3 we noted that our definition of a tame congruence could be extended to intervals in a congruence lattice. For $\mathcal{A}$ a finite algebra and α and β in $Con\,\mathcal{A}$ with $\alpha \subset \beta$, the interval $[\alpha, \beta] \subseteq Con\,\mathcal{A}$ is called tame in $\mathcal{A}$ if the congruence relation β/α of the quotient algebra $\mathcal{A}/\alpha$ is tame. Then β/α-minimal sets are called $[\alpha, \beta]$-minimal, and the corresponding minimal algebras of the quotient algebra $\mathcal{A}/\alpha$ are called $[\alpha, \beta]$-minimal.

Definition 11.4.2 Let C be $[\theta, \Theta]$- minimal and let $[\alpha, \beta]$ be tame in $\mathcal{A}$. A $[\theta, \Theta]$- *trace* is a set $N \subseteq C$ such that N is a Θ equivalence class $[x]_\Theta$ for which $[x]_\Theta \neq [x]_\theta$. An $[\alpha, \beta]$- *trace* N is a set $N \subseteq A$ such that there is some $[\alpha, \beta]$-minimal set U such that $N \subseteq U$ and N is an $[\alpha/U, \beta/U]$- trace of $\mathcal{A}/U$ (so $N = [x]_{\beta \cap U}$ for some $x \in U$ with $[x]_{\beta \cap U} \cap U \not\subseteq [x]_\alpha$).

The following theorem was proved by Hobby and McKenzie in [57].

Theorem 11.4.3 *Let* $[\alpha, \beta]$ *be a tame interval for the algebra* $\mathcal{A}$ *and let* N *be an* $[\alpha, \beta]$-*trace. Then* $(\mathcal{A}|_N)/(\alpha|_N)$ *is a minimal algebra. Moreover for all* $[\alpha, \beta]$-*traces* N *the minimal algebras* $(\mathcal{A}|_N)/(\alpha|_N)$ *have the same type 1–5.* ∎

This result allows us to extend our definition of type from minimal and tame algebras to tame intervals in a congruence lattice.

Definition 11.4.4 Let $[\alpha, \beta]$ be a tame interval of the finite algebra $\mathcal{A}$. The *type* of the interval is the type of the minimal algebra $(\mathcal{A}|_N)/(\alpha|_N)$ for an arbitrary $[\alpha, \beta]$-trace N.

We saw earlier that all two–element lattices are tight. In a congruence lattice $Con\,\mathcal{A}$, a two-element interval is called a *prime interval*. It follows that on a finite algebra $\mathcal{A}$ all prime intervals $[\alpha, \beta] \subseteq Con\,\mathcal{A}$ are tame. In this case, for any $[\alpha, \beta]$-minimal set N, the induced algebra $\mathcal{A}|_N$ is minimal with respect to $(\alpha|_N, \beta|_N)$.

Definition 11.4.5 The *type of a prime interval* $[\alpha, \beta]$ is the type of the $((\alpha|_N)/(\beta|_N))$-minimal algebra $\mathcal{A}|_N$, where N is an $[\alpha, \beta]$-minimal set.

Finally, our definition of type can be extended to arbitrary intervals and varieties. If $[\gamma, \lambda]$ is an arbitrary interval in $Con\,\mathcal{A}$, we define type $\{\gamma, \lambda\} :=$ $\{\text{type}(\alpha, \beta) \mid \gamma \leq \alpha, \beta \leq \lambda\}$, and type $\mathcal{A} = \{\text{type}\{\Delta_A, A \times A\}\}$. The type of a variety V of algebras is the set of all the types of finite algebras in the variety:

$$\text{type}\{V\} = \bigcup \; \{\text{type } \mathcal{A} \mid \mathcal{A} \in V \text{ and } \mathcal{A} \text{ is finite}\}.$$

We now present some examples. For the first one, we recall that a finite algebra $\mathcal{A} = (A; F^A)$ is called primal if its term clone is all of O_A.

Theorem 11.4.6 *Every primal algebra has type 3 (Boolean type).*

Proof: Let $\mathcal{A}$ be primal. We proved in Proposition 10.5.7 that any primal algebra is simple. By the primality, for any two elements $a, b \in A$ there is a polynomial operation f with $f(A) = \{a, b\}$. Therefore every two–element subset of A is a minimal set. It follows that if U is such a minimal set, the algebra $\mathcal{A}|_U$ is two–element. Let us denote the two elements by 0 and 1. Then there are term operations $\wedge$, $\vee$ and $\neg$ on $\mathcal{A}|_U$ which satisfy

$$
\begin{aligned}
0 \wedge 0 &= 0 \wedge 1 = 1 \wedge 0 = 0, \quad 1 \wedge 1 = 1, \\
0 \vee 1 &= 1 \vee 0 = 1 \vee 1 = 1, \quad 0 \vee 0 = 0, \\
\neg 0 &= 1, \quad \neg 1 = 0.
\end{aligned}
$$

Then the algebra $\mathcal{A}|_U$ is polynomially equivalent to the two-element Boolean algebra, which has type 3. ∎

Example 11.4.7 1. Let $\mathcal{A}$ be a finite algebra such that $clone\mathcal{A} = Pol\,\varrho$ for some partial order ϱ on A which has a least element 0 and a greatest element 1. Let

$$
g(x) = \begin{cases} 0 & \text{if} \quad x = 0 \\ 1 & \text{if} \quad x \neq 0. \end{cases}
$$

Then g is monotone, and the image set $U = \{0, 1\}$ gives a two-element minimal algebra $\mathcal{A}|_U$. Using

$$
\varphi(x, y) = inf(g(x), y) = \begin{cases} 0 & \text{if} \quad x = 0 \\ y & \text{if} \quad x \neq y, \end{cases}
$$

$$
\varphi'(x, y) = sup(g(x), y) = \begin{cases} 1 & \text{if} \quad x \neq 0 \\ y & \text{if} \quad x = 0, \end{cases}
$$

we see that $\mathcal{A}$ has type 4.

2. It can be shown that like primal algebras, functionally complete algebras have type 3.

We conclude this section with some results on the types possible for tame algebras.

Theorem 11.4.8 *Let $\mathcal{A}$ be a finite tame algebra.*

(i) If $Con\,\mathcal{A}$ has more than two elements, then type $\mathcal{A} \in \{1,2\}$.

(ii) If there is no homomorphic image of the lattice $Con\,\mathcal{A}$ which is isomorphic to a congruence lattice of a vector space with more than one element, then type $\mathcal{A} = \{1\}$.

Proof: (i) Suppose that type $\mathcal{A} \notin \{1,2\}$. Then any minimal set U on $\mathcal{A}$ has $|U| = 2$, and the induced algebra $\mathcal{A}|_U$ has only two elements and is simple. We saw in Theorem 11.1.2 that the mapping $\varphi_U(\theta) := \theta|_U$ is a lattice homomorphism from $Con\,\mathcal{A}$ onto $Con(\mathcal{A}|_U)$. By the simplicity of $\mathcal{A}|_U$, for any congruence θ on $\mathcal{A}$ we have $\theta|_U \in \{\Delta_U,\ U \times U\}$. But now the definition of a tame algebra, Definition 11.2.1, shows that $Con\,\mathcal{A}$ can only contain the two congruences Δ_A and $A \times A$.

(ii) Since there are vector spaces with two-element congruence lattices, we may assume that $|Con\,\mathcal{A}| = 2$. But then by (i) only type $\mathcal{A} \in \{1,2\}$ is possible. But the type of $\mathcal{A}$ cannot be 2, since for every $U \in Min(\mathcal{A})$ the lattice $Con(\mathcal{A}|_U)$ is a homomorphic image of $Con\,\mathcal{A}$ by Theorem 11.1.2 again, and $\mathcal{A}|_U$ would be equivalent to a vector space with the congruence lattice $Con(\mathcal{A}|_U)$. ∎

Lemma 11.4.9 *Let $\mathcal{A}$ be a finite algebra such that $Con\,\mathcal{A}$ is isomorphic to the lattice $\mathcal{M}_n$ for some $n \geq 3$. If $n-1$ is not a prime power, then $\mathcal{A}$ is tame and* type $\mathcal{A} = 1$.

Proof: By example 11.2.9 we know that the algebra $\mathcal{A}$ is tame. Moreover, for $n \geq 3$ the lattices $\mathcal{M}_n$ are simple. To see this, let $\theta \neq \Delta_A$ and $(a,b) \in \theta$ with $a \neq b$, $b \neq 1$ and $a \neq 0$. Then $(a \wedge a, a \wedge b) = (a,0)$, if $b \neq 1$ and $(a \vee a, a \vee 1)$ if $b = 1$, and we have $\theta = A \times A$. Therefore, all homomorphic images with more than one element are isomorphic to $\mathcal{M}_n$. For every finite field $\mathcal{K}$, the cardinality $|K|$ is a prime power, and the congruence lattice of a two-dimensional vector space over $\mathcal{K}$ has the form $\mathcal{M}_{|K|+1}$. By 11.4.8 (ii) we have type $\mathcal{A} = 1$. ∎

Now we consider properties of tame algebras of type 1 or type 2.

Definition 11.4.10 An algebra $\mathcal{A}$ satisfies the *strong term condition* if for all $n \in I\!N$, all n-ary term operations t^A of $\mathcal{A}$ and all $b, c_1, \ldots, c_n, d_1, \ldots, d_n \in A$ the following implication is satisfied:

$$t^A(c_1, c_2, \ldots, c_n) = t^A(d_1, d_2, \ldots, d_n) \quad \Rightarrow$$
$$t^A(b, c_2, \ldots, c_n) = t^A(b, d_2, \ldots, d_n).$$

An algebra satisfying the strong term condition is called *strongly abelian*.

An algebra A satisfies the *term condition* if for all $n \in \mathbb{N}$, for all n-ary term operations t^A of A and all a, b, c_2, ..., c_n, d_2, ..., d_n the following implication holds:

$$t^A(a, c_2, \ldots, c_n) = t^A(a, d_2, \ldots, d_n) \quad \Rightarrow$$
$$t^A(b, c_2, \ldots, c_n) = t^A(b, d_2, \ldots, d_n).$$

An algebra which satisfies the term condition is said to be *abelian*.

It is clear that if an algebra A satisfies the strong term condition then it also satisfies the term condition, so that strongly abelian algebras are abelian.

The concept of an abelian algebra generalizes the usual concept of an abelian group. A group $\mathcal{G} = (G; \cdot, {}^{-1}, e)$ is called abelian in the group-theoretical sense if it satisfies the identity $x \cdot y \approx y \cdot x$. We show that a group is abelian in this sense iff it is abelian as an algebra. If $\mathcal{G}$ is an abelian algebra then $t^G(x, y, z) := yxz$ satisfies $t^{\mathcal{G}}(e, e, a) = a = t^{\mathcal{G}}(e, a, e) \Rightarrow t^{\mathcal{G}}(b, e, a) = t^{\mathcal{G}}(b, a, e)$ and thus $ba = ab$, making $\mathcal{G}$ an abelian group. Conversely when $\mathcal{G}$ is an abelian group, the n-ary term operations of $\mathcal{G}$ for $n \geq 1$ are exactly those of the form $t^{\mathcal{G}}(x_1, \ldots, x_n) = x_1^{k_1} \cdots x_n^{k_n}$ with $k_1, \ldots, k_n \in \mathbb{Z}$, and it is easy to show that these terms satisfy the term condition.

Example 11.4.11 Consider the binary operations f and g given by the Cayley tables below.

f	0	1	2	3
0	0	0	1	1
1	0	0	1	1
2	0	0	1	1
3	2	2	3	3

g	0	1	2	3
0	0	0	1	1
1	0	0	1	1
2	0	0	1	1
3	2	2	0	0.

It can be verified that for all evaluations f satisfies the implication of the strong term condition, while g satisfies the implication of the term condition.

Abelian and strongly abelian algebras will be studied further in the next chapter. For the moment, we characterize the types of such algebras.

Theorem 11.4.12 *Let A be a finite tame algebra. Then*

(i) $\mathcal{A}$ *is strongly abelian iff* type $\mathcal{A} = 1$, *and*

(ii) $\mathcal{A}$ *is abelian iff* type $\mathcal{A} \in \{1, 2\}$. ∎

Corollary 11.4.13 *Every finite algebra $\mathcal{A}$ whose congruence lattice $Con\,\mathcal{A}$ is isomorphic to the lattice $\mathcal{M}_n$ for some $n \geq 3$ is abelian. If in addition $n - 1$ is not a prime power, then $\mathcal{A}$ is also strongly abelian.*

Proof: We have shown that the lattices $\mathcal{M}_n$ with $m \geq 3$ are tight, and therefore every finite algebra with $Con\,\mathcal{A} \cong \mathcal{M}_n$ is tame. By Theorem 11.4.8 (i) we have type $\mathcal{A} \in \{1, 2\}$, and then Theorem 11.4.12 (ii) shows that $\mathcal{A}$ is abelian. If $n - 1$ is not a prime power then by Lemma 11.4.9 and Theorem 11.4.12 $\mathcal{A}$ is strongly abelian. ∎

11.5 Mal'cev Conditions and Omitting Types

Hobby and McKenzie proved in [57] that there is a close connection between the types of finite algebras $\mathcal{A}$ and Mal'cev-type conditions for the varieties $V(\mathcal{A})$ they generate. To present their results in this section, we will use our Mal'cev condition results from Chapter 9 along with the following new properties of congruences.

We recall from Section 6.7 that an algebra $\mathcal{A}$ is called *meet-semidistributive* iff whenever congruences θ_1, θ_2 and θ_3 in $Con\,\mathcal{A}$ satisfy $\theta_1 \wedge \theta_2 = \theta_1 \wedge \theta_3$, they also satisfy $\theta_1 \wedge \theta_2 = \theta_1 \wedge (\theta_2 \vee \theta_3)$.
Dually, $\mathcal{A}$ is called *join-semidistributive* iff whenever congruences θ_1, θ_2 and θ_3 in $Con\,\mathcal{A}$ satisfy $\theta_1 \vee \theta_2 = \theta_1 \vee \theta_3$, they also satisfy $\theta_1 \vee \theta_2 = \theta_1 \vee (\theta_2 \wedge \theta_3)$.

An algebra which is both meet- and join-semidistributive is called *semidistributive*. A variety is called semidistributive if all the algebras in it are semidistributive.

Hobby and McKenzie ([57]) proved the following "omitting type" theorem, which characterizes the types which cannot occur for a variety by means of the Mal'cev-condition properties of the variety.

Theorem 11.5.1 *Let $\mathcal{A}$ be a finite algebra, with $V(\mathcal{A})$ the variety generated by $\mathcal{A}$. Then:*

(i) *If $V(\mathcal{A})$ is congruence distributive, then type $\{V(\mathcal{A})\} \cap \{1,2,5\} = \emptyset$;*

(ii) *If $V(\mathcal{A})$ is congruence permutable, then type $\{V(\mathcal{A})\} \subseteq \{2,3\}$;*

(iii) *$V(\mathcal{A})$ is n-permutable for some $n \geq 2$ iff type $\{V(\mathcal{A})\} \subseteq \{2,3\}$;*

(iv) *type $\{V(\mathcal{A})\} = \{3\}$ iff in $V(\mathcal{A})$ there exist terms $f_0(x,y,z,u), \ldots, f_n(x,y,z,u)$, for $n \geq 2$, such that the following identities hold in V:*

$$f_0(x,y,y,z) \approx x,$$
$$f_i(x,x,y,x) \approx f_{i+1}(x,y,y,x), \quad \text{for all } i < n,$$
$$f_i(x,x,y,y) \approx f_{i+1}(x,y,y,y), \quad \text{for all } i < n,$$
$$f_n(x,x,y,z) \approx z;$$

(v) *type $\{V(\mathcal{A})\} \cap \{1,2\} = \emptyset$ iff $V(\mathcal{A})$ is meet-semidistributive iff the class of all lattices isomorphic to a sublattice of $\operatorname{Con}\mathcal{B}$ for some $\mathcal{B} \in V(\mathcal{A})$ does not contain the lattice $\mathcal{M}_3$.* ∎

As an easy corollary we obtain the following result.

Corollary 11.5.2 *Let $\mathcal{A}$ be a finite algebra. If $V(\mathcal{A})$ is both congruence distributive and n-permutable for some $n \geq 2$ then type $\{V(\mathcal{A})\} = 3$.* ∎

To formulate the next theorem we need the concept of a 1-*snag*.

Definition 11.5.3 *A 1-snag of an algebra $\mathcal{A}$ is a pair (a,b) of distinct elements of A such that for some $f \in P^2(\mathcal{A})$, the equations $f(a,b) = f(b,a)$ and $f(b,b) = b$ are satisfied. We denote the set of all 1-snags of $\mathcal{A}$ by $S_{n_1}(\mathcal{A})$.*

Hobby and McKenzie used 1-snags to characterize locally finite varieties, that is, varieties in which every finitely generated algebra is finite, of type 1.

Lemma 11.5.4 *Let $V(\mathcal{A})$ be a locally finite variety. Then type $\{V(\mathcal{A})\} = \{1\}$ iff for every algebra $\mathcal{B}$ in $V(\mathcal{A})$ the set $S_{n_1}(\mathcal{B}) = \emptyset$.* ∎

For algebras which generate congruence modular varieties we have the following useful lemma, due to E.W. Kiss and E. Pröhle ([62]).

Lemma 11.5.5 *Let $\mathcal{A}$ be a finite algebra. If $V(\mathcal{A})$ is congruence modular then* type $\{V(\mathcal{A})\}$ = type $\{Sub(\mathcal{A})\}$. ∎

We saw in Chapter 9 that for a two-element algebra $\mathcal{A}$, the variety $V(\mathcal{A})$ is congruence modular iff $V(\mathcal{A})$ is congruence distributive or congruence permutable. Combining this with Lemma 11.5.5 gives a characterization of the type of $V(\mathcal{A})$ for any two-element algebra $\mathcal{A}$.

Theorem 11.5.6 *Let $\mathcal{A}$ be a two-element algebra of type i, for $i \in \{1,2,3,4,5\}$. Then* type $\{V(\mathcal{A})\} = \{i\}$.

Proof: We consider two cases, depending on whether $V(\mathcal{A})$ is congruence modular or not.

Case 1: $V(\mathcal{A})$ is congruence modular. In this case we apply Lemma 11.5.5., and the fact that $\mathcal{A}$ has no non-trivial subalgebras and is simple, to conclude that type $\{V(\mathcal{A})\}$ = type$\{Sub(\mathcal{A})\}$ = type$\{\mathcal{A}\} = \{i\}$.

Case 2: $V(\mathcal{A})$ is not congruence modular. In this case $V(\mathcal{A})$ is neither congruence distributive nor congruence permutable. From our classification of two-element algebras in Section 10.3, we know that exactly the following two-element algebras have this property:

$\mathcal{P}_1 = (\{0,1\}; \wedge)$, $\mathcal{P}_3 = (\{0,1\}; \wedge, c_0^1)$, $\mathcal{P}_5 = (\{0,1\}; \wedge, c_1^1)$,
$\mathcal{P}_6 := (\{0,1\}; \wedge c_0^1, c_1^1)$, $\mathcal{O}_1 = (\{0,1\}; c_1^1)$, $\mathcal{O}_4 = (\{0,1\}; \neg)$,
$\mathcal{O}_8 = (\{0,1\}; c_0^1, c_1^1)$, $\mathcal{O}_9 = (\{0,1\}; \neg, c_0^1)$

(and the dually isomorphic algebras).

The algebras $\mathcal{O}_1$, $\mathcal{O}_8$, $\mathcal{O}_9$ have type 1. For each of these algebras $\mathcal{A}$ there is no algebra $\mathcal{B}$ in $V(\mathcal{A})$ with an essentially binary term operation, so we have $Sn_1(\mathcal{B}) = \emptyset$ for any $\mathcal{B} \in V(\mathcal{A})$. Lemma 11.5.4 then shows that type $\{V(\mathcal{A})\} = \{1\}$ for these algebras $\mathcal{A}$.

The algebras $\mathcal{O}_1$, $\mathcal{P}_3$, $\mathcal{P}_5$ and $\mathcal{P}_6$ are all the two-element algebras of type 5. The varieties generated by these algebras are the varieties of semilattices, semilattices with zero, semilattices with unit, and bounded semilattices. It was proved by G. Czedli in [14] that these varieties are all congruence meet-semidistributive; so by Theorem 11.5.1 (v) the type of $V(\mathcal{A})$

cannot contain 1 or 2. If 3 or 4 is in type$\{V(\mathcal{A})\}$ then there must exist a finite algebra $\mathcal{B} \in V(\mathcal{A})$ and a prime interval $(\delta, \theta) \in Con\, \mathcal{B}$ such that $(N; q|_N,\ p|_N)$ is a two-element lattice, where N is the uniquely determined (δ, θ)-trace and q, p are binary polynomial operations of $\mathcal{B}$ which satisfy $p(x, 1) = p(1, x) = p(x, x) = x$ and $q(x, 0) = q(0, x) = q(x, x)$ for all $x \in B$ and $N = \{0, 1\}$ (see [57]). But this contradicts the fact that $\mathcal{B}$ is a finite semilattice (with unit or bounded). Therefore, type $\{V(\mathcal{A})\} = \{5\}$.   ∎

For minimal algebras $\mathcal{A}$ with $|A| \geq 3$ we have the following results.

Lemma 11.5.7 *Let $\mathcal{A}$ be a minimal algebra with $|A| \geq 3$. If* type $\mathcal{A} = 1$ *then* type $\{V(\mathcal{A})\} = \{1\}$.

Proof: If all polynomial operations of $\mathcal{A}$ are essentially unary, then any algebra $\mathcal{B} \in V(\mathcal{A})$ has the same property. By Lemma 11.5.4 we have type $\{V(\mathcal{A})\} = \{1\}$.   ∎

Theorem 11.5.8 *Let $\mathcal{M}$ be a minimal algebra. Then we have:*

$\mathcal{M}$ has type 3 $\Leftrightarrow$ $V(\mathcal{M})$ is congruence distributive and there is an
$\qquad\qquad\qquad\qquad n \geq 2$ such that $V(\mathcal{M})$ is n-permutable;

$\mathcal{M}$ has type 4 $\Leftrightarrow$ $V(\mathcal{M})$ is congruence distributive, and there is no
$\qquad\qquad\qquad\qquad n \geq 2$ such that $V(\mathcal{M})$ is n-permutable;

$\mathcal{M}$ has type 5 $\Leftrightarrow$ $V(\mathcal{M})$ is not congruence distributive and not
$\qquad\qquad\qquad\qquad n$-permutable for a certain natural number $n \geq 2$,
$\qquad\qquad\qquad\qquad$ but $V(\mathcal{M})$ is meet-semidistributive;

$\mathcal{M}$ has type 2 $\Leftrightarrow$ $V(\mathcal{M})$ is not congruence distributive and not
$\qquad\qquad\qquad\qquad$ meet-semidistributive, and $V(\mathcal{M}^+)$ is
$\qquad\qquad\qquad\qquad n$-permutable for some n;

$\mathcal{M}$ has type 1 $\Leftrightarrow$ $V(\mathcal{M})$ is not congruence distributive and not
$\qquad\qquad\qquad\qquad$ meet-semidistributive, and $V(\mathcal{M}^+)$ is not
$\qquad\qquad\qquad\qquad n$-permutable for a certain natural number $n \geq 2$.

Proof: If $\mathcal{M}$ has type 3, then $\mathcal{M}$ is polynomially equivalent to the two-element Boolean algebra and therefore is one of the following two-element algebras (up to isomorphism):

$\mathcal{C}_1, \mathcal{C}_3, \mathcal{C}_4, \mathcal{D}_3, \mathcal{D}_1, \mathcal{F}_5^n, \mathcal{F}_8^n, \mathcal{F}_5^\infty, \mathcal{F}_8^\infty$ (for $n \geq 2$).

As we saw in Section 9.6, it was shown by M. Reschke, O. Lüders and K. Denecke in [100] that these algebras generate congruence distributive and n-permutable varieties, for $n = 2$ or $n = 3$.

If $\mathcal{M}$ has type 4, then $\mathcal{M}$ is polynomially equivalent to the two-element lattice, and $\mathcal{M}$ is one of the following two-element algebras (up to isomorphism):

$\mathcal{A}_1$, $\mathcal{A}_3$, $\mathcal{A}_4$, $\mathcal{D}_2$, $\mathcal{F}_6^n$, $\mathcal{F}_7^n$, $\mathcal{F}_6^\infty$, $\mathcal{F}_7^\infty$ (for $n \geq 2$).

We saw in Section 9.6 that these algebras generate varieties which are congruence distributive, but not n-permutable for $n \geq 2$.

If $\mathcal{M}$ has type 5, then $\mathcal{M}$ is (up to isomorphism) one of the two-element algebras $\mathcal{P}_1$, $\mathcal{P}_3$, $\mathcal{P}_5$, $\mathcal{P}_6$. In this case (see [100]) the variety $V(\mathcal{M})$ is neither congruence distributive nor congruence permutable, but was shown to be meet-semidistributive by D. Papert in [85].

If $\mathcal{M}$ has type 1, then $\mathcal{M}$ has no essentially at least binary term operations. Since the Mal'cev-type conditions for congruence distributivity and n-permutability require essentially at least ternary term operations, we see that $V(\mathcal{M})$ is not congruence distributive and not n-permutable for some $n \geq 2$. Also by Theorem 11.5.1 (v) we see that $V(\mathcal{M})$ cannot be meet-semidistributive.

If $\mathcal{M}$ has type 2, then by Theorem 11.5.1 parts (i) and (v) $V(\mathcal{M})$ cannot be congruence distributive or meet-semidistributive. Hobby and McKenzie showed in [57] that $\mathcal{M}$ has a Mal'cev operation as a polynomial operation, so that $\mathcal{M}^+$ has a Mal'cev operation as a term operation and thus $V(\mathcal{M}^+)$ is congruence permutable (that is, 2-permutable).

Conversely, if $V(\mathcal{M})$ is congruence distributive and n-permutable for some $n \geq 2$ then by Corollary 11.5.2 type $\{V(\mathcal{M})\} = \{3\}$ and type $\mathcal{M} = 3$. If $V(\mathcal{M})$ is congruence distributive but there is no $n \geq 2$ such that $V(\mathcal{M})$ is n-permutable, then by Theorem 11.5.1 (i) type $\{V(\mathcal{M})\} \cap \{1, 2, 5\} = \emptyset$, so type $\{V(\mathcal{M})\} \subseteq \{3, 4\}$. By the proof of Theorem 11.5.6 and Lemma 11.5.7 type $\{V(\mathcal{M})\} = \{3, 4\}$ is impossible for a minimal algebra $\mathcal{M}$. If type $\{V(\mathcal{M})\} = \{3\}$, then type $\mathcal{M} = 3$ and $V(\mathcal{M})$ would be n-permutable for some $n \geq 2$. Therefore type $\{V(\mathcal{M})\} = \{4\}$.

If $V(\mathcal{M})$ is not congruence distributive, not n-permutable for some $n \geq 2$, but is meet–semidistributive, then type $\{V(\mathcal{M})\} \cap \{1, 2\} = \emptyset$, so we have type $\{V(\mathcal{M})\} \subseteq \{3, 4, 5\}$. Using Theorem 11.5.6 and Lemma 11.5.7 again, this means that type $\{V(\mathcal{M})\}$ must be one of $\{3\}$, $\{4\}$ or $\{5\}$. The first two of these are impossible, since otherwise $V(\mathcal{M})$ would be congruence distributive or n-permutable for some $n \geq 2$. Therefore we have type $\{5\}$.

Next, if $V(\mathcal{M})$ is not congruence distributive and not meet- semidistributive, then $1 \in \text{type}\{V(\mathcal{M})\}$ or $2 \in \text{type}\{V(\mathcal{M})\}$. Having $V(\mathcal{M}^+)$ n-permutable means that type $\{V(\mathcal{M}^+)\} \subseteq \{2, 3\}$. Thus type $\mathcal{M}^+$ must be 2 or 3. But if type $\mathcal{M}^+ = 3$, then type $\mathcal{M} = 3$ and type $\{V(\mathcal{M})\} = \{3\}$, making $V(\mathcal{M})$ congruence distributive. We are left with type $\mathcal{M}^+ = $ type $\mathcal{M} = 2$.

Finally, if $V(\mathcal{M})$ is not congruence distributive and not meet-semidistributive, and $V(\mathcal{M}^+)$ is not n-permutable for some $n \geq 2$, then type $\mathcal{M} \notin \{2, 3, 4, 5\}$, and therefore type $\mathcal{M} = 1$.   ∎

11.6   Residually Small Varieties

To illustrate the applications of tame congruence theory, we present in this section two results on residually small varieties. We recall from Section 6.7 that a variety V is called *residually small* if there is a cardinal number λ such that every subdirectly irreducible algebra in V has at most λ elements.

Varieties which are not residually small are called *residually large*. A variety is called *locally finite* if every finitely generated algebra in the variety is finite.

Example 11.6.1 The variety of distributive lattices is generated by the two-element distributive lattice $2_D = (\{0, 1\}; \wedge, \vee)$. It is wellknown that every distributive lattice is isomorphic to a subdirect power of this algebra 2_D. Therefore 2_D is the only subdirectly irreducible algebra in $V(2_D)$, and $V(2_D)$ is residually small. Similarly it can be shown that the variety of Boolean algebras and the variety generated by a primal algebra are residually small.

Hobby and McKenzie proved the following result in [57].

Proposition 11.6.2 *Let $\mathcal{A}$ be a finite algebra such that $Con\mathcal{A}$ has a sublat-*

tice of the form given by the Hasse diagram below (the pentagon) in which β covers α (there are no congruences properly between them), and such that type $[\alpha, \beta] = 2$ and type $[\Delta_A, \delta] \in \{3, 4\}$.

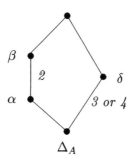

Then $V(\mathcal{A})$ is residually large. ∎

Theorem 11.6.3 *Every locally finite variety which omits the types 1 and 5, and is residually small, is congruence-modular.* ∎

11.7 Exercises

11.7.1. Let $\mathcal{L}$ be the lattice given by the following Hasse diagram:

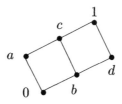

Let $\mu : L \mapsto L$ be the mapping which maps 0 and a to c and all other elements to 1. Show that μ is strictly extensive and a meet-endomorphism.

11.7.2. Verify that the lattices from Example 11.2.9 are tight.

11.7.3. Verify that the two lattices shown below are not tight.

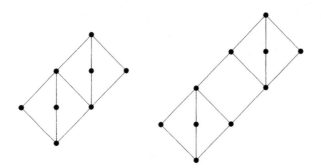

11.7.4. Prove that any functionally complete algebra has type 3.

11.7.5. A lattice $\mathcal{L}$ is called *order polynomially complete* iff every monotone mapping $f : L^n \to L$, for any $n \in \mathbb{N}$, is a polynomial operation of $\mathcal{L}$. Prove that a lattice $\mathcal{L}$ is order polynomially complete iff $\mathcal{L}$ is both tight and simple.

11.7.6. Let $(S, \cdot)$ be a finite semigroup. Show that for each element $a \in S$ there is some integer $k \geq 1$ such that $e = a^k$ is an idempotent, that is, $a^{2k} = a^k$. Moreover, there is an integer k such that $a^{2k} = a^k$ holds for every $a \in S$.

11.7.7. Prove that every finite quasigroup generates a congruence permutable variety.

Chapter 12

Term Condition and Commutator

In this chapter we show how concepts such as an abelian group, the commutator of a group and related concepts like solvable groups may be generalized to arbitrary universal algebras. The *commutator* of two elements a and b in a group $\mathcal{G}$ is the element $[a, b] := a^{-1}b^{-1}ab$. The *commutator group* of $\mathcal{G}$ is the normal subgroup of $\mathcal{G}$ which is generated by the set $\{[a, b] \mid a, b \in G\}$ of all commutators of $\mathcal{G}$. In a sense, the commutator subgroup of a group $\mathcal{G}$ measures how far the group is from being commutative. More generally, if $\mathcal{M}$ and $\mathcal{N}$ are two normal subgroups of $\mathcal{G}$ then the commutator group of $\mathcal{M}$ and $\mathcal{N}$, written as $[\mathcal{M}, \mathcal{N}]$, is the normal subgroup of $\mathcal{G}$ generated by the set $\{[a, b] \mid a \in M, b \in N\}$. It is well known that $[\mathcal{M}, \mathcal{N}]$ is the least normal subgroup of $\mathcal{G}$ with the property that $gh = hg$ for all $g \in M/_{[\mathcal{M},\mathcal{N}]}, h \in N/_{[\mathcal{M},\mathcal{N}]}$.

12.1 The Term Condition

We begin by generalizing the concept of an abelian group to arbitrary algebras. If $\mathcal{G}$ is an abelian group then the n-ary term operations of $\mathcal{G}$ are exactly operations induced by the terms $t(x_1, \ldots, x_n) = x_1^{k_1} \cdot \ldots \cdot x_n^{k_n}$, for $k_1, \ldots, k_n \in \mathbb{Z}$. These terms are normal forms for arbitrary n-ary terms over the variety of all abelian groups, meaning that any n-ary term is equivalent to one of these terms modulo IdV, where V is the variety of all abelian groups. We saw in Section 11.4 that a group $\mathcal{G}$ is abelian iff its term operations satisfy the so-called *term condition* (TC) of Definition 11.4.10: for any n-ary term t (with induced term operation $t^{\mathcal{G}}$) and any elements $a, b, c_2, \ldots, c_n$,

289

$d_2, \ldots, d_n \in G$

$$(\text{TC}) \quad t^{\mathcal{G}}(a, c_2, \ldots, c_n) = t^{\mathcal{G}}(a, d_2, \ldots, d_n)$$
$$\Rightarrow t^{\mathcal{G}}(b, c_2, \ldots, c_n) = t^{\mathcal{G}}(b, d_2, \ldots, d_n).$$

This gives us an equivalent characterization of the abelian property for groups, by means of term operations, and suggests how to generalize the concept of abelian to arbitrary algebras.

Definition 12.1.1 An algebra $\mathcal{A}$ satisfies the *term condition* (TC) if for all $n \in \mathbb{N}\backslash\{0\}$, for all n-ary term operations $t^{\mathcal{A}}$ of $\mathcal{A}$ and for all elements $a, b, c_2, \ldots, c_n, d_2, \ldots, d_n \subset A$ the following implication is satisfied:

$$(\text{TC}) \quad t^{\mathcal{A}}(a, c_2, \ldots, c_n) = t^{\mathcal{A}}(a, d_2, \ldots, d_n)$$
$$\Rightarrow t^{\mathcal{A}}(b, c_2, \ldots, c_n) = t^{\mathcal{A}}(b, d_2, \ldots, d_n).$$

An algebra $\mathcal{A}$ is called *abelian* if it satisfies the term condition (TC).

Example 12.1.2 1. Every abelian group is an abelian algebra.

2. Zero-semigroups, right-zero semigroups and left-zero-semigroups are semigroups satisfying, respectively, the following identities: $x_1 x_2 \approx x_3 x_4$, $x_1 x_2 \approx x_2$ and $x_1 x_2 \approx x_1$. (Here we follow the convention of writing the semigroup operation as juxtaposition.) These are all abelian algebras. (See Exercise 12.3.1.)

A rectangular band is a semigroup which satisfies the identities $x_1 x_2 x_3 \approx x_1 x_3$ and $x_1^2 \approx x_1$. Since in a rectangular band any at least binary term can be written as a product of the first and last variable used in it, it is easy to show that any rectangular band is an abelian algebra: we have $ac_2 \cdots c_n = ad_2 \cdots d_n \Rightarrow ac_n = ad_n \Rightarrow bac_n = bad_n \Rightarrow bc_n = bd_n \Rightarrow bc_2 \cdots c_n = bd_2 \cdots d_n$.

3. Every algebra which has only unary fundamental operations is abelian, since then every term operation is also unary, and the term condition TC is always satisfied.

4. Every subalgebra of an abelian algebra is abelian and every direct product of abelian algebras is abelian. (See Exercise 12.3.2.)

The next theorem shows that the property of being abelian can be characterized by properties of the congruence lattice of the algebra $\mathcal{A}^2$.

Theorem 12.1.3 *An algebra $\mathcal{A}$ is abelian if and only if there is a congruence relation θ on the algebra $\mathcal{A}^2$ such that the diagonal $\triangle_A = \{(a, a) \mid a \in A\}$ is a congruence class of θ.*

Proof: We show that $\triangle_A$ is a class or block of the congruence $\langle \triangle_A \rangle_{Con\mathcal{A}^2}$ which is generated by $\triangle_A$. We must show that for arbitrary elements $a, b \in A$ we have $((a, a), (b, b)) \in \langle \triangle_A \rangle_{Con\mathcal{A}^2}$, while for any $b \neq c$ in A, with $u = (b, c)$, we have $((a, a), u) \notin \langle \triangle_A \rangle_{Con\mathcal{A}^2}$. We use the characterization from Lemma 5.3.3 of the congruence generated by $\triangle_A$ on $\mathcal{A}^2$. Thus we have to show that for all $a, b \in A$ and all unary polynomial operations $p^{\mathcal{A} \times \mathcal{A}}$ of $\mathcal{A} \times \mathcal{A}$, the implication

$$p^{\mathcal{A} \times \mathcal{A}}((a, a)) \in \triangle_A \Rightarrow p^{\mathcal{A} \times \mathcal{A}}((b, b)) \in \triangle_A \quad (*)$$

is satisfied. Every unary polynomial operation $p^{\mathcal{A}}$ of $\mathcal{A}$ arises from some n-ary term operation $t^{\mathcal{A}}$ of $\mathcal{A}$, by substitution of constants from A for $n - 1$ of the variables in $t^{\mathcal{A}}$. Therefore we have:

$$p^{\mathcal{A} \times \mathcal{A}}((a, a)) = (p^{\mathcal{A}}(x), p^{\mathcal{A}}(y))$$
$$= (t^{\mathcal{A}}(a, c_2, \ldots, c_n), t^{\mathcal{A}}(a, d_2, \ldots, d_n)) \in \triangle_A$$
$$\Rightarrow t^{\mathcal{A}}(a, c_2, \ldots, c_n) = t^{\mathcal{A}}(a, d_2, \ldots, d_n)$$
$$\Rightarrow t^{\mathcal{A}}(b, c_2, \ldots, c_n) = t^{\mathcal{A}}(b, d_2, \ldots, d_n)$$
$$\Rightarrow p^{\mathcal{A} \times \mathcal{A}}(b, b) \in \triangle_A,$$

using the term condition (TC).

Since in this sense the condition $(*)$ is equivalent to the term condition, and $(*)$ is also equivalent to the fact that $\triangle_A$ is a block of the congruence $\langle \triangle_A \rangle_{Con\mathcal{A}^2}$, we have proved our proposition. ∎

The basic term condition (TC) can be generalized to a *term condition for congruences*, which we shall denote by (TCC).

Definition 12.1.4 Let $\mathcal{A}$ be an algebra and let θ_1 and θ_2 be congruences on $\mathcal{A}$. The algebra $\mathcal{A}$ satisfies the *term condition (TCC) for congruences* with respect to the pair (θ_1, θ_2) if for all $n \in \mathbb{N} \setminus \{0\}$, for all n-ary term operations

t^A of A and for all $(a, b) \in \theta_2, (c_2, d_2) \in \theta_1, \ldots, (c_n, d_n) \in \theta_1$ the implication

$$t^A(a, c_2, \ldots, c_n) = t^A(a, d_2, \ldots, d_n) \quad \Rightarrow$$
$$t^A(b, c_2, \ldots, c_n) = t^A(b, d_2, \ldots, d_n) \quad \text{(TCC)}$$

is satisfied.

This generalized term condition (TCC) also has an interpretation in group theory. There is a well known one-to-one correspondence between normal subgroups of a group G and congruence relations of the group: if N is a normal subgroup of G, the relation θ_N defined by

$$(a, b) \in \theta_N \quad \Leftrightarrow \quad ab^{-1} \in N \quad \Leftrightarrow \quad \exists h \in N(a = hb)$$

is a congruence. Conversely, if θ is a congruence on a group G, the congruence class of the identity element of G forms a normal subgroup of G. From this correspondence we have the following equivalence.

Lemma 12.1.5 *The following two statements are equivalent for normal subgroups M and N of a group G:*

(i) $ab = ba$ for all $a \in M$ and $b \in N$,
(ii) G satisfies the term condition (TCC) with respect to (θ_M, θ_N).

Proof: (ii) $\Rightarrow$ (i): Suppose that (ii) is satisfied, and let $a \in M$ and $b \in N$. Consider the term operation $t^G(x, y, z) := yxz$ over G. Since $(e, b) \in \theta_N$ and $(e, a), (a, e) \in \theta_M$, for e the identity element of the group, we have $t^G(e, e, a) = a = t^G(e, a, e)$; using the term condition (TCC) on this gives $t^G(b, e, a) = t^G(b, a, e)$, meaning $ab = ba$.

(i) $\Rightarrow$ (ii): As we have seen, in a group G the n-ary term operations t, for $n \geq 1$, have the form $t(x_1, \ldots, x_n) = x_{i_1}^{k_1} \cdots x_{i_m}^{k_m}$, for some $i_1, \ldots, i_m \in \{1, \ldots, n\}$ and $k_1, \ldots, k_m \in \mathbb{Z}$. For $(a, b) \in \theta_N$, $(c_2, d_2), \ldots, (c_n, d_n) \in \theta_M$ and $t(a, c_2, \ldots, c_n) = t(a, d_2, \ldots, d_n)$, we have to show that we can replace every occurrence of a in $t(a, c_2, \ldots, c_n)$ and in $t(a, d_2, \ldots, d_n)$ by b, and still have the equality preserved. This can be done step by step, for every occurrence of a, using the commutativity assumption and the properties of normal subgroups. We illustrate with an example. Suppose that $ac_2ac_3 = ad_2ad_3$, with $(c_2, d_2) \in \theta_M$, $(c_3, d_3) \in \theta_M$ and $(a, b) \in \theta_N$. Then there exist elements g_1 and $g_2 \in M$ and $h \in N$ such that $c_2 = g_1 d_2$, $c_3 = g_2 d_3$, and $a = hb$. This gives us $ac_2 ac_3 = hbg_1 d_2 hbg_2 d_3 = hbhg_1 d_2 bg_2 d_3 = hbhc_2 bc_3$, and similarly $ad_2 ad_3 = hbd_2 hbd_3 = hbhd_2 bd_3$. Thus if $ac_2 ac_3 = ad_2 ad_3$, we obtain $hbhc_2 bc_3 = hbhd_2 bd_3$, with $hbh \in N$. Left multiplication by $b(hbh)^{-1}$ then gives $bc_2 bc_3 = bd_2 bd_3$, as required. ∎

The generalized term condition (TCC) can also be characterized using the diagonal relation $\triangle_A$. Let θ_1 and θ_2 be in $\mathit{Con}A$. We define $D_{\theta_2} := \{((a,a),(b,b)) \mid (a,b) \in \theta_2\}$, and denote by $\langle D_{\theta_2} \rangle_{\mathit{Con}\theta_1}$ the congruence relation of the subalgebra $\theta_1 \subseteq A \times A$ which is generated by D_{θ_2}. Then we have the following equivalence.

Theorem 12.1.6 *Let θ_1 and θ_2 be congruences on an algebra A. Then A satisfies the term condition (TCC) with respect to (θ_1, θ_2) if and only if the diagonal $\triangle_A$ is a union of congruence classes of $\langle D_{\theta_2} \rangle_{\mathit{Con}\theta_1}$, that is, iff*
$$[\triangle_A]_{\langle D_{\theta_2} \rangle_{\mathit{Con}\theta_1}} = \triangle_A.$$

Proof: By Lemma 5.3.3, it is enough to show that for all $a, b \in A$ and every unary polynomial operation p^θ on $A \times A$ we have

$$p^\theta((a,a)) \in \triangle_A \ \Rightarrow \ p^\theta((b,b)) \in \triangle_A. \quad (*)$$

Every unary polynomial operation p^A arises from an n-ary term operation t^A, by substitution of constants for $n-1$ variables. Therefore we have $p^{\,\theta}((a,b)) = (p^A(a), p^A(b)) = (t^A(a, c_2, \ldots, c_n), t^A(a, d_2, \ldots, d_n)) \in \triangle_A$. Then the implication $(*)$ is equivalent to the implication $t^A(a, c_2, \ldots, c_n) = t^A(a, d_2, \ldots, d_n) \Rightarrow t^A(b, c_2, \ldots, c_n) = t^A(b, d_2, \ldots, d_n)$, for $(c_2, d_2), \ldots, (c_n, d_n) \in \theta_1$ and $(a, b) \in \theta_2$. ∎

12.2 The Commutator

Now we define the following generalization of the concept of the term condition. Let A be an algebra and let α, β, δ be congruence relations on A. We will consider the following condition (1):

(1)   For all $n \in \mathbb{N}$, for all n-ary term operations t^A of A and all $(a, b) \in \beta$ and $(c_2, d_2), \ldots, (c_n, d_n) \in \alpha$,

$$(t^{\mathcal{A}}(a, c_2, \ldots, c_n), t^{\mathcal{A}}(a, d_2, \ldots, d_n)) \in \delta \quad \Rightarrow$$
$$(t^{\mathcal{A}}(b, c_2, \ldots, c_n), t^{\mathcal{A}}(b, d_2, \ldots, d_n)) \in \delta.$$

For a given α and β in $Con\mathcal{A}$, we will consider the set of all $\delta \in Con\mathcal{A}$ satisfying (1). We note first that this set is non-empty, since for instance $\delta = A \times A$ satisfies (1). It is also true that this set is closed under intersection. If the congruences δ_1 and δ_2 satisfy the implication of condition (1), for fixed congruences α and β, then for any $(a, b) \in \beta$ and $(c_2, d_2), \ldots, (c_n, d_n) \in \alpha$ we have

$$(t^{\mathcal{A}}(a, c_2, \ldots, c_n), t^{\mathcal{A}}(a, d_2, \ldots, d_n)) \in \delta_1 \cap \delta_2$$
$$\Rightarrow (t^{\mathcal{A}}(a, c_2, \ldots, c_n), t^{\mathcal{A}}(a, d_2, \ldots, d_n)) \in \delta_1 \quad \text{and}$$
$$(t^{\mathcal{A}}(a, c_2, \ldots, c_n), t^{\mathcal{A}}(a, d_2, \ldots, d_n)) \in \delta_2$$
$$\Rightarrow (t^{\mathcal{A}}(b, c_2, \ldots, c_n), t^{\mathcal{A}}(b, d_2, \ldots, d_n)) \in \delta_1 \quad \text{and}$$
$$(t^{\mathcal{A}}(b, c_2, \ldots, c_n), t^{\mathcal{A}}(b, d_2, \ldots, d_n)) \in \delta_2$$
$$\Rightarrow (t^{\mathcal{A}}(b, c_2, \ldots, c_n), t^{\mathcal{A}}(b, d_2, \ldots, d_n)) \in \delta_1 \cap \delta_2.$$

These observations motivate the following definition:

Definition 12.2.1 Let α and β be congruences on an algebra $\mathcal{A}$. The smallest congruence relation $\delta \in Con\mathcal{A}$ satisfying condition (1) is called the *commutator* of α and β, and is denoted by $[\alpha, \beta]$.

Lemma 12.2.2 *Let α and β be congruences on an algebra $\mathcal{A}$. Then $[\alpha, \beta] \subseteq \alpha \cap \beta$.*

Proof: We shall show that condition (1) is satisfied for both $\delta = \alpha$ and $\delta = \beta$; since $[\alpha, \beta]$ is the smallest congruence to satisfy condition (1) we then must have $[\alpha, \beta] \subseteq \alpha$ and $[\alpha, \beta] \subseteq \beta$, and therefore $[\alpha, \beta] \subseteq \alpha \cap \beta$.

Let (a, b) be a pair in β and let (c_i, d_i) be in α for $i = 2, \ldots, n$. Then $(t^{\mathcal{A}}(b, c_2, \ldots, c_n), t^{\mathcal{A}}(b, d_2, \ldots, d_n)) \in \alpha$. From $(t^{\mathcal{A}}(a, c_2, \ldots, c_n), t^{\mathcal{A}}(a, d_2, \ldots, d_n)) \in \beta$ and $(a, b) \in \beta$ we get

$$(t^{\mathcal{A}}(b, c_2, \ldots, c_n), t^{\mathcal{A}}(b, d_2, \ldots, d_n)) \in \beta$$

and

$$(t^{\mathcal{A}}(a, c_2, \ldots, c_n), t^{\mathcal{A}}(a, d_2, \ldots, d_n)) \in \beta,$$

and by transitivity

$$(t^{\mathcal{A}}(b, c_2, \ldots, c_n), t^{\mathcal{A}}(b, d_2, \ldots, d_n)) \in \beta. \qquad \blacksquare$$

The following theorem connects the commutator with the term condition for congruences.

Theorem 12.2.3 *For any congruences α and β on an algebra $\mathcal{A}$, the commutator $[\alpha, \beta]$ is the least congruence relation $\delta \subseteq \alpha \cap \beta$ for which $\mathcal{A}/\delta$ satisfies the term condition with respect to $(\alpha/\delta, \beta/\delta)$.*

Proof: For $\delta \in Con\mathcal{A}$ with $\delta \subseteq \alpha \cap \beta$, satisfaction of condition (1) is equivalent to having $\mathcal{A}/\delta$ satisfy the term condition with respect to $(\alpha/\delta, \beta/\delta)$. The claim follows. ∎

Let $\mathcal{G}$ be a group. Using the correspondence between normal subgroups $\mathcal{N}$ and congruences θ_N, and the equivalence from Lemma 12.1.5, it is straightforward to verify that the usual group-theoretic commutator of two normal subgroups $\mathcal{M}$ and $\mathcal{N}$ of a group $\mathcal{G}$ agrees with the commutator congruence $[\theta_M, \theta_N]$ as defined in 12.2.1.

In group theory the operation of forming the commutator of a group $\mathcal{G}$ can also be iterated. That is, we form the i-th iterated commutator (also called the i-th derivation) of $\mathcal{G}$, inductively by

$$D^0\mathcal{G} = \mathcal{G}, \; D^1\mathcal{G} = [\mathcal{G}, \mathcal{G}], \; \ldots, D^{i+1}\mathcal{G} = [D^i\mathcal{G}, D^i\mathcal{G}].$$

This gives a series

$$\mathcal{G} = D^0\mathcal{G} \supseteq D^1\mathcal{G} \cdots \supseteq D^i\mathcal{G} \supseteq \cdots \qquad (N),$$

in which each $D^{i+1}\mathcal{G}$ is a normal subgroup of $D^i\mathcal{G}$ and each quotient (or factor) group $D^i\mathcal{G}/D^{i+1}\mathcal{G}$ is abelian. If this chain stops at some finite stage n with $D^n\mathcal{G}$ equal to the trivial subgroup $\mathcal{E}$ of $\mathcal{G}$, then the series (N) is called a *normal series* with abelian factors, and $\mathcal{G}$ is said to be *solvable*. Thus a group $\mathcal{G}$ is solvable if and only if there is a natural number n such that $D^n\mathcal{G} = \mathcal{E}$.

To generalize this process for arbitrary algebras, we make the following definition.

Definition 12.2.4 For any algebra $\mathcal{A}$, we define for $\nabla_A = A \times A$ and the diagonal $\triangle_A$ the following congruences:

$$\nabla^\circ_A := \nabla_A, \quad \nabla^{k+1}_A := [\nabla^k_A, \nabla^k_A],$$

$$\nabla^{(\circ)}_A := \nabla_A, \quad \nabla^{(k+1)}_A := [\nabla^{(k)}_A, \nabla^{(k)}_A].$$

The algebra $\mathcal{A}$ is said to be *solvable of degree n* if $\nabla^{(n)}_A = \triangle_A$.

An algebra $\mathcal{A}$ is called *nilpotent* if it is solvable of degree 1, so that $[\nabla_A, \nabla_A] = \triangle_A$. In fact the following conditions are equivalent for an algebra $\mathcal{A}$:

(i) $\mathcal{A}$ is abelian;
(ii) $\mathcal{A}$ satisfies the term condition (1) from the beginning of Section 12.2;
(iii) The diagonal $\triangle_A$ is a block of a congruence relation on $\mathcal{A} \times \mathcal{A}$;
(iv) $[\nabla_A, \nabla_A] = \triangle_A$;
(v) $\mathcal{A}$ is nilpotent of degree 1.

In Chapter 11 we introduced the concept of polynomially equivalent algebras, as algebras having the same universe and the same polynomial operations. The following theorem of H. P. Gumm ([52]) uses polynomial equivalence to characterize abelian algebras in congruence permutable varieties.

Theorem 12.2.5 *Let V be a congruence permutable variety. For every algebra $\mathcal{A} \in V$ the following are equivalent:*

(i) $\mathcal{A}$ is abelian,
(ii) $\mathcal{A}$ is polynomially equivalent to a module over a ring.

Proof: (ii) $\Rightarrow$ (i): it is an easy exercise to show that the polynomial operations of a module satisfy the term condition (TC) from Definition 12.1.1. Therefore modules are abelian. This is also true for algebras which are polynomially equivalent to a module over a ring.

(i) $\Rightarrow$ (ii): Let $\mathcal{A}$ be an abelian algebra in V, so that $\mathcal{A}$ satisfies the term condition (TC). Since V is congruence permutable, there is a Mal'cev term p over V satisfying the identities $p(x, x, y) \approx y$ and $p(x, y, y) \approx x$ in the variety V. We now use the term operation p^A to define a module structure on the set A, and show that $\mathcal{A}$ is polynomially equivalent to this module. The rather lengthy proof will be broken into six steps.

Step 1. We fix an element $0 \in A$ (since A is non-empty) and define a binary operation $+$ and a unary operation $-$ on A, by

$$
\begin{aligned}
x + y : &= p(x, 0, y), \\
-x : &= p(0, x, 0).
\end{aligned}
$$

Then $(A; +, -)$ is an abelian group.

Proof: We note first that 0 is a neutral element for the operation $+$, since for any $a \in A$ we have $a + 0 = p(a, 0, 0) = a = p(0, 0, a) = 0 + a$.

Next we use the fact that the term condition is satisfied for all polynomial operations, including all operations built up from $+$, $-$ and 0. Then for any $a \in A$, we have $p(0, 0, -a) = p(0, 0, p(0, a, 0)) = p(0, a, 0)$, and by applying the term condition (TC) to the equation $p(0, 0, -a) = p(0, a, 0)$ we can replace the first 0 by any element from A. Replacing 0 by a we obtain $p(a, 0, -a) = p(a, a, 0)$, and therefore $a + (-a) = 0$. Similarly we have $p(-a, 0, 0) = p(p(0, a, 0), 0, 0) = p(0, a, 0)$, and using the term condition to replace 0 by a gives $p(-a, 0, a) = p(0, a, a)$, and $(-a) + a = 0$. Here we are also using commutativity, which follows from $b = p(a, a, b) = p(b, a, a)$ by the term condition if we replace a by 0.

Finally, we have associativity, which follows from $(a + 0) + (b + 0) = (a + b) + (0 + \underline{0})$, using the term condition to replace 0 by c at the position indicated by underlining.

Step 2. Every polynomial operation of $\mathcal{A}$ is affine with respect to $(A; +, -, 0)$; that is, for every n-ary polynomial operation f^A of $\mathcal{A}$ and all $a_1, \ldots, a_n, b_1, \ldots, b_n \in A$ we have

$$
f^A(a_1 + b_1, \ldots, a_n + b_n) = f^A(a_1, \ldots, a_n) + f^A(b_1, \ldots, b_n) - f^A(0, \ldots, 0).
$$

Proof: Using the term condition on the equation

$$
\begin{aligned}
f^A(a_1 + \underline{0}, \ldots, a_n + 0) + f^A(0, \ldots, 0) &= \\
f^A(a_1, \ldots, a_n) + f^A(0 + \underline{0}, \ldots, 0 + 0) &
\end{aligned}
$$

to replace 0 by b_1 (in the two underlined places) we get

$$
\begin{aligned}
f^A(a_1 + b_1, \ldots, a_n + 0) + f^A(0, \ldots, 0) &= \\
f^A(0 + b_1, \ldots, 0 + 0) + f^A(a_1, \ldots, a_n), &
\end{aligned}
$$

and continuing in this way we finally get

$$f^A(a_1 + b_1, \ldots, a_n + b_n) + f^A(0, \ldots, 0) =$$
$$f^A(a_1, \ldots, a_n) + f^A(b_1, \ldots, b_n).$$

Step 3. Every unary polynomial operation r of A with $r(0) = 0$ is an endomorphism of $(A; +, -, 0)$.

Proof: This is a direct consequence of Step 2.

Step 4. Let R be the set of all unary polynomial operations of A mapping 0 to 0. We define operations

$$\begin{aligned}
(r + s)(a) &:= r(a) + s(a), \\
(-r)(a) &:= -r(a), \\
\overline{0}(a) &:= 0, \text{ and} \\
(r \circ s)(a) &:= r(s(a))
\end{aligned}$$

on R. Then $(R; +, -, \overline{0}, \circ)$ forms a ring.

Proof: Since the set of all endomorphisms of $(A; +, -, 0)$ forms a ring, we only have to show that $(R; +, -, \overline{0}, \circ)$ is a subring of the full endomorphism ring of $(A; +, -, 0)$. This is clear since if r and s are unary polynomial operations of A which fix 0, then so are $r + s$, $-r$, $\overline{0}$ and $r \circ s$.

Step 5. $(A; +, -, 0, R)$ is a module over the ring $(R; +, -, \overline{0}, \circ)$.

Proof: This follows from the fact that $(R; +, -, \overline{0}, \circ)$ is a ring of endomorphisms of $(A; +, -, 0)$.

Step 6. The polynomial operations of the algebra A are exactly the operations of the form $f(x_1, \ldots, x_n) = c + r_1 x_1 + \cdots + r_n x_n$, for $n \in \mathbb{N}$, $c \in A$ and $r_1, \ldots, r_n \in R$.

Proof: It is clear that all operations of the given form are polynomial operations of A. Assume that f^A is a polynomial operation of A. From Step 2 we have

$$\begin{aligned}
f^{\mathcal{A}}(a_1,\ldots,a_n) &= f^{\mathcal{A}}(a_1+0,0+a_2,\ldots,0+a_n) \\
&= (f^{\mathcal{A}}(a_1,0,\ldots,0) - f^{\mathcal{A}}(0,\ldots,0)) + \\
&\quad f^{\mathcal{A}}(0,a_2,\ldots,a_n) \\
&\;\;\vdots \\
&= (f^{\mathcal{A}}(a_1,0,\ldots,0) - f^{\mathcal{A}}(0,\ldots,0)) + \cdots \\
&\quad + (f^{\mathcal{A}}(0,\ldots,0,a_n) - f^{\mathcal{A}}(0,\ldots,0)) + \\
&\quad f^{\mathcal{A}}(0,\ldots,0).
\end{aligned}$$

Now the polynomial operations induced by the polynomials

$$r_1(x) := f(x,0,\ldots,0) - f(0,\ldots,0)$$
$$\vdots$$
$$r_n(x) := f(0,\ldots,0,x) - f(0,\ldots,0)$$

are elements of R, so we have

$$f^{\mathcal{A}}(a_1,\ldots,a_n) = f^{\mathcal{A}}(0,\ldots,0) + r_1^{\mathcal{A}}(a_1) + \cdots + r_n^{\mathcal{A}}(a_n). \qquad \blacksquare$$

12.3 Exercises

12.3.1. Verify that any left-zero, right-zero or zero-semigroup is abelian, as claimed in Example 12.1.2 (2).

12.3.2. Prove that any subalgebra of an abelian algebra is abelian, and that every direct product of abelian algebras is abelian.

12.3.3. Prove that the collection of all abelian algebras in a congruence permutable variety forms a subvariety.

12.3.4. Does the algebra $\mathcal{A} = (\{a,b,c,d\}; f)$, with

f	a	b	c	d
a	b	a	c	d
b	a	d	b	c
c	d	c	a	b
d	c	b	d	a

satisfy the term condition?

12.3.5. Complete the proof given in this chapter that a group is abelian iff it is commutative. Prove that a ring is abelian iff it satisfies $xy = 0$.

12.3.6. Prove that every module over a ring is abelian.

12.3.7. Let $\mathcal{A}$ be a module, and let $p(x, y, z)$ be a ternary polynomial satisfying the identities $p(x, x, y) \approx p(y, x, x) \approx y$ in $\mathcal{A}$. Show that $p(x, y, z)$ can be nothing other than $x - y + z$.

Chapter 13

Complete Sublattices

We have seen that the collection of all varieties of a given type forms a complete lattice, as does the collection of all clones of operations defined on a fixed set. These two lattices play an important role in universal algebra, but their study is made difficult by the fact that the lattices are large (usually uncountably infinite) and very complex. Thus we look for new approaches or tools to use in their study. One such approach is to try to study some smaller parts of the large lattice. Such smaller parts should have the same algebraic structure, so we are interested in the study of complete sublattices of a complete lattice.

In this chapter we describe some new methods for producing complete sublattices of a given complete lattice. As we saw in Chapter 2, the two complete lattices we are interested in both arise as the lattice of closed sets under a closure operation, which can be obtained via a Galois-connection. This leads us to the study of new closure operators and Galois-connections, which produce sublattices of the original lattice of closed sets. We develop the theory of such sublattices in this chapter. In the next chapter we will apply this general theory to our two specific examples of the lattices of varieties and clones.

13.1 Conjugate Pairs of Closure Operators

Our basic concepts of closure operators and Galois-connections were developed in Chapter 2. We saw there that any closure operator γ defined on a set A gives us a closure system, the set $\mathcal{H}_\gamma$ of all γ-closed subsets of A, and that any such closure system forms a complete lattice. In this lattice, the meet operation, also the greatest lower bound or infimum with respect to the partial order of set inclusion, is the operation of intersection. The join operation however is not usually just the union: we have

$$\bigvee \mathcal{B} = \bigcap \{H \in \mathcal{H}_\gamma \mid H \supseteq \bigcup \mathcal{B}\}$$

for every $\mathcal{B} \subseteq \mathcal{H}_\gamma$. One situation when we do have the join operation equal to union is the following.

Definition 13.1.1 A closure operator γ defined on a set A is said to be *additive* if for all $T \subseteq A, \gamma(T) = \bigcup_{a \in T} \gamma(a)$. (Note that we write $\gamma(a)$ for $\gamma(\{a\})$).

We can show easily that when γ is an additive closure operator, the least upper bound operation on the lattice $\mathcal{H}_\gamma$ agrees with $\bigcup \mathcal{B}$ (see M. Reichel, [98] or D. Dikranjan and E. Giuli, [39]). We always have $\bigcup \mathcal{B} \subseteq \gamma(\bigcup \mathcal{B})$ because of the extensivity of γ. Conversely, if $a \in \bigcup \mathcal{B}$ then $a \in B \in \mathcal{B}$ for some set $B \in \mathcal{B}$, and since $B \in \mathcal{H}_\gamma$ we have $\gamma(a) \subseteq \bigcup \mathcal{B}$ and $\gamma(\bigcup \mathcal{B}) = \bigcup_{a \in \bigcup \mathcal{B}} \gamma(a) \subseteq \bigcup_{a \in \bigcup \mathcal{B}} \bigcup \mathcal{B} = \bigcup \mathcal{B}$. This means that $\bigcup \mathcal{B}$ is γ-closed and $\bigvee \mathcal{B} = \bigcup \mathcal{B}$. In other words, when γ is an additive closure operator on A, the corresponding closure system forms a complete sublattice of the lattice $(\mathcal{P}(A); \wedge = \bigcap, \vee = \bigcup)$ of all subsets of A.

Definition 13.1.2 Let γ_1 be a closure operator defined on the set A and let γ_2 be a closure operator defined on the set B. Let $R \subseteq A \times B$ be a relation between A and B. Then γ_1 and γ_2 are called *conjugate* with respect to R if for all $t \in A$ and all $s \in B, \gamma_1(t) \times \{s\} \subseteq R$ iff $\{t\} \times \gamma_2(s) \subseteq R$.

This property of conjugacy of two closure operators is defined in terms of individual elements. When the two operators are also additive, we can extend this to sets of elements. Thus when (γ_1, γ_2) is a pair of additive closure operators, γ_1 on A and γ_2 on B, and they are conjugate with respect to a relation $R \subseteq A \times B$, then for all $X \subseteq A$ and all $Y \subseteq B$ we have $X \times \gamma_2(Y) \subseteq R$ if and only if $\gamma_1(X) \times Y \subseteq R$.

Examples of conjugate pairs of additive closure operators will be given in the next chapter. In the rest of this section we develop the general theory of such operators. We assume throughout that we have two sets A and B, and that R is a relation from A to B. We know from Chapter 2 that this relation induces a Galois-connection (μ, ι) between A and B, for which the two maps $\mu\iota$ and $\iota\mu$ are closure operators. Moreover, the pair $(\mu\iota, \iota\mu)$ is always conjugate with respect to the original relation R. But $\mu\iota$ and $\iota\mu$ need not be additive in general.

Our goal is to construct, given a relation R and the induced Galois-connection, a new relation which is connected to R, but gives a smaller lattice of closed sets. One way to do this is by using conjugate pairs of closure operators.

Definition 13.1.3 Let $\gamma := (\gamma_1, \gamma_2)$ be a conjugate pair of additive closure operators, with respect to a relation $R \subseteq A \times B$. Let R_γ be the following relation between A and B:

$$R_\gamma := \{(t, s) \in A \times B \mid \gamma_1(t) \times \{s\} \subseteq R\}.$$

We now have two relations and Galois-connections between A and B. We have the original relation R, with induced Galois-connection (μ, ι) between A and B, and corresponding lattices of closed sets. We also have the new relation R_γ and its induced Galois-connection, which we shall denote by $(\mu_\gamma, \iota_\gamma)$. The following theorem gives some properties relating the two Galois-connections.

Theorem 13.1.4 Let $\gamma = (\gamma_1, \gamma_2)$ be a conjugate pair of additive closure operators with respect to $R \subseteq A \times B$. Then for all $T \subseteq A$ and $S \subseteq B$, the following properties hold:

(i) $\mu_\gamma(T) = \mu(\gamma_1(T))$,

(ii) $\mu_\gamma(T) \subseteq \mu(T)$,

(iii) $\gamma_2(\mu_\gamma(T)) = \mu_\gamma(T)$,

(iv) $\gamma_1(\iota(\mu_\gamma(T))) = \iota(\mu_\gamma(T))$,

(v) $\mu_\gamma(\iota_\gamma(S)) = \mu(\iota(\gamma_2(S)))$;   and dually, *(i')* $\iota_\gamma(S) = \iota(\gamma_2(S))$,

(ii') $\iota_\gamma(S) \subseteq \iota(S)$,

(iii') $\gamma_1(\iota_\gamma(S)) = \iota_\gamma(S)$,

(iv') $\gamma_2(\mu(\iota_\gamma(S))) = \mu(\iota_\gamma(S))$,

(v') $\iota_\gamma(\mu_\gamma(T)) = \iota(\mu(\gamma_1(T)))$.

Proof: We will prove only (i)-(v), the proofs of the other propositions being dual.

(i) By definition,
$$\begin{aligned}
\mu_\gamma(T) &= \{b \in B \mid \forall a \in T \; ((a,b) \in R_\gamma)\} \\
&= \{b \in B \mid \forall a \in T \; (\gamma_1(a) \times \{b\} \subseteq R)\} \\
&= \{b \in B \mid \forall a \in \gamma_1(T) \; ((a,b) \in R)\} \quad = \quad \mu(\gamma_1(T)).
\end{aligned}$$

(ii) Since γ_1 is a closure operator, we have $T \subseteq \gamma_1(T)$; and thus, since μ reverses inclusions, $\mu(T) \supseteq \mu(\gamma_1(T))$. Using (i) we obtain $\mu_\gamma(T) \subseteq \mu(T)$.

(iii) Extensivity of γ_2 implies $\mu_\gamma(T) \subseteq \gamma_2(\mu_\gamma(T))$. Now let $S \subseteq \mu_\gamma(T)$. Then for all $s \in S$ and for all $t \in T$, $(t,s) \in R_\gamma$, and by definition of R_γ we get $\{t\} \times \gamma_2(s) \subseteq R$. Idempotency of γ_2 gives $\{t\} \times \gamma_2(\gamma_2(s)) \subseteq R$ and thus $\gamma_2(s) \subseteq \mu_\gamma(T)$ for all $s \in S$. By additivity of γ_2 we get $\gamma_2(S) = \bigcup_{s \in S} \gamma_2(s) \subseteq \mu_\gamma(T)$; and taking $S = \mu_\gamma(T)$ we obtain $\gamma_2(\mu_\gamma(T)) \subseteq \mu_\gamma(T)$. Altogether we have the equality $\mu_\gamma(T) = \gamma_2(\mu_\gamma(T))$.

(iv) $\gamma_1(\iota(\mu_\gamma(T))) = \gamma_1(\iota(\gamma_2(\mu_\gamma(T)))) = \gamma_1(\iota_\gamma(\mu_\gamma(T)))$
$= \iota_\gamma(\mu_\gamma(T)) = \iota(\gamma_2(\mu_\gamma(T))) = \iota(\mu_\gamma(T))$, by parts (i) and (i').

(v) $\mu_\gamma(\iota_\gamma(S)) = \mu(\gamma_1(\iota_\gamma(S))) = \mu(\iota_\gamma(S)) = \mu(\iota(\gamma_2(S)))$.   ■

The next theorem is our "Main Theorem for Conjugate Pairs of Closure Operators." It shows that when we consider sets which are closed under the original Galois-connection from R, there are four equivalent conditions for such sets to also be closed under the new connection from R_γ.

Theorem 13.1.5 *(Main Theorem for Conjugate Pairs of Additive Closure Operators) Let R be a relation between sets A and B, with corresponding Galois-connection (μ, ι). Let $\gamma = (\gamma_1, \gamma_2)$ be a conjugate pair of additive* *closure operators with respect to the relation R. Then for all sets $T \subseteq A$ with $\iota(\mu(T)) = T$ the following propositions (i) - (iv) are equivalent; and dually, for all sets $S \subseteq B$ with $\mu(\iota(S)) = S$, propositions (i') - (iv') are equivalent:*

(i) $T = \iota_\gamma(\mu_\gamma(T))$,

(ii) $\gamma_1(T) = T$,

(iii) $\mu(T) = \mu_\gamma(T)$,

(iv) $\gamma_2(\mu(T)) = \mu(T)$;   and dually,

(i') $S = \mu_\gamma(\iota_\gamma(S))$,

(ii') $\gamma_2(S) = S$,

(iii') $\iota(S) = \iota_\gamma(S)$,

(iv') $\gamma_1(\iota(S)) = \iota(S)$.

Proof: We prove the equivalence of (i), (ii), (iii) and (iv); the equivalence of the four dual statements can be proved dually.

(i) $\Rightarrow$ (ii): We always have $T \subseteq \gamma_1(T)$, since γ_1 is a closure operator. Since $\iota\mu$ is a closure operator we also have $\gamma_1(T) \subseteq \iota\mu(\gamma_1(T)) = \iota_\gamma(\mu_\gamma(T)) = T$, by 13.1.4 (v') and (i).

(ii) $\Rightarrow$ (iii): We have $\mu(T) = \mu(\gamma_1(T)) = \mu_\gamma(T)$ by (ii) and 13.1.4 (i).

(iii) $\Rightarrow$ (iv): We have $\gamma_2(\mu(T)) = \gamma_2(\mu_\gamma(T)) = \mu_\gamma(T)$, using 13.1.4 (iv) and (iii).

(iv) $\Rightarrow$ (i): Since the $\iota_\gamma\mu_\gamma$-closed sets are exactly the sets of the form $\iota_\gamma(S)$, we have to find a set $S \subseteq B$ with $T = \iota_\gamma(S)$. But we have $\iota_\gamma(\mu(T)) = \iota(\gamma_2(\mu(T))) = \iota(\mu(T)) = T$, by 13.1.4 (i') and our assumption that T is $\iota\mu$-closed. ∎

Before we use this Main Theorem to produce our complete sublattices, we need the following additional properties.

Theorem 13.1.6 *Let R be a relation between sets A and B, with Galois-connection (μ, ι). Let $\gamma = (\gamma_1, \gamma_2)$ be a conjugate pair of additive closure operators with respect to R. Then for all sets $T \subseteq A$ and $S \subseteq B$, the following properties hold:*

(i) $\gamma_1(T) \subseteq \iota(\mu(T))$ $\Leftrightarrow$ $\iota(\mu(T)) = \iota_\gamma(\mu_\gamma(T))$;

(ii) $\gamma_1(T) \subseteq \iota(\mu(T))$ $\Leftrightarrow$ $\gamma_1(\iota(\mu(T))) = \iota(\mu(T))$;

(i') $\gamma_2(S) \subseteq \mu(\iota(S))$ $\Leftrightarrow$ $\mu(\iota(S)) = \mu_\gamma(\iota_\gamma(S))$;

(ii') $\gamma_2(S) \subseteq \mu(\iota(S))$ $\Leftrightarrow$ $\gamma_2(\mu(\iota(S))) = \mu(\iota(S))$.

Proof: We prove only (i') and (ii'); the others are dual.

(i') Suppose that $\gamma_2(S) \subseteq \mu(\iota(S))$. Since $\mu\iota$ is a closure operator we have $\mu(\iota(S)) = \mu(\iota(\mu(\iota(S)))) \supseteq \mu(\iota(\gamma_2(S))) = \mu_\gamma(\iota_\gamma(S))$, by our assumption and by 13.1.4 (v). Also $S \subseteq \gamma_2(S)$, and hence we have $\mu(\iota(S)) \subseteq \mu(\iota(\gamma_2(S))) = \mu_\gamma(\iota_\gamma(S))$, again by 13.1.4 (v). For the converse we have $\gamma_2(S) \subseteq \mu(\iota(\gamma_2(S))) = \mu_\gamma(\iota_\gamma(S)) = \mu(\iota(S))$, using the extensivity of $\mu\iota$, 13.1.4 (v) and our assumption.

(ii') Let $\gamma_2(S) \subseteq \mu(\iota(S))$. Then $S \subseteq \gamma_2(S)$ implies $\gamma_2(\mu(\iota(S))) \subseteq \gamma_2(\mu(\iota(\gamma_2(S))))$. We also have $\gamma_2(\mu(\iota(\gamma_2(S)))) = \gamma_2(\mu(\iota_\gamma(S)))$ by Theorem 13.1.4 (i'), and $\gamma_2(\mu(\iota(\gamma_2(S)))) = \mu(\iota_\gamma(S))$ by 13.1.4 (iv'). In addition, $\mu(\iota_\gamma(S)) = \mu(\iota(\gamma_2(S))) \subseteq \mu(\iota(\mu(\iota(S)))) = \mu(\iota(S))$. Altogether we obtain $\gamma_2(\mu(\iota(S))) \subseteq \mu(\iota(S))$. The opposite inclusion is always true, since γ_2 is a closure operator. Conversely, $S \subseteq \mu(\iota(S))$ implies $\gamma_2(S) \subseteq \gamma_2(\mu(\iota(S))) = \mu(\iota(S))$, by the extensivity of $\mu\iota$, the monotonicity of γ_2 and our assumption. ∎

Now we are ready to produce our complete sublattices. We know that from the original relation R and Galois-connection (μ, ι) we have two (dually isomorphic) complete lattices of closed sets, the lattices $\mathcal{H}_{\mu\iota}$ and $\mathcal{H}_{\iota\mu}$. We also get two complete lattices of closed sets from the new Galois-connection $(\mu_\gamma, \iota_\gamma)$ induced by R_γ. Our result is that each new complete lattice is in fact a complete sublattice of the corresponding original complete lattice.

Theorem 13.1.7 *Let R be a relation from A to B, with induced Galois-connection (μ, ι). Let $\gamma = (\gamma_1, \gamma_2)$ be a conjugate pair of additive closure* *operators with respect to R. Then the lattice $\mathcal{H}_{\mu_\gamma \iota_\gamma}$ of sets closed under $\mu_\gamma \iota_\gamma$ is a complete sublattice of the lattice $\mathcal{H}_{\mu\iota}$, and dually the lattice $\mathcal{H}_{\iota_\gamma \mu_\gamma}$ is a complete sublattice of the lattice $\mathcal{H}_{\iota\mu}$.*

Proof: As a closure system $\mathcal{H}_{\mu_\gamma \iota_\gamma}$ is a complete lattice, and we have to prove that it is a complete sublattice of the complete lattice $\mathcal{H}_{\mu\iota}$. We begin by showing that it is a subset. Let $S \in \mathcal{H}_{\mu_\gamma \iota_\gamma}$, so that $\mu_\gamma(\iota_\gamma(S)) = S$. Then $\mu(\iota(S)) = \mu(\iota(\mu_\gamma(\iota_\gamma(S)))) = \mu(\iota(\mu(\iota(\gamma_2(S))))) = \mu(\iota(\gamma_2(S))) = \mu_\gamma(\iota_\gamma(S)) = S$ by 13.1.4 (v), and thus $S \in \mathcal{H}_{\mu\iota}$. This shows $\mathcal{H}_{\mu_\gamma \iota_\gamma} \subseteq \mathcal{H}_{\mu\iota}$. Since every S in $\mathcal{H}_{\mu_\gamma \iota_\gamma}$ satisfies $\mu(\iota(S)) = S$, we can apply Theorem 13.1.5 (ii'), to get

$$S \in \mathcal{H}_{\mu_\gamma \iota_\gamma} \iff S = \mu_\gamma(\iota_\gamma(S)) \iff S = \gamma_2(S) \iff S \in \mathcal{H}_{\gamma_2}.$$

As we remarked after Definition 13.1.1, the fact that γ_2 is an additive closure operator means that the corresponding closure system is a complete sublattice of the lattice $(\mathcal{P}(B); \cap, \cup)$ of all subsets of B; that is, on our lattice $\mathcal{H}_{\mu_\gamma \iota_\gamma}$ the meet operation agrees with ordinary set-intersection and the join agrees with union. We already know that the meet operation in $\mathcal{H}_{\mu\iota}$ also agrees with intersection, so we only need to show that $\mathcal{H}_{\mu_\gamma \iota_\gamma}$ is closed under the join operation of $\mathcal{H}_{\mu\iota}$. Let $(S_k)_{k \in J}$ be an indexed family of subsets of B. Then

$$\mu_\gamma(\iota_\gamma(\bigvee_{k \in J} S_k)) = \gamma_2(\bigvee_{k \in J} S_k) = \gamma_2(\mu(\iota(\bigcup_{k \in J} S_k)))$$
$$= \gamma_2(\mu(\iota_\gamma(\bigcup_{k \in J} S_k))) = \mu(\iota_\gamma(\bigcup_{k \in J} S_k)) = \mu(\iota(\bigcup_{k \in J} S_k)) = \bigvee_{k \in J} S_k,$$

by Theorem 13.1.4 (iv'); and then using 13.1.4 (iii) we have

$$\iota(\bigcup_{k \in J} S_k) = \bigcap_{k \in J} \iota(S_k) = \bigcap_{k \in J} \iota_\gamma(S_k) = \iota_\gamma(\bigcup_{k \in J} S_k). \qquad \blacksquare$$

Thus conjugate pairs of additive closure operators give us a way to construct complete sublattices of a given closure lattice. We may also define an order relation on the set of all conjugate pairs of additive closure operators: for $\alpha = (\gamma_1, \gamma_1')$ and $\beta = (\gamma_2, \gamma_2')$ we set

$$\alpha \leq \beta :\Leftrightarrow (\forall T \subseteq A)(\forall S \subseteq B)[\gamma_1(T) \subseteq \gamma_2(T) \text{ and } \gamma_1'(S) \subseteq \gamma_2'(S)].$$

When $\alpha \leq \beta$, it can be shown that the lattice $\mathcal{H}_{\mu_\beta \iota_\beta}$ is a sublattice of $\mathcal{H}_{\mu_\alpha \iota_\alpha}$, and dually that $\mathcal{H}_{\iota_\beta \mu_\beta}$ is a sublattice of $\mathcal{H}_{\iota_\alpha \mu_\alpha}$.

The following additional properties may also be verified:

(i) $\gamma_1(\iota(\mu(T))) = \iota(\mu(T)) \iff T = \iota(\mu(\gamma_1(T)))$,   and

(i´) $\gamma_2(\mu(\iota(S))) = \mu(\iota(S)) \iff S = \mu(\iota(\gamma_2(T)))$.

S. Arworn in [1] has generalized the theory of conjugate pairs of additive closure operators to the situation of conjugate pairs of extensive, additive operators.

13.2   Galois Closed Subrelations

In the previous section we developed a method to produce complete sublattices of a given complete lattice. We started with a relation R which induced a Galois-connection (μ, ι) between two sets A and B, and used the two closure operators $\mu\iota$ and $\iota\mu$ to produce our (dually isomorphic) lattices of closed sets. We then used two new closure operators on our sets, which are additive and conjugate with respect to our relation R, to determine a new relation R_γ, which in turn induces a Galois-connection and closure operators. We showed that the sets closed under these new operators form complete sublattices of the original lattices of closed sets.

In this section we examine in more detail the relations such as R_γ which determine complete sublattices of our original lattices. As our starting point, we assume as before that we have a relation R from A to B, which induces a Galois-connection (μ, ι) and from which we obtain complete lattices $\mathcal{H}_{\iota\mu}$ and $\mathcal{H}_{\mu\iota}$ of closed subsets of A and of B respectively. Then we consider a subrelation R' of the initial relation R, from which we obtain a new Galois-connection and two new complete lattices. We describe a property of the subrelation R' which is sufficient to guarantee that the new complete lattices will be complete sublattices of the original lattices. This property is called the *Galois-closed subrelation property*. Moreover, we show that any complete sublattices of our original lattices arise in this way.

Definition 13.2.1 Let R and R' be relations between sets A and B, and let (μ, ι) and (μ', ι') be the Galois-connections between A and B induced by R and R', respectively. The relation R' is called a *Galois-closed subrelation of R* if:

1) $R' \subseteq R$,   and

2) $\forall T \subseteq A$, $\forall S \subseteq B$ $(\mu'(T) = S$ and $\iota'(S) = T \Rightarrow \mu(T) = S$ and $\iota(S) = T)$.

Directly from this definition we can prove the following equivalent characterizations of Galois-closed subrelations, as shown by B. Ganter and R. Wille in [47] and by S. Arworn, K. Denecke and R. Pöschel in [4] and [22].

Proposition 13.2.2 Let $R' \subseteq R$ be relations between sets A and B. Then the following are equivalent:

(i) R' is a Galois-closed subrelation of R;

(ii) For any $T \subseteq A$, if $\iota'\mu'(T) = T$ then $\mu(T) = \mu'(T)$, and for any $S \subseteq B$, if $\mu'\iota'(S) = S$ then $\iota(S) = \iota'(S)$;

(iii) For all $T \subseteq A$ and for all $S \subseteq B$ the equations $\iota'\mu'(T) = \iota\,\mu'(T)$ and $\mu'\iota'(S) = \mu\,\iota'(S)$ are satisfied.

Proof: (i) $\Rightarrow$ (ii): Let R' be a Galois-closed subrelation of R, and let $T \subseteq A$ with $\iota'\mu'(T) = T$. Define S to be the set $\mu'(T)$. Then we have $\mu'(T) = S$ and $\iota'(S) = T$, and applying the second part of the definition of a Galois-closed subrelation gives us $\mu(T) = S$ and $\iota(S) = T$. In particular, $\mu(T) = S = \mu'(T)$. The claim for subsets S of B is proved similarly.

(ii) $\Rightarrow$ (iii): Assume that $T \subseteq A$ and $S \subseteq B$. Then $\mu'(T)$ is a closed subset of B under the closure operator $\mu'\iota'$, and $\iota'(S)$ is a closed subset of A under the closure operator $\iota'\mu'$. This means that

$$\mu'\iota'\mu'(T) = \mu'(T) \text{ and } \iota'\mu'\iota'(S) = \iota'(S).$$

Then by condition (ii) we get

$$\iota\,\mu'(T) = \iota'\mu'(T) \text{ and } \mu\,\iota'(S) = \mu'\iota'(S).$$

(iii) $\Rightarrow$ (i): Assume now that $T \subseteq A$ and $S \subseteq B$ such that $\mu'(T) = S$ and $\iota'(S) = T$. It follows that

$$\begin{aligned}
\mu'\iota'(S) &= \mu'(T), & \iota'\mu'(T) &= \iota'(S), \text{ and} \\
\mu\,\iota'(S) &= \mu(T), & \iota\,\mu'(T) &= \iota(S).
\end{aligned}$$

Then by condition (iii) we get

$$\mu(T) = \mu'(T) = S \text{ and } \iota(S) = \iota'(S) = T.$$

This shows that R' is a Galois-closed subrelation of R.   ∎

We leave it as an exercise for the reader to verify, using the results of Section 13.1, that if $\gamma := (\gamma_1, \gamma_2)$ is a pair of additive closure operators which are conjugate with respect to a relation $R \subseteq A \times B$, then the relation R_γ of Definition 13.1.3 is a Galois-closed subrelation of R.

Before we can prove our main theorem, we need the following well-known result for Galois-connections. The proof is straightforward, and is left as an exercise for the reader (see Exercise 2.4.2).

Lemma 13.2.3 *Let $R \subseteq A \times B$ be a relation between the sets A, B and let (μ, ι) be the Galois-connection between A and B induced by R. Then for any families $\{T_j \subseteq A \mid j \in J\}$ and $\{S_j \subseteq B \mid j \in J\}$, the following equalities hold:*

$$(i)\ \mu(\bigcup_{j \in J} T_j) = \bigcap_{j \in J} \mu(T_j),$$

$$(ii)\ \iota(\bigcup_{j \in J} S_j) = \bigcap_{j \in J} \iota(S_j).$$ ∎

Ganter and Wille showed in [47] that there is a one-to-one correspondence between Galois-closed subrelations of a relation $R \subseteq A \times B$ and complete sublattices of the corresponding lattices $\mathcal{H}_{\iota\mu}$ and $\mathcal{H}_{\mu\iota}$ of closed sets. (Note however that the terminology in [47] is different from ours.) The remainder of this section is devoted to the proof of this claim. We will show that any Galois-closed subrelation R' of the relation R yields a lattice of closed subsets of A which is a complete sublattice of the corresponding lattice $\mathcal{H}_{\iota\mu}$ for R. Conversely, we also show that any complete sublattice of the lattice $\mathcal{H}_{\iota\mu}$ occurs as the lattice of closed sets induced from some Galois-closed subrelation of R. Dual results of course hold for the set B.

Theorem 13.2.4 *([47], [22], and [4]) Let $R \subseteq A \times B$ be a relation between sets A and B, with induced Galois-connection (μ, ι). Let $\mathcal{H}_{\iota\mu}$ be the corresponding lattice of closed subsets of A.*

(i) If $R' \subseteq A \times B$ is a Galois-closed subrelation of R, then the class $\mathcal{U}_{R'} := \mathcal{H}_{\iota'\mu'}$ is a complete sublattice of $\mathcal{H}_{\iota\mu}$.

(ii) If $\mathcal{U}$ is a complete sublattice of $\mathcal{H}_{\iota\mu}$, then the relation

$$R_{\mathcal{U}} := \bigcup \{T \times \mu(T) \mid T \in \mathcal{U}\}$$

is a Galois-closed subrelation of R.

(iii) For any Galois-closed subrelation R' of R and any complete sublattice $\mathcal{U}$ of $\mathcal{H}_{\iota\mu}$, we have

$$\mathcal{U}_{R_{\mathcal{U}}} = \mathcal{U} \quad \text{and} \quad R_{\mathcal{U}_{R'}} = R'.$$

Proof: (i) We begin by verifying that any subset of A which is closed under the operator $\iota'\mu'$ is also closed under $\iota\mu$, so that the lattice $\mathcal{H}_{\iota'\mu'}$ is at least a subset of $\mathcal{H}_{\iota\mu}$. Let $T \in \mathcal{H}_{\iota'\mu'}$, so that $\iota'\mu'(T) = T$. By Proposition 13.2.2, parts (ii) and (iii), we have

$$\iota\,\mu(T) = \iota\,\mu'(T) = \iota'\mu'(T) = T.$$

Therefore, $\mathcal{H}_{\iota'\mu'} \subseteq \mathcal{H}_{\iota\mu}$.

Next we have to show that this subset is in fact a sublattice. This means showing that for any family $\{T_j \mid j \in J\}$ of sets in $\mathcal{H}_{\iota'\mu'}$, both the sets $\bigwedge_{\mathcal{H}_{\iota\mu}} \{T_j \mid j \in J\}$ and $\bigvee_{\mathcal{H}_{\iota\mu}} \{T_j \mid j \in J\}$ are in $\mathcal{H}_{\iota'\mu'}$.

We start with the meet operation. We know from above that the collection $\{T_j \mid j \in J\}$ is also a family of sets in $\mathcal{H}_{\iota\mu}$. Since

$$\bigwedge_{\mathcal{H}_{\iota\mu}} \{T_j \mid j \in J\} = \bigcap_{j\in J} T_j = \iota\mu(\bigcap_{j\in J} T_j),$$

we have

$$\bigcap_{j\in J} T_j = \iota\mu(\bigcap_{j\in J} \iota'\mu'(T_j)).$$

Applying Lemma 13.2.3 to this, and then using Proposition 13.2.2 (iii) twice, we get

$$\bigcap_{j\in J} T_j = \iota\mu\iota'(\bigcup_{j\in J} \mu'(T_j)) = \iota\mu'\iota'(\bigcup_{j\in J} \mu'(T_j)) = \iota'\mu'\iota'(\bigcup_{j\in J} \mu'(T_j)).$$

Using the closure operator properties, and Lemma 13.2.3 once more, we get

$$\iota'\mu'(\bigcap_{j\in J} T_j) = \iota'\mu'(\bigcap_{j\in J} \iota'\mu'(T_j)) = \iota'\mu'\iota'(\bigcup_{j\in J} \mu'(T_j))$$
$$= \iota'(\bigcup_{j\in J} \mu'(T_j)) \subseteq \iota'\mu'(T_j) = T_j,$$

for all $j \in J$.

Thus we have $\iota'\mu'(\bigcap_{j \in J} T_j) \subseteq \bigcap_{j \in J} T_j$. The reverse inclusion is always true for a closure operator, so altogether we have $\iota'\mu'(\bigcap_{j \in J} T_j) = \bigcap_{j \in J} T_j$. This shows that $\bigcap_{j \in J} T_j \in \mathcal{H}_{\iota'\mu'}$.

Now we consider the join,

$$\bigvee_{\mathcal{H}_{\iota\mu}} \{T_j \mid j \in J\} = \iota\mu(\bigcup_{j \in J} T_j).$$

By repeated use of Lemma 13.2.3 and Proposition 13.2.2, we see that

$$
\begin{aligned}
\iota'\mu'\iota\mu(\bigcup_{j \in J} T_j) &= \iota'\mu'\iota(\bigcap_{j \in J} \mu(T_j)) && (13.2.3) \\
&= \iota'\mu'\iota(\bigcap_{j \in J} \mu'(T_j)) && (13.2.2 \text{ (ii)}) \\
&= \iota'\mu'\iota\mu'(\bigcup_{j \in J} T_j) && (13.2.3) \\
&= \iota'\mu'\iota'\mu'(\bigcup_{j \in J} T_j) && (13.2.2 \text{ (iii)}) \\
&= \iota'\mu'(\bigcup_{j \in J} T_j) && (\text{by closure properties}) \\
&= \iota\mu'(\bigcup_{j \in J} T_j) && (13.2.2 \text{ (iii)}) \\
&= \iota(\bigcap_{j \in J} \mu'(T_j)) && (13.2.3) \\
&= \iota(\bigcap_{j \in J} \mu(T_j)) && (13.2.2 \text{ (ii)}) \\
&= \iota\mu(\bigcup_{j \in J} T_j) && (13.2.3).
\end{aligned}
$$

This shows that $\iota\mu(\bigcup_{j \in J} T_j)$ is also a fixed point under $\iota'\mu'$, so that it too is an element of $\mathcal{H}_{\iota'\mu'}$.

(ii) Now let $\mathcal{U}$ be any complete sublattice of $\mathcal{H}_{\iota\mu}$. We define the relation

$$R_{\mathcal{U}} := \bigcup \{T \times \mu(T) \mid T \in \mathcal{U}\},$$

which we will prove is a Galois-closed subrelation of R. First, for each non-empty $T \in \mathcal{U}$ we have $\mu(T) = \{s \in B \mid \forall t \in T, \ (t, s) \in R\}$, so that $T \times \mu(T) \subseteq R$. Therefore $R_{\mathcal{U}} \subseteq R$. To show that the second condition of the definition of a Galois-closed subrelation is met, we let (μ', ι') be the Galois-connection between sets A and B induced by $R_{\mathcal{U}}$, and assume that $\mu'(T) = S$ and $\iota'(S) = T$ for some $T \subseteq A$ and $S \subseteq B$. Our goal is to prove that

$$\mu(T) = S \quad \text{and} \quad \iota(S) = T. \quad (*)$$

The proof that (*) holds will be divided into a number of steps. We begin with two facts we shall need.

Fact 1: For any set $T \in \mathcal{U}$, we have $\mu'(T) = \mu(T)$.

Proof of Fact 1: Let $T \in \mathcal{U}$. By definition we have

$$\mu'(T) = \{s \in B \mid \forall t \in T, \ (t, s) \in R_{\mathcal{U}}\}.$$

This means that $\mu'(T)$ is the greatest subset of B with $T \times \mu'(T) \subseteq R_{\mathcal{U}}$. But from the definition of $R_{\mathcal{U}}$ we have $T \times \mu(T) \subseteq R_{\mathcal{U}}$. Therefore we have $\mu(T) \subseteq \mu'(T)$. The opposite inclusion also holds since $R_{\mathcal{U}} \subseteq R$. Altogether we have $\mu'(T) = \mu(T)$.

Fact 2: For any set T in $\mathcal{U}$, if $\mu(T) = S$ then $\iota(S) = \iota'(S)$.

Proof of Fact 2: Let $T \in \mathcal{U}$ and let $\mu(T) = S$. This means that $T \times S \subseteq R_{\mathcal{U}}$. Since $\iota'(S) = \{t \in A \mid \forall s \in S, \ (t, s) \in R_{\mathcal{U}}\}$, the set $\iota'(S)$ is the greatest subset of A with $\iota'(S) \times S \subseteq R_{\mathcal{U}}$. This shows that $T \subseteq \iota'(S)$. But we also have $T = \iota\mu(T) = \iota(S)$, so we now have $\iota(S) \subseteq \iota'(S)$. The opposite inclusion also holds since, as we showed just above, $R_{\mathcal{U}} \subseteq R$. Altogether we get $\iota'(S) = \iota(S)$.

Returning now to the proof of (*), we let $T \subseteq A$ and $S \subseteq B$, with $\mu'(T) = S$ and $\iota'(S) = T$. If T is the empty set, we use Facts 1 and 2 to conclude that (*) holds; so we may now assume that T is non-empty. For each $t \in T$, we define

$$D_t = \bigcap \ \{T' \in \mathcal{U} \mid t \in T' \ \text{and} \ S \subseteq \mu(T')\}.$$

We will show the following facts:

(a) $D_t \neq \emptyset$.

(b) $\mu'(\{t\}) = \mu'(D_t)$.

(c) $\iota'\mu'(\{t\}) = D_t$.

(d) $T = \bigcup\limits_{t \in T} D_t.$

(e) $\mu(T) = \mu'(T) = S$ and $\iota(S) = \iota'(S) = T$, and (*) holds.

Proof of (a): Since $t \in T$ and $\mu'(T) = S$, we have $(t, s) \in R_{\mathcal{U}}$ for all $s \in S$. From the definition of $R_{\mathcal{U}}$ we see that for each s in S there exists a set $T_s \in \mathcal{U}$ such that $(t, s) \in T_s \times \mu(T_s)$. Therefore $t \in \bigcap\limits_{s \in S} T_s$, $\bigcap\limits_{s \in S} T_s \in \mathcal{U}$, and $S \subseteq \mu(\bigcap\limits_{s \in S} T_s)$, which shows that $D_t \neq \emptyset$.

Proof of (b): If $s \in \mu'(\{t\})$, then $(t, s) \in R_{\mathcal{U}}$ and there is a set $T' \in \mathcal{U}$ containing t for which $(t, s) \in T' \times \mu(T')$. Fact 1 tells us that $\mu'(T') = \mu(T')$. By definition we have $D_t \subseteq T'$, and applying μ' reverses this inclusion to $\mu'(D_t) \supseteq \mu'(T')$. Since s is in $\mu'(T')$ we now have $s \in \mu'(D_t)$. This shows that $\mu'(\{t\}) \subseteq \mu'(D_t)$. Conversely, $t \in D_t$ and so $\{t\} \subseteq D_t$, and then applying μ' gives $\mu'(\{t\}) \supseteq \mu'(D_t)$. Altogether we have the equality $\mu'(\{t\}) = \mu'(D_t)$.

Proof of (c): From (b) we have $\mu'(\{t\}) = \mu'(D_t)$. Since $D_t \in \mathcal{U}$, we have $\iota'\mu'(\{t\}) = \iota'\mu'(D_t) = \iota'\mu(D_t) = \iota\mu(D_t) = D_t$, by Facts 1 and 2.

Proof of (d): It is clear from the definition of D_t that $T \subseteq \bigcup\limits_{t \in T} D_t$. For the opposite inclusion we have

$$
\begin{aligned}
T = \iota'\mu'(T) &= \iota'\mu'(\bigcup\limits_{t \in T}\{t\}) \\
&\supseteq \iota'\mu'(\{t\}) \quad \text{for all } t \in T \\
&= D_t \quad \text{for all } t \in T,
\end{aligned}
$$

using the result of (c).

Proof of (e): We start with the fact that $T = \bigcup\limits_{t \in T} D_t$, from (d). Since $D_t \in \mathcal{U}$, we can apply Fact 1, to get

$$
\begin{aligned}
\mu(T) &= \mu(\bigcup\limits_{t \in T} D_t) = \bigcap\limits_{t \in T} \mu(D_t) \\
&= \bigcap\limits_{t \in T} \mu'(D_t) = \mu'(\bigcup\limits_{t \in T} D_t) = \mu'(T) = S.
\end{aligned}
$$

From Fact 2, we have $\iota(S) = \iota'(S) = \iota'\mu'(T) = T$. This shows that (*) holds, completing the proof of (ii) that $R_{\mathcal{U}}$ is a Galois-closed subrelation of R.

(iii) Now we must show that for any complete sublattice $\mathcal{U}$ of $\mathcal{H}_{\iota\mu}$, and any Galois-closed subrelation R' of R, we have $\mathcal{U}_{R_\mathcal{U}} = \mathcal{U}$ and $R_{\mathcal{U}_{R'}} = R'$.

We know that $\mathcal{U}_{R_\mathcal{U}} := \mathcal{H}_{\iota'\mu'}$, the lattice of subsets of A closed under the closure operator $\iota'\mu'$ induced from the relation $R_\mathcal{U}$. This means that $T \in \mathcal{U}_{R_\mathcal{U}}$ iff $\iota'\mu'(T) = T$. First let $T \in \mathcal{U}_{R_\mathcal{U}}$, and let S be the set $\mu'(T)$. Then we have $\iota'(S) = T$, and since $R_\mathcal{U}$ is a Galois-closed subrelation of R we conclude that

$$\mu(T) = S \quad \text{and} \quad \iota(S) = T.$$

If $T = \emptyset$ then $T \in \mathcal{U}$, and for $T \neq \emptyset$ we use the same argument as before to show that $T = \bigcup_{t \in T} D_t$ (see (d) above). But now $T = \iota\mu(T) = \iota\mu(\bigcup_{t \in T} D_t) = \sup\{D_t \mid t \in T\} \in \mathcal{U}$. This shows one direction, that $\mathcal{U}_{R_\mathcal{U}} \subseteq \mathcal{U}$.

For the opposite inclusion, let $T \in \mathcal{U}$. Then using the fact that $\mathcal{U}$ is a sublattice of $\mathcal{H}_{\iota\mu}$, along with Fact 2, we have

$$\iota'\mu'(T) = \iota'\mu(T) = \iota\mu(T) = T.$$

This shows that $T \in \mathcal{H}_{\iota'\mu'}$, which is equal to $\mathcal{U}_{R_\mathcal{U}}$. We now have the required equality $\mathcal{U}_{R_\mathcal{U}} = \mathcal{U}$.

Now let R' be a Galois-closed subrelation of R, and set

$$\begin{aligned}
\mathcal{U}_{R'} &:= \mathcal{H}_{\iota'\mu'} = \{T \subseteq A \mid \iota'\mu'(T) = T\}, \quad \text{and} \\
R_{\mathcal{U}_{R'}} &:= \cup\{T \times \mu(T) \mid T \in \mathcal{U}_{R'}\}.
\end{aligned}$$

We will show that $R_{\mathcal{U}_{R'}} = R'$.
First, if $(t, s) \in R'$ then $s \in \mu'(\{t\})$. Setting $S := \mu'(\{t\})$, we have $s \in S$ and $\iota'(S) = \iota'\mu'(\{t\})$. Now taking $T := \iota'\mu'(\{t\})$, we have $\iota'\mu'(T) = T$, so $T \in \mathcal{U}_{R'}$ and $\mu'(T) = S$ and $\iota'(S) = T$. Therefore $\mu(T) = S$ and $\iota(S) = T$.

Since $t \in \iota'\mu'(\{t\}) = T$ and $s \in S = \mu(T)$, we get $(t, s) \in T \times \mu(T)$ and $T \in \mathcal{U}_{R'}$. Hence $(t, s) \in R_{\mathcal{U}_{R'}}$, and we have shown that $R' \subseteq R_{\mathcal{U}_{R'}}$.

To show the opposite inclusion, let $T \in \mathcal{U}_{R'}$, and let $S = \mu(T)$. Then from Facts 1 and 2 we have

$$\mu'(T) = \mu(T) = S \quad \text{and} \quad \iota'(S) = \iota(S) = \iota\mu(T) = T.$$

Therefore $T \times \mu(T) \subseteq R'$, and $R_{\mathcal{U}_{R'}} \subseteq R'$. Altogether, we have $R_{\mathcal{U}_{R'}} = R'$. This completes the proof of part (iii), and of Theorem 13.2.4. ∎

13.3   Closure Operators on Complete Lattices

In this section we will describe one more method to produce complete sub-lattices of a given complete lattice. We will do so by consideration of the fixed points of a certain kind of closure operator defined on the complete lattice.

As before, we start with a relation R between two sets A and B, with induced Galois-connection (μ, ι) and corresponding closure operators $\iota\mu$ on A and $\mu\iota$ on B. We denote by $\mathcal{H}_{\iota\mu}$ and $\mathcal{H}_{\mu\iota}$ the corresponding complete lattices of closed sets on A and B respectively. Now we assume that $\gamma_1 : \mathcal{P}(A) \to \mathcal{P}(A)$ and $\gamma_2 : \mathcal{P}(B) \to \mathcal{P}(B)$ are additive closure operators which are conjugate with respect to the relation R. As we saw in Section 13.1, this conjugate pair determines a new relation R_γ from A to B, with its own induced Galois-connection $(\iota_\gamma, \mu_\gamma)$ and closure operators $\iota_\gamma\mu_\gamma$ and $\mu_\gamma\iota_\gamma$. We now have three lattices of closed sets on A, corresponding to closure under the operators $\iota\mu$, $\iota_\gamma\mu_\gamma$ and γ_1, and dually three lattices on B. From Theorems 13.1.5 and 13.1.7 we have the following connection between these lattices:

$$\mathcal{H}_{\iota_\gamma\mu_\gamma} = \mathcal{H}_{\iota\mu} \cap \mathcal{H}_{\gamma_1} \quad \text{and} \quad \mathcal{H}_{\mu_\gamma\iota_\gamma} = \mathcal{H}_{\mu\iota} \cap \mathcal{H}_{\gamma_2}.$$

For notational convenience, we shall henceforth denote the lattice $\mathcal{H}_{\iota_\gamma\mu_\gamma}$ by $\mathcal{S}_{\gamma_1}$, and dually the lattice $\mathcal{H}_{\mu_\gamma\iota_\gamma}$ by $\mathcal{S}_{\gamma_2}$.

In this section we examine how the lattice $\mathcal{S}_{\gamma_1}$ is situated in the lattice $\mathcal{H}_{\iota\mu}$. In particular, for any set $T \in \mathcal{H}_{\iota\mu}$ we can look for the least γ_1-closed class $^{\gamma_1}T$ containing T and the greatest γ_1-closed class $_{\gamma_1}T$ contained in T. Thus we consider two operators $\Gamma_1 : T \mapsto {}^{\gamma_1}T$ and $\Gamma_2 : T \mapsto {}_{\gamma_1}T$, whose properties will be studied in more detail. We will present our definitions and results for the lattices $\mathcal{H}_{\iota\mu}$ and $\mathcal{S}_{\gamma_1}$ on A, but of course these results can be dualized for the corresponding lattices on B as well.

Definition 13.3.1 Let T be an arbitrary subset of A. Then we define:

$$^{\gamma_1}T := \bigcap \{T' \in \mathcal{S}_{\gamma_1} \mid T' \supseteq T\}$$

$$_{\gamma_1}T := \overset{\mathcal{S}_{\gamma_1}}{\bigvee} \{T' \in \mathcal{S}_{\gamma_1} \mid T' \subseteq T\} = \overset{\mathcal{H}_{\iota\mu}}{\bigvee} \{T' \in \mathcal{S}_{\gamma_1} \mid T' \subseteq T\}.$$

(Note that the last equality holds because $\mathcal{S}_{\gamma_1}$ is a complete sublattice of $\mathcal{H}_{\iota\mu}$, as we proved in Theorem 13.1.7.)

We shall need the following preliminary lemma.

Lemma 13.3.2 *For any sets T and T' in $\mathcal{H}_{\iota\mu}$ and any set S in $\mathcal{H}_{\mu\iota}$,*

(i) $T' \subseteq T$ *iff* $\mu(T') \supseteq \mu(T)$.

(ii) $S = \mu(T) \in \mathcal{S}_{\gamma_2}$ *iff* $T = \iota(S) \in \mathcal{S}_{\gamma_1}$.

Proof: (i) This follows directly from the definition of a Galois-connection.

(ii) Let $T \in \mathcal{H}_{\iota\mu}$ and $S \in \mathcal{H}_{\mu\iota}$. If $S = \mu(T) \in \mathcal{S}_{\gamma_2}$, then $\iota(S) = \iota\mu(T) = T$ and also $\iota\gamma_2(S) = \iota(S)$. This gives $T = \iota\gamma_2(S)$. Applying μ to both sides and using 13.1.4 (i') and 13.2.2 (iii), we get

$$\mu(T) = \mu\iota\gamma_2(S) = \mu\iota_\gamma(S) = \mu_\gamma\iota_\gamma(S).$$

Now we apply ι to both sides of this result, to get $\iota\mu_\gamma\iota_\gamma(S) = \iota(S)$. Finally we have

$$\iota_\gamma(S) = \iota_\gamma\mu_\gamma\iota_\gamma(S) = \iota\mu_\gamma\iota_\gamma(S) = \iota(S) = \iota\mu(T) = T,$$

so $T = \iota(S) \in \mathcal{H}_{\iota_\gamma\mu_\gamma} = \mathcal{S}_{\gamma_1}$. The other direction can be proved similarly. ∎

Now we can prove our first properties of the sets ${}^{\gamma_1}T$ and ${}_{\gamma_1}T$.

Proposition 13.3.3 *Let $T \subseteq A$. Then:*

(i) ${}^{\gamma_1}T = \iota_\gamma\mu_\gamma(T) = \iota\mu_\gamma(T) = \iota\mu\,\gamma_1(T)$; *and in particular,* ${}^{\gamma_1}T$ *is the $\iota_\gamma\mu_\gamma$-closed set generated by T.*

(ii) *If $T = \iota\mu(T)$, then*

(a) ${}_{\gamma_1}T = T$ *iff* ${}^{\gamma_1}T = T$ *iff* $\gamma_1(T) = T$,

(b) ${}_{\gamma_1}T$ *is the greatest $\iota_\gamma\mu_\gamma$-closed set contained in T,* *and*

(c) ${}_{\gamma_1}T = \iota_\gamma\mu(T) = \iota\,\gamma_2\,\mu(T)$.

Proof: (i) This follows from the definition of ${}^{\gamma_1}T$, Proposition 13.2.2 (iii) and Proposition 13.1.4 (i).

(ii) (a) The first equivalence follows from the definitions, while the second one follows directly from Theorem 13.1.5 and the fact that $\mathcal{S}_{\gamma_1} = \mathcal{H}_{\iota_\gamma\mu_\gamma}$.

(ii) (b) By definition, $_{\gamma_1}T$ is a join of elements in the lattice $\mathcal{S}_{\gamma_1}$, so it is in $\mathcal{S}_{\gamma_1}$. Moreover, this lattice is equal to $\mathcal{H}_{\iota_\gamma\mu_\gamma}$, which is a complete sublattice of $\mathcal{H}_{\iota\mu}$, so $_{\gamma_1}T$ is $\iota_\gamma\mu_\gamma$-closed. It is also contained in T, since all the sets in the join are contained in T and T itself is assumed to be $\iota\mu$-closed. Moreover, every $\iota_\gamma\mu_\gamma$-closed set T' is also in $\mathcal{S}_{\gamma_1}$ and therefore contained in $_{\gamma_1}T$ by definition; so $_{\gamma_1}T$ is the largest such set.

(ii) (c) By Lemma 13.3.2 we have

$$
\begin{aligned}
_{\gamma_1}T &= \overset{\mathcal{H}_{\iota\mu}}{\bigvee}\{T' \in \mathcal{S}_{\gamma_1} \mid T' \subseteq T\} = \iota\mu\bigcup\{T' \in \mathcal{S}_{\gamma_1} \mid T' \subseteq T\} \\
&= \iota\bigcap\{\mu(T') \in \mathcal{S}_{\gamma_2} \mid T' \subseteq T\} \\
&= \iota\bigcap\{\mu(T') \in \mathcal{S}_{\gamma_2} \mid \mu(T') \supseteq \mu(T)\} \\
&= \iota\bigcap\{S' \in \mathcal{S}_{\gamma_2} \mid S' \supseteq \mu(T)\} = \iota\mu_\gamma\iota_\gamma\mu(T) \\
&= \iota_\gamma\mu_\gamma\iota_\gamma\mu(T), \qquad \text{by 13.2.2} \\
&= \iota_\gamma\mu(T).
\end{aligned}
$$
∎

For any set T in the lattice $\mathcal{H}_{\iota\mu}$, let us denote by $[_{\gamma_1}T,\ ^{\gamma_1}T]$ the interval between $_{\gamma_1}T$ and $^{\gamma_1}T$ in $\mathcal{H}_{\iota\mu}$. Such intervals will be called γ_1-intervals in $\mathcal{H}_{\iota\mu}$. It is possible that different sets T may produce the same γ_1-interval. This suggests that we define an equivalence relation $\sim$ on $\mathcal{H}_{\iota\mu}$, by $T_1 \sim T_2$ $: \iff [_{\gamma_1}T_1,\ ^{\gamma_1}T_1] = [_{\gamma_1}T_2,\ ^{\gamma_1}T_2]$.

A set T in $\mathcal{H}_{\iota\mu}$ will be called $\iota_\gamma\mu_\gamma$-*collapsing* if the γ_1-interval $[_{\gamma_1}T,\ ^{\gamma_1}T] = \{T\}$, that is, if the interval "collapses" to the singleton set containing T. Thus collapsing sets are uniquely characterized by their $\iota_\gamma\mu_\gamma$-closure.

Proposition 13.3.4 *Let T, T_1 and T_2 be $\iota\mu$-closed subsets of A. Then the following properties hold:*

(i) $T \in \mathcal{S}_{\gamma_1}$ *iff* $[_{\gamma_1}T,\ ^{\gamma_1}T] = \{T\}$,

(ii) $T_1 \subseteq T_2$ *implies* $_{\gamma_1}T_1 \subseteq {}_{\gamma_1}T_2$ *and* $^{\gamma_1}T_1 \subseteq {}^{\gamma_1}T_2$,

(iii) $_{\gamma_1}T_1 \overset{\mathcal{H}_{\iota\mu}}{\wedge} {}_{\gamma_1}T_2 = {}_{\gamma_1}(T_1 \overset{\mathcal{H}_{\iota\mu}}{\wedge} T_2)$,

(iv) $^{\gamma_1}T_1 \overset{\mathcal{H}_{\iota\mu}}{\vee} {}^{\gamma_1}T_2 = {}^{\gamma_1}(T_1 \overset{\mathcal{H}_{\iota\mu}}{\vee} T_2)$.

Proof: (i) and (ii) follow directly from Definition 13.3.1.

(iii) Since $\mathcal{S}_{\gamma_1}$ is a complete sublattice of $\mathcal{H}_{\iota\mu}$ by Theorem 13.1.7, the set on the left hand side is an element of $\mathcal{S}_{\gamma_1}$, and it is contained in both T_1 and T_2. Therefore it is also contained in the greatest set from $\mathcal{S}_{\gamma_1}$ to contain T_1 and T_2, which is the set on the right hand side of (iii). Conversely, the set on the right hand side of (iii) is by definition an element of $\mathcal{S}_{\gamma_1}$, and by part (ii) it is contained in both $_{\gamma_1}T_1$ and $_{\gamma_1}T_2$. Thus it is also contained in the set on the left hand side.

(iv) This is dual to (iii). ∎

Proposition 13.3.5 (i) *The mapping* $\Gamma_1 : \mathcal{H}_{\iota\mu} \to \mathcal{H}_{\iota\mu}$ *defined by* $T \mapsto {}^{\gamma_1}T$ *is a closure operator on* $\mathcal{H}_{\iota\mu}$, *and satisfies*

$$
{}^{\gamma_1}\left(\overset{\mathcal{H}_{\iota\mu}}{\bigvee} \{T_j \mid j \in J\} \right) = \overset{\mathcal{H}_{\iota\mu}}{\bigvee} \{{}^{\gamma_1}T_j \mid j \in J\}.
$$

(ii) *The mapping* $\Gamma_2 : \mathcal{H}_{\iota\mu} \to \mathcal{H}_{\iota\mu}$ *defined by* $T \mapsto {}_{\gamma_1}T$ *is a kernel operator on* $\mathcal{H}_{\iota\mu}$, *and satisfies*

$$
{}_{\gamma_1}\left(\bigcap \{T_j \mid j \in J\} \right) = \bigcap \{{}_{\gamma_1}T_j \mid j \in J\}.
$$

Proof: By part (i) of Proposition 113.3.3, the new operator Γ_1 coincides with the closure operator $\iota_\gamma\mu_\gamma$; so it is a closure operator. The operator Γ_2 is isotone by 13.3.4 part (ii). It is intensive, that is, $\Gamma_2(T) \subseteq T$ for any T in $\mathcal{H}_{\iota\mu}$, by 13.3.3 part (ii)(b). It is also idempotent, since $_{\gamma_1}T \in \mathcal{S}_{\gamma_1}$ is γ_1-closed and therefore $_{\gamma_1}(_{\gamma_1}T) = {}_{\gamma_1}T$ by 13.3.2 part (ii)(a). Thus Γ_2 is a kernel operator.

The two equalities are generalizations of parts (iv) and (iii) of Proposition 13.3.4, and may be proved in the same manner. Note that $\bigcap$ equals $\overset{\mathcal{H}_{\iota\mu}}{\bigwedge}$ in the lattice $\mathcal{H}_{\iota\mu}$. ∎

Remark 13.3.6 From the definitions and Proposition 13.3.4 parts (iii) and (iv), we conclude that the operators Γ_1 and Γ_2 have the following properties:

(i) The mapping Γ_1 is a join-retraction from $\mathcal{H}_{\iota\mu}$ onto $\mathcal{S}_{\gamma_1} \subseteq \mathcal{H}_{\iota\mu}$; that is, it is an idempotent join-homomorphism which is the identity map on $\mathcal{S}_{\gamma_1}$.

(ii) Analogously, the mapping Γ_2 is a meet-retraction from $\mathcal{H}_{\iota\mu}$ onto $\mathcal{S}_{\gamma_1}$; that is, it is an idempotent meet-homomorphism which is the identity map on $\mathcal{S}_{\gamma_1}$.

(iii) Note that, in general, Γ_1 does not preserve meets and Γ_2 does not preserve joins.

In Chapter 2 we showed that there is a 1-1 correspondence between closure operators $\varphi : \mathcal{L} \to \mathcal{L}$ on a complete lattice $\mathcal{L}$ and closure systems $\mathcal{S}$ on $\mathcal{L}$ (that is, subsets of $\mathcal{L}$ closed under arbitrary meets). This correspondence occurs via the following maps. For any closure operator $\varphi : \mathcal{L} \to \mathcal{L}$, we get the closure system

$$S := Fix(\varphi) := \mathcal{H}_\varphi = \{T \in \mathcal{L} \mid \varphi(T) = T\}$$

of all fixed points of φ; and for any closure system $\mathcal{S}$ we have the closure operator φ defined by

$$\varphi(T) = \varphi_S(T) := \overset{\mathcal{L}}{\bigwedge}\{T' \in \mathcal{S} \mid T \leq T'\} \text{ for } T \in \mathcal{L}.$$

Moreover, for any closure system $\mathcal{S}$ and any closure operator φ, we have $\mathcal{H}_{\varphi_S} = \mathcal{S}$ and $\varphi_{\mathcal{H}_\varphi} = \varphi$.

There is also a dual 1-1 correspondence between kernel operators and kernel systems on a complete lattice $\mathcal{L}$. A kernel system on $\mathcal{L}$ is a family $\mathcal{S}$ of subsets of $\mathcal{L}$ which is closed under arbitrary joins. Then for kernel operators $\psi : \mathcal{L} \to \mathcal{L}$ on $\mathcal{L}$ and kernel systems $\mathcal{S}$ on $\mathcal{L}$, we set

$$S \quad := \quad Fix(\psi) := \{T \in \mathcal{L} \mid \psi(T) = T\};$$
$$\psi(T) \quad := \quad \psi_S(T) := \overset{\mathcal{L}}{\bigvee}\{T' \in \mathcal{S} \mid T' \leq T\} \text{ for } T \in \mathcal{L}.$$

We have $Fix(\psi_S) = \mathcal{S}$ and $\psi_{Fix(\psi)} = \psi$, for any kernel system $\mathcal{S}$ and any kernel operator ψ on $\mathcal{L}$.

A result of A. Tarski ([111]) shows that for any closure operator φ on a complete lattice $\mathcal{L}$, the closure system (fixed-point set) $\mathcal{H}_\varphi$ is always a complete lattice with respect to $\leq$. However, it is not necessarily a sublattice of $\mathcal{L}$. Thus we look for some additional condition under which a complete sublattice is obtained. The answer (and its dual for kernel operators) is given in the following theorem.

Theorem 13.3.7 *Let $\mathcal{L}$ be a complete lattice.*

(i) *If φ is a closure operator on $\mathcal{L}$ which satisfies*

$$\varphi(\overset{\mathcal{L}}{\bigvee}\{T_j \mid j \in J\}) = \overset{\mathcal{L}}{\bigvee}\{\varphi(T_j) \mid j \in J\} \qquad (*)$$

for every index set J, then the set of all fixed points under φ,

$$\mathcal{H}_\varphi = \{T \in \mathcal{L} \mid \varphi(T) = T\},$$

is a complete sublattice of $\mathcal{L}$ and $\varphi(\mathcal{L}) = \mathcal{H}_\varphi$.

(ii) *Conversely, if $\mathcal{H}$ is a complete sublattice of $\mathcal{L}$, then the function $\varphi_{\mathcal{H}}$ which is defined by*

$$\varphi_{\mathcal{H}}(T) := \overset{\mathcal{L}}{\bigwedge}\{T' \in \mathcal{H} \mid T \leq T'\}$$

is a closure operator on $\mathcal{L}$ with $\varphi_{\mathcal{H}}(\mathcal{L}) = \mathcal{H}$, and $\varphi_{\mathcal{H}}$ satisfies the condition ($*$). Moreover, $\mathcal{H}_{\varphi_{\mathcal{H}}} = \mathcal{H}$ and $\varphi_{\mathcal{H}_\varphi} = \varphi$.

(iii) *If ψ is a kernel operator on $\mathcal{L}$ which satisfies*

$$\psi(\overset{\mathcal{L}}{\bigwedge}\{T_j \mid j \in J\}) = \overset{\mathcal{L}}{\bigwedge}\{\psi(T_j) \mid j \in J\} \qquad (**)$$

for every index set J, then the set of all fixed points under ψ,

$$\mathcal{H}_\psi = \{T \in \mathcal{L} \mid \psi(T) = T\},$$

is a complete sublattice of $\mathcal{L}$ and $\psi(\mathcal{L}) = \mathcal{H}_\psi$.

(iv) *Conversely, if $\mathcal{H}$ is a complete sublattice of $\mathcal{L}$ then the function $\psi_{\mathcal{H}}$ which is defined by*

$$\psi_{\mathcal{H}}(T) := \overset{\mathcal{L}}{\bigvee}\{T' \in \mathcal{H} \mid T' \leq T\}$$

is a kernel operator on $\mathcal{L}$ with $\psi_{\mathcal{H}}(\mathcal{L}) = \mathcal{H}_\psi$, and ψ satisfies the condition ($**$). Moreover, $\mathcal{H}_{\psi_{\mathcal{H}}} = \mathcal{H}$ and $\psi_{\mathcal{H}_\psi} = \psi$.

Proof: (i) Let φ be a closure operator on $\mathcal{L}$ which satisfies the condition (*). We have to prove that the set of all fixed points under φ is a complete sublattice of $\mathcal{L}$, that is, that for any index set J, both

$$\overset{\mathcal{L}}{\bigwedge}\{T_j \in \mathcal{H}_\varphi \mid j \in J\} \in \mathcal{H}_\varphi \quad \text{and} \quad \overset{\mathcal{L}}{\bigvee}\{T_j \in \mathcal{H}_\varphi \mid j \in J\} \in \mathcal{H}_\varphi.$$

It is clear that $\varphi(\overset{\mathcal{L}}{\bigwedge}\{T_j \in \mathcal{H}_\varphi \mid j \in J\}) \geq \overset{\mathcal{L}}{\bigwedge}\{T_j \in \mathcal{H}_\varphi \mid j \in J\}$. For each $j \in J$ we have $T_j = \varphi(T_j) \geq \varphi(\overset{\mathcal{L}}{\bigwedge}\{T_j \in \mathcal{H}_\varphi \mid j \in J\})$, and from this we obtain $\overset{\mathcal{L}}{\bigwedge}\{T_j \in \mathcal{H}_\varphi \mid j \in J\} \geq \varphi(\overset{\mathcal{L}}{\bigwedge}\{T_j \in \mathcal{H}_\varphi \mid j \in J\})$. Altogether this gives equality, and $\overset{\mathcal{L}}{\bigwedge}\{T_j \in \mathcal{H}_\varphi \mid j \in J\} \in \mathcal{H}_\varphi$. The fact that φ satisfies the join condition (*) gives $\overset{\mathcal{L}}{\bigvee}\{T_j \in \mathcal{H}_\varphi \mid j \in J\} \in \mathcal{H}_\varphi$. Thus we have a sublattice of $\mathcal{L}$. It is clear that $\mathcal{H}_\varphi \subseteq \varphi(\mathcal{L})$. Since φ is idempotent, $\varphi(T)$ is in $\mathcal{H}_\varphi$ for all $T \in \mathcal{L}$. This shows that $\varphi(\mathcal{L}) = \mathcal{H}_\varphi$.

(ii) By Remark 13.3.6, we need only show that the closure operator defined by

$$\varphi_\mathcal{H}(T) := \overset{\mathcal{L}}{\bigwedge}\{T' \in \mathcal{H} \mid T \leq T'\}$$

satisfies condition (*) and that $\varphi_\mathcal{H}(\mathcal{L}) = \mathcal{H}$. We prove the latter fact first. Since $\mathcal{H}$ is a complete sublattice of $\mathcal{L}$, we have $\varphi_\mathcal{H}(\mathcal{L}) \subseteq \mathcal{H}$. For the opposite inclusion, we see that for any $T \in \mathcal{H}$

$$\varphi_\mathcal{H}(T) = \overset{\mathcal{L}}{\bigwedge}\{T' \in \mathcal{H} \mid T \leq T'\} = T.$$

Thus $\mathcal{H} \subseteq \varphi_\mathcal{H}(\mathcal{L})$, and altogether we have $\mathcal{H} = \varphi_\mathcal{H}(\mathcal{L})$.

Since for each $j \in J$ we have $T_j \leq \varphi_\mathcal{H}(T_j)$ and $\varphi_\mathcal{H}(T_j) \in \mathcal{H}$, the set

$$\overset{\mathcal{L}}{\bigvee}\{\varphi_\mathcal{H}(T_j) \mid j \in J\}$$

is an upper bound of the set $\{T_j \in \mathcal{L} \mid j \in J\}$. Therefore

$$\overset{\mathcal{L}}{\bigvee}\{T_j \in \mathcal{L} \mid j \in J\} \leq \overset{\mathcal{L}}{\bigvee}\{\varphi_\mathcal{H}(T_j) \mid j \in J\}.$$

Since the set on the right hand side of this inequality is an element of $\mathcal{H}$, applying $\varphi_\mathcal{H}$ on both sides gives

$$\varphi_\mathcal{H}(\overset{\mathcal{L}}{\bigvee}\{T_j \in \mathcal{L} \mid j \in J\}) \leq \overset{\mathcal{L}}{\bigvee}\{\varphi_\mathcal{H}(T_j) \in \mathcal{L} \mid j \in J\}.$$

Since $\overset{\mathcal{L}}{\bigvee}\{T_j \in \mathcal{L} \mid j \in J\} \geq T_j$, we have $\varphi_{\mathcal{H}}(\overset{\mathcal{L}}{\bigvee}\{T_j \in \mathcal{L} \mid j \in J\}) \geq \varphi_{\mathcal{H}}(T_j)$ for all $j \in J$. Thus also $\varphi_{\mathcal{H}}(\overset{\mathcal{L}}{\bigvee}\{T_j \in \mathcal{L} \mid j \in J\}) \geq \overset{\mathcal{L}}{\bigvee}\{\varphi_{\mathcal{H}}(T_j) \mid j \in J\}$, giving the required equality.

(iii), (iv) These proofs are analogous to those of (i) and (ii).

The equations follow from Remark 13.3.6, by restricting the one-to-one mapping between closure operators and complete lattices to closure operators satisfying condition (*) and to complete sublattices.   ∎

We can apply this Theorem to the special case of conjugate pairs of closure operators studied in Section 13.1. Using the notation from 13.1 and applying 13.3.5 part (i), we take $\mathcal{L} = \mathcal{H}_{\iota\mu}$, and $\varphi = \Gamma_1$ ($= \iota_\gamma\mu_\gamma$, as in 13.3.3 (i)) and $\psi = \Gamma_2$ ($= \iota_\gamma\mu$, as in 13.3.3 (ii)). This gives an additional proof of the fact that $\mathcal{S}_{\gamma_1} = \mathcal{H}_{\iota_\gamma\mu_\gamma} = Fix(\varphi) = Fix(\psi)$ is a complete sublattice of $\mathcal{H}_{\iota\mu}$.

Closure and kernel operators on complete lattices have been studied by K. P. Shum and A. Yang in [107] and K. Denecke in [22]. The techniques of this section have also been used by K. Denecke and S. L. Wismath in [38] in a general construction to produce complete sublattices. The application of this construction to the Galois-connection (Id, Mod) encompasses several well-known results on regular and normal identities, as well as some new families of identities and varieties.

13.4   Exercises

13.4.1. Let R be a relation between sets A and B. Prove that the closure operators $\mu\iota$ and $\iota\mu$ obtained from the Galois-connection (μ, ι) induced by R are conjugate with respect to R.

13.4.2. Prove the additional properties for conjugate pairs of additive closure operators listed at the end of Section 13.1.

13.4.3. Let R be a relation between sets A and B, with induced Galois-connection (μ, ι). Let $\gamma := (\gamma_1, \gamma_2)$ be a pair of additive closure operators which are conjugate with respect to R. Verify that the relation R_γ defined in 13.1.3 is a Galois-closed subrelation of R.

13.4.4. Prove that for an additive closure operator γ, the least upper bound operation on the lattice $\mathcal{H}_\gamma$ agrees with the union operation.

13.4.5. This exercise investigates partial closure operators. Let A be a non-empty set. A partial mapping $C : P(A) \to P(A)$ is called a partial closure operator on A if it satisfies the following conditions, for every $X, Y \subseteq A$:

(i) if $C(X)$ is defined, then $X \subseteq C(X)$,
(ii) if $C(X)$ and $C(Y)$ are defined, then $X \subseteq Y$ implies $C(X) \subseteq C(Y)$,
(iii) if $C(X)$ is defined, then $C(C(X)) = C(X)$,   and
(iv) $C(\{x\})$ is defined for every $x \in A$.

If $X \subseteq A$ and $C(X) = X$, then X is said to be a closed set. A family $\mathcal{F}$ of subsets of A is called a partial closure system on A if it satisfies the following two conditions:

(i) $\bigcup \mathcal{F} = A,$ and
(ii) for every $x \in A$, $\bigcap \{X \in \mathcal{F} \mid x \in X\} \in \mathcal{F}$.

Prove that the family of closed sets of a partial closure operator on a set A is a partial closure system on A, and conversely that for every partial closure system $\mathcal{F}$ on A there is a partial closure operator on A whose family of closed sets is exactly $\mathcal{F}$. (See B. Šešelja and A. Tepavčević, [106].)

13.4.6. Prove that every partially ordered set $(P; \leq)$ is isomorphic to a partial closure system on P, ordered by inclusion.

Chapter 14

G-Clones and M-Solid Varieties

In Chapter 13 we studied methods of producing complete sublattices of a complete lattice. We now apply these methods to our two chief examples of complete lattices, the lattice of all clones on a fixed set and the lattice of all varieties of algebras of a given type. We have seen, in Chapters 2 and 6, that both of these lattices arise as the lattices of closed sets from a Galois-connection.

14.1 G-Clones

In this section we apply our theory of Galois-closed subrelations to the lattice of clones on a fixed set. We assume a fixed base set A, and denote by $O(A)$ the set of all finitary operations on A and by $R(A)$ the set of all finitary relations on set A. As our basic relation between these two sets we have the relation R of preservation:

$$R = \{(f, \rho) \mid f \in O(A),\ \rho \in R(A)\ \text{and}\ f\ \text{preserves}\ \rho\}.$$

We saw in Chapter 2 that this relation induces a Galois-connection of the form (Pol_A, Inv_A), between sets of operations and sets of relations. From this we obtain two lattices of closed sets, the lattice of all *clones* on the set A and the lattice of all *relational clones* on A. Thus clones on A are sets C of operations for which $Pol_A Inv_A C = C$, and dually relational clones are sets Q of relations for which $Inv_A Pol_A Q = Q$. Now we want to produce a

Galois-closed subrelation of the relation R, with a corresponding complete sublattice of the lattice of clones on A.

To form this subrelation, we focus on certain kinds of operations on A. We let $\mathcal{S}_A$ be the symmetric group of all permutations defined on the set A. Then for every n-ary operation $f \in O(A)$ and any permutation $s \in \mathcal{S}_A$ we can define a new operation f^s, of the same arity as f, by

$$f^s(a_1, \ldots, a_n) := s(f(s^{-1}(a_1), s^{-1}(a_2), \ldots, s^{-1}(a_n)),$$

for all $a_1, \ldots, a_n \in A$.

We use this to define, for any fixed subgroup $\mathcal{G} \subseteq \mathcal{S}_A$ of permutations, a mapping γ_G^O on operations and sets of operations. For any operation f and any set $F \subseteq O(A)$, we set

$$\gamma_G^O(f) := \{f^s \mid s \in G\} \text{ and } \gamma_G^O(F) := \bigcup_{f \in F} \gamma_G^O(f).$$

This gives a map γ_G^O on the power set of $O(A)$, which is our first candidate for a closure operator.

Lemma 14.1.1 *For every subgroup $\mathcal{G} \subseteq \mathcal{S}_A$ the operator*

$$\gamma_G^O : \mathcal{P}(O(A)) \to \mathcal{P}(O(A))$$

defined by $F \mapsto \gamma_G^O(F)$ is an additive closure operator on $O(A)$.

Proof: By definition our mapping γ_G^O is additive and therefore monotone, so that

$$F_1 \subseteq F_2 \Rightarrow \gamma_G^O(F_1) \subseteq \gamma_G^O(F_2).$$

Since the subgroup $\mathcal{G}$ contains the identity permutation φ_{id} and $f^{\varphi_{id}} = f$ for every $f \in F$, we have $F \subseteq \gamma_G^O(F)$, making the operator γ_G^O extensive. From extensivity and monotonicity it follows that $\gamma_G^O(F) \subseteq \gamma_G^O(\gamma_G^O(F))$. For the other inclusion for idempotency, we see that for any two permutations s and s' in G we have

$$
\begin{aligned}
(f^s)^{s'}(a_1, \ldots, a_n) &= s'(f^s(s'^{-1}(a_1), \ldots, s'^{-1}(a_n))) \\
&= s'(s(f(s^{-1}(s'^{-1}(a_1)), \ldots, s^{-1}(s'^{-1}(a_n))))) \\
&= (s' \circ s)(f((s' \circ s)^{-1}(a_1), \ldots, (s' \circ s)^{-1}(a_n))),
\end{aligned}
$$

and since $s' \circ s \in G$ we have $\gamma_G^O(\gamma_G^O(F)) \subseteq \gamma_G^O(F)$.   ∎

Definition 14.1.2 Let $G \subseteq S_A$ be a permutation group on the set A. A clone C on A is called a *G-clone* if $\gamma_G^O(C) = C$; so C is closed with respect to the operator γ_G^O.

G-clones have been studied by several authors: see for instance Gorlov and Pöschel, [50], and N. van Hoa, [56]. The special case where $f^s = f$ for each element f of the clone and a permutation $s \in S_A$ was considered by Demetrovics and Hannák in [17], Demetrovics, Hannák and Marchenkov in [18] and [19], by Marchenkov in [76] and by Csákány and Gavalcová in [13].

We can use the closure operator γ_G^O to define another relation R_G between operations and relations on A, by setting

$$R_G := \{(f, \rho) \mid f \in O(A), \ \rho \in R(A) \text{ and } \gamma_G^O(f) \times \{\rho\} \subseteq R\}.$$

To show that R_G is a Galois-closed subrelation of R, we look for another additive closure operator, this time on the set of relations on A, in order to make a pair of operators conjugate with respect to the original relation R of preservation. For any h-ary relation $\rho \in R^h(A)$ and any $s \in G$, we define (as in Rosenberg, [103])

$$\rho^s := \{(s(x_1), \ldots, s(x_h)) \mid (x_1, \ldots x_h) \in \rho\}.$$

As before, we use this to define an operator on individual relations and on sets Q of relations on A, with $\gamma_G^R(\rho) := \{\rho^s \mid s \in G\}$ and $\gamma_G^R(Q) := \bigcup_{\rho \in Q} \gamma_G^R(\rho)$. Then it is straightforward to verify, as in Lemma 14.1.1, that the mapping γ_G^R is a closure operator on $R(A)$.

Lemma 14.1.3 *For every subgroup $G \subseteq S_A$ the operator*

$$\gamma_G^R : \mathcal{P}(R(A)) \to \mathcal{P}(R(A))$$

defined by $Q \mapsto \gamma_G^R(Q)$ is an additive closure operator on $R(A)$. ∎

Definition 14.1.4 A relational clone $Q \subseteq R(A)$ is called a *G-relational clone* if $\gamma_G^R(Q) = Q$.

Now we have a pair (γ_G^O, γ_G^R) of additive closure operators, between sets of operations and sets of relations, and we can verify that these operators are conjugate with respect to the relation R of preservation.

Lemma 14.1.5 *For any $f \in O(A)$, for any $\rho \in R(A)$ and for any subgroup $\mathcal{G} \subseteq \mathcal{S}_A$, we have:*

$$\gamma_G^O(f) \text{ preserves } \rho \quad \text{iff} \quad f \text{ preserves } \gamma_G^R(\rho).$$

Proof: We prove first that for every $s \in \mathcal{S}_A$, f preserves ρ iff f^s preserves ρ^s (see I. G. Rosenberg, [103]). If $(a_{11}, \ldots, a_{1h}), \ldots, (a_{n1}, \ldots, a_{nh})$ are h-tuples from ρ then

$$(f^s(s(a_{11}), \ldots, s(a_{n1})), \ldots, f^s(s(a_{1h}), \ldots, s(a_{nh})))$$
$$= (s(f(s^{-1}(s(a_{11})), \ldots, s^{-1}(s(a_{n1})))), \ldots,$$
$$s(f(s^{-1}(s(a_{1h})), \ldots, s^{-1}(s(a_{nh})))))$$
$$= (s(f(a_{11}, \ldots, a_{n1})), \ldots, s(f(a_{1h}, \ldots, a_{nh}))) \in \rho^s,$$

and thus $(f(a_{11}, \ldots, a_{n1}), \ldots, f(a_{1h}, \ldots, a_{nh})) \in \rho$, and conversely. Now if $\gamma_G^O(f)$ preserves ρ then $f^{s^{-1}}$ preserves ρ for every $s^{-1} \in G$. Using the result just proved, we have that $(f^{s^{-1}})^s = f$ preserves ρ^s, for every $s \in G$. This means that f preserves $\gamma_G^R(\rho)$. The converse can be shown in a similar way. ∎

Combining the three previous Lemmas gives the following conclusion.

Theorem 14.1.6 *Let $\mathcal{G}$ be a subgroup of $\mathcal{S}_A$. Then the pair $\gamma_G := (\gamma_G^O, \gamma_G^R)$ is a conjugate pair of additive closure operators with respect to the relation*

$$R = \{(f, \rho) \mid f \in O(A), \rho \in R(A) \text{ and } f \text{ preserves } \rho\}. \quad \blacksquare$$

Our conjugate pair of closure operators induces the relation

$$R_G = \{(f, \rho) \mid f \in O(A), \rho \in R(A) \text{ and } \{f\} \times \gamma_G^R(\rho) \subseteq R\}$$

between $O(A)$ and $R(A)$. The Galois-connection induced by this relation R_G is denoted by $(GInv_A, GPol_A)$. We know from Section 13.2 that this relation is a Galois-closed subrelation of the original relation R. We also have a number of properties of our closure operators and closed sets, from the theorems of Section 13.1.

Theorem 14.1.7 *Let $\mathcal{G} \subseteq \mathcal{S}_A$ be a subgroup of the full permutation group $\mathcal{S}_A$ on a set A. Then the set of all G-clones of operations defined on A forms*

a complete sublattice of the lattice $\mathcal{L}_A$ of all clones of operations defined on
A. We denote this sublattice of $\mathcal{L}_A$ by $\mathcal{L}_A^G$. If $\mathcal{G}$ is a subgroup of $\mathcal{G}'$, then
$\mathcal{L}_A^{G'} \subseteq \mathcal{L}_A^G$, and $\mathcal{L}_A^{G'}$ is a complete sublattice of $\mathcal{L}_A^G$.

Dually, the set of all G-relational clones forms a complete sublattice of the
lattice of all relational clones.

Proof: The fact that we get complete sublattices, of the lattices of clones and
of relational clones respectively, comes from Theorem 13.1.7. If $\mathcal{G} \subseteq \mathcal{G}' \subseteq \mathcal{S}_A$
are subgroups, then clearly $\gamma_G^O(F) \subseteq \gamma_{G'}^O(F)$ for every $F \subseteq O(A)$, and if
$\gamma_{G'}^O(F) = F$ then also $\gamma_G^O(F) = F$, so $\mathcal{L}_A^{G'} \subseteq \mathcal{L}_A^G$. Moreover $R_{G'}$ is a Galois-
closed subrelation of R_G, so that $\mathcal{L}_A^{G'}$ is a complete sublattice of $\mathcal{L}_A^G$. ∎

Theorem 13.1.4 gives us a number of interactions between the various closed
sets, which we restate here for the G-clone setting.

Proposition 14.1.8 *For all $F \subseteq O(A)$ and all $Q \subseteq R(A)$, the following
properties hold:*

(i) $GPol_AQ = Pol_A\gamma_G^R(Q)$,
(ii) $GPol_AQ \subseteq Pol_AQ$,
(iii) $\gamma_G^O(GPol_AQ) = GPol_AQ$,
(iv) $GPol_AGInv_AF = Pol_AInv_A\gamma_G^O(F)$;   *and dually,*

(i') $GInv_AF = Inv_A\gamma_G^O(F)$,
(ii') $GInv_AF \subseteq Inv_AF$,
(iii') $\gamma_G^R(GInv_AF) = GInv_AF$,
(iv') $GInv_AGPol_AQ = Inv_APol_A\gamma_G^R(Q)$. ∎

The Main Theorem for Conjugate Pairs of additive closure operators, Theorem 13.1.5, gives us a characterization of G-clones and G-relational clones.

Theorem 14.1.9 *Let $\mathcal{G} \subseteq \mathcal{S}_A$ be a subgroup of the full symmetric group on
set A. Let $F \subseteq O(A)$ be a clone and let $Q \subseteq R(A)$ be a relational clone. Then
the following conditions (i) - (v) are equivalent for F, and dually conditions
(i') -(v') are equivalent for Q:*

(i) $F = GPol_AGInv_AF$,   (i') $Q = GInv_AGPol_AQ$,
(ii) $\gamma_G^O(F) = F$,   (ii') $\gamma_G^R(Q) = Q$,
(iii) $Inv_AF = GInv_AF$,   (iii') $Pol_AQ = GPol_AQ$,
(iv) $\gamma_G^R(Inv_AF) = Inv_AF$,   (iv') $\gamma_G^O(Pol_AQ) = Pol_AQ$,
(v) $F = Pol_AGInv_AF$;   *and dually,*   (v') $Q = Inv_AGPol_AQ$.

Proof: These conditions all come from Theorem 13.1.5, except for (iv) and (iv$'$), which are simply applications of (ii$'$) and (ii). ∎

We also have the following conditions which can be derived from Theorem 13.1.6 (see also K. Denecke and M. Reichel, [35]).

Proposition 14.1.10 *Let* $\mathcal{G} \subseteq \mathcal{S}_A$ *be a subgroup of the full permutation group on set* A. *Then for any* $F \subseteq O(A)$ *and* $Q \subseteq R(A)$ *the following properties hold:*

(i) $\gamma_G^O(F) \subseteq \ Pol_A Inv_A F \ \Leftrightarrow \ Pol_A Inv_A F = \\ GPol_A GInv_A F,$

(ii) $\gamma_G^O(F) \subseteq \ Pol_A Inv_A F \ \Leftrightarrow \ \gamma_G^O(Pol_A Inv_A F) = \\ Pol_A Inv_A F,$

(i$'$) $\gamma_G^R(Q) \subseteq \ Inv_A Pol_A Q \ \Leftrightarrow \ Inv_A Pol_A Q = \\ GInv_A GPol_A Q,$

(ii$'$) $\gamma_G^R(Q) \subseteq \ Inv_A Pol_A Q \ \Leftrightarrow \ \gamma_G^R(Inv_A Pol_A Q) = \\ Inv_A Pol_A Q.$ ∎

Condition (ii) is a useful tool in checking whether a clone is a G-clone. Suppose that we have a generating set or basis F for a clone, so $\langle F \rangle :=$ $Pol_A Inv_A F$. To test if this clone is a G-clone, by (ii) it is enough to check whether the set $\gamma_G^O(F)$ is included in $\langle F \rangle$.

As an example we will apply this method to $\mathcal{L}_2$, the lattice of all clones on the two-element set $A = \{0,1\}$. As we saw in Section 10.3, this lattice was first completely described by E. L. Post, in [95]. We will use here the notation of Jablonskij, Gawrilow and Kudrjawzew in [60]. It is clear that the group S_2 of all permutations defined on the set $\{0,1\}$ contains only one non-trivial function, namely the negation $\neg : x \mapsto \neg x$. This means that S_2-clones are those clones F which are self-dual as sets, that is, sets with $F^{\neg} = F$. We list here the clones we shall need, with the notational convention that we denote a clone by the name used in Section 10.3 for the corresponding two-element algebra.

$\mathcal{O}_1 = J_A$,   the clone of projections,

$\mathcal{O}_4 = \langle \neg \rangle$, the clone of projections and their negations,

$\mathcal{O}_8 = \langle c_0^2, c_1^2 \rangle$, the clone of constants (c_0^2, c_1^2 are the constants with value 0 and 1, respectively),

$\mathcal{O}_9 = \langle c_0^2, \neg \rangle$, the clone of essentially unary operations,

$\mathcal{L}_4 = \langle g \rangle$, $g(x, y, z) = x + y + z$, the clone of linear
   idempotent operations,

$\mathcal{L}_5 = \langle g, \neg \rangle$, the clone of linear self-dual operations,

$\mathcal{L}_1 = \langle c_1^2, + \rangle$, the clone of all linear operations,

$\mathcal{D}_2 = \langle h \rangle$, $h(x, y, z) := (x \wedge y) \vee (x \wedge z) \vee (y \wedge z)$,
   the clone of self-dual monotone operations,

$\mathcal{D}_1 = \langle g, h \rangle$, the clone of self-dual idempotent operations,

$\mathcal{D}_3 = \langle g, h, \neg \rangle$, the clone of self-dual operations,

$\mathcal{A}_4 = \langle \wedge, \vee \rangle$, the clone of monotone idempotent operations,

$\mathcal{A}_1 = \langle c_0^2, c_1^2, \wedge, \vee \rangle$, the clone of monotone operations,

$\mathcal{C}_4 = \langle \vee, t \rangle$, for $t(x, y, z) = x \wedge (y + z + 1)$, the clone of idempotent
   operations,

$\mathcal{C}_1 = O(A)$, the clone of all operations on $\{0, 1\}$.

By checking the generating systems, it can be verified that all of these clones
are S_2-clones. That these are all the S_2-clones was proved by Gorlov and
Pöschel in [50].

Theorem 14.1.11 *([50]) There are exactly fourteen S_2-clones on the two-
element set $A = \{0, 1\}$:*
$\mathcal{O}_1$, $\mathcal{O}_4$, $\mathcal{O}_8$, $\mathcal{O}_9$, $\mathcal{L}_1$, $\mathcal{L}_4$, $\mathcal{L}_5$, $\mathcal{D}_2$, $\mathcal{D}_1$, $\mathcal{D}_3$, $\mathcal{A}_4$, $\mathcal{A}_1$, $\mathcal{C}_4$ *and* $\mathcal{C}_1$. ∎

The complete sublattice $\mathcal{L}_2^{S_2}$ of $\mathcal{L}_2$ is given by the Hasse diagram below.
Gorlov and Pöschel also proved (see [50]) that there are forty-eight S_3-clones
of operations defined on the three-element set $A = \{0, 1, 2\}$.

14.2   *H*-clones

In [50], V. V. Gorlov and R. Pöschel described several generalizations of
G-clones to *H*-clones, where *H* is a transformation monoid. We will use a
new approach here, by applying a closure operator on the lattice of all clones
which is different from that given in Section 14.1.

Let $\mathcal{T}_A = (O^1(A), \circ, \varphi_{id})$ be the monoid of all unary mappings or *transformations* on *A*, where $\circ$ is the composition of unary operations and φ_{id} is the
identity mapping on *A*.

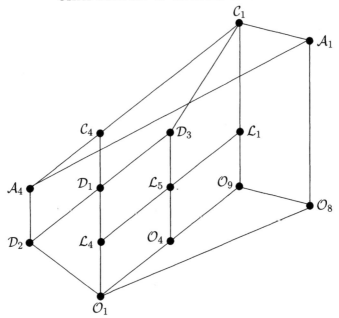

For every unary mapping $\varphi \in O^1(A)$ and every n-ary mapping $f \in O^n(A)$, we define a new mapping f^φ, by setting

$$f^\varphi(a_1, \ldots, a_n) = \varphi(f(\varphi(a_1), \ldots, \varphi(a_n))),$$

for all $(a_1, \ldots, a_n) \in A^n$. We use this to define, for any set H of unary mappings, an operator on individual mappings and on sets $F \subseteq O(A)$ of mappings, by

$$\gamma_H^O(f) := \{f^\varphi \mid \varphi \in H\} \quad \text{and} \quad \gamma_H^O(F) := \bigcup_{f \in F} \gamma_H^O(f).$$

When H is the base set of a submonoid of $\mathcal{T}_A$, we get the following result.

Lemma 14.2.1 *For every submonoid $\mathcal{H}$ of $\mathcal{T}_A$, the mapping γ_H^O is an additive closure operator on $O(A)$.*   ■

We define the following subrelation of the preservation relation R:

$$R_H := \{(f, \rho) \mid f \in O(A), \rho \in R(A) \text{ and } \gamma_H^O(f) \times \{\rho\} \subseteq R\}.$$

This subrelation induces a Galois-connection $(HPol_A, HInv_A)$ between sets of relations and sets of operations on A. The sets of operations which are closed under this Galois-connection, that is the clones F such that $\gamma_H^O(F) = F$, are called *H*-clones. We will denote by $\mathcal{L}_A^H$ the lattice of all *H*-clones on A. Since we do not have a conjugate pair of additive closure operators here, it is not clear whether this lattice forms a complete sublattice of the lattice $\mathcal{L}_A$. It is always at least a meet-subsemilattice of $\mathcal{L}_A$.

We conclude this section by examining in more detail the case that A is the two-element set $\{0, 1\}$. In this case we have exactly four unary operations: the identity operation φ_{id}, the two constant operations c_0 and c_1, and the negation operation $\neg$. It is easy to check that the four-element monoid $\mathcal{T}(\{0, 1\})$ then has the following proper submonoids:

$H_1 = \{\varphi_{id}\}$, $H_2 = \{\varphi_{id}, c_0\}$, $H_3 = \{\varphi_{id}, c_1\}$,
$H_4 = \{\varphi_{id}, c_0, c_1\}$, $H_5 = \{\varphi_{id}, \neg\}$.

The submonoid lattice of $\mathcal{T}_{\{0,1\}}$ has the form

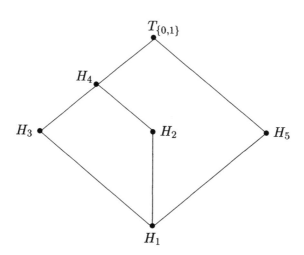

We obtain the following complete lattices of *H*-closed sets:

(i) $\gamma_{H_1}^O(F) = F$ holds for all clones $F \subseteq O(A)$, so that $\mathcal{L}_A^{H_1} = \mathcal{L}_A$.

(ii) $\gamma_{H_2}^O(F) = F \cup F^{c_0} = F \cup \{c_0^n \mid n \in \mathbb{N}\} = F$ iff $c_0 \in F$. This means that $\mathcal{L}_A^{H_2}$ is the set of all clones containing the constant c_0. (Here c_0^n is the *n*-ary constant 0 operation.)

(iii) $\mathcal{L}_A^{H_3}$ is the set of all clones containing the constant c_1.

(iv) $\mathcal{L}_A^{H_4}$ is the set of all clones containing both constants c_0 and c_1.

(v) $\mathcal{L}_A^{H_5} = \{\mathcal{O}_1, \mathcal{O}_4, \mathcal{O}_8, \mathcal{O}_9, \mathcal{L}_4, \mathcal{L}_5, \mathcal{L}_1, \mathcal{D}_2, \mathcal{D}_1, \mathcal{D}_3, \mathcal{A}_4, \mathcal{A}_1, \mathcal{C}_4, \mathcal{C}_1\}$ is the set of all self-dual clones.

It turns out that all of these lattices are complete sublattices of $\mathcal{L}_A$, for $A = \{0, 1\}$. The following picture shows the structure of the set of all these lattices.

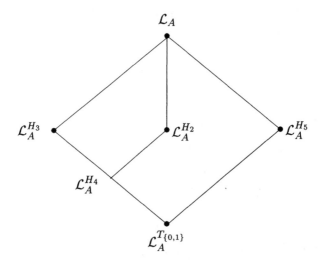

14.3 *M*-Solid Varieties

In this section we apply our theory of conjugate pairs of closure operators to the lattice of all varieties of a given type. We assume a fixed type τ of operation symbols, and consider the sets $A = W_\tau(X)^2$ of all identities of type τ and $B = Alg(\tau)$ of all algebras of type τ. Between these two sets we have the basic relation R of satisfaction: that is, a pair $(s \approx t, \mathcal{A}) \in R$ iff the algebra $\mathcal{A}$ satisfies the identity $s \approx t$. As we saw in Section 6.1, this relation induces the Galois-connection (Id, Mod) between sets of identities and sets of algebras. On the algebra side, the closed sets are the equational classes or varieties, and we have the complete lattice $\mathcal{L}(\tau)$ of all varieties of type τ. Dually, the closed sets on the identity side are the equational theories, and we have the complete lattice $\mathcal{E}(\tau)$ of all equational theories of type τ. These two lattices $\mathcal{L}(\tau)$ and $\mathcal{E}(\tau)$ are dually isomorphic, and in general are very large and complex. Thus it is important to find some means of studying at least portions of these lattices, such as complete sublattices.

Our goal is to introduce two new closure operators on our sets A and B, which we shall show form a conjugate pair of additive closure operators. The results of Section 13.1 then give us complete sublattices of our two lattices. The new operators we use are based on the concept of hypersatisfaction of an identity by a variety. We begin with the definition of a hypersubstitution, as introduced by Denecke, Lau, Pöschel and Schweigert in [32]. A complete study of hypersubstitutions and hyperidentities may be found in [37].

A hypersubstitution of type τ is a mapping which associates to every operation symbol f_i a term $\sigma(f_i)$ of type τ, of the same arity as f_i. Any hypersubstitution σ can be uniquely extended to a map $\hat{\sigma}$ on the set $W_\tau(X)$ of all terms of type τ inductively as follows:

(i) if $t = x_j$ for some $j \geq 1$, then $\hat{\sigma}[t] = x_j$;
(ii) if $t = f_i(t_1, \ldots, t_{n_i})$ for some n_i-ary operation symbol f_i and some terms $t_1, \ldots, t_{n_i}$, then $\hat{\sigma}[t] = \sigma(f_i)(\hat{\sigma}[t_1], \ldots, \hat{\sigma}[t_{n_i}])$.

Here the left side of (ii) means the composition of the term $\sigma(f_i)$ and the terms $\hat{\sigma}[t_1], \ldots, \hat{\sigma}[t_{n_i}]$.

We can define a binary operation $\circ_h$ on the set $Hyp(\tau)$ of all hypersubstitutions of type τ, by taking $\sigma_1 \circ_h \sigma_2$ to be the hypersubstitution which maps each fundamental operation symbol f_i to the term $\hat{\sigma}_1[\sigma_2(f_i)]$. That is,

$$\sigma_1 \circ_h \sigma_2 := \hat{\sigma}_1 \circ \sigma_2,$$

where $\circ$ denotes ordinary composition of functions. We will show that this operation is associative, and that the set of all hypersubstitutions forms a monoid. The identity element is the *identity hypersubstitution* σ_{id}, which maps every f_i to $f_i(x_1, \ldots, x_{n_i})$.

Proposition 14.3.1 *Let τ be any fixed type.*

(i) For any two hypersubstitutions σ and ρ of type τ, we have $(\sigma \circ_h \rho)\check{} = (\hat{\sigma} \circ \rho)\check{} = \hat{\sigma} \circ \hat{\rho}$.
(ii) The binary operation $\circ_h$ is associative.
(iii) $(Hyp(\tau); \circ_h, \sigma_{id})$ is a monoid.

Proof: (i) This can be proved by induction on the complexity of terms, using the definition of the extension of a hypersubstitution; we leave the details as an exercise for the reader. (See Exercise 14.5.2.)

(ii) Let σ_1, σ_2 and σ_3 be any three elements of $Hyp(\tau)$. Then from (i) we have $\sigma_1 \circ_h (\sigma_2 \circ_h \sigma_3) = \hat{\sigma}_1 \circ (\hat{\sigma}_2 \circ \sigma_3) = (\hat{\sigma}_1 \circ \hat{\sigma}_2) \circ \sigma_3 = (\hat{\sigma}_1 \circ \sigma_2)\check{} \circ \sigma_3 = (\sigma_1 \circ_h \sigma_2) \circ_h \sigma_3$.

(iii) This follows directly from (ii).   ∎

Definition 14.3.2 Let $\mathcal{M}$ be any submonoid of $Hyp(\tau)$. An algebra $\mathcal{A}$ is said to M-hypersatisfy an identity $u \approx v$ if for every hypersubstitution $\sigma \in M$, the identity $\hat{\sigma}[u] \approx \hat{\sigma}[v]$ holds in $\mathcal{A}$. In this case we say that the identity $u \approx v$ is an *M-hyperidentity* of $\mathcal{A}$. An identity is called an M-hyperidentity of a variety V if it holds as an M-hyperidentity in every algebra in V. A variety V is called *M-solid* if every identity of V is an M-hyperidentity of V. When M is the whole monoid $Hyp(\tau)$, an M-hyperidentity is called a *hyperidentity*, and an M-solid variety is called a *solid* variety.

Let M be any submonoid of $Hyp(\tau)$. Since M contains the identity hypersubstitution, any M-hyperidentity of a variety V is an identity of V. This means that the relation of M-hypersatisfaction, defined between $Alg(\tau)$ and $W_\tau(X)^2$, is a subrelation of the relation of satisfaction from which we induced our Galois-connection (Id, Mod). The new Galois-connection induced by the relation of M-hypersatisfaction is $(H_M Mod, H_M Id)$, defined on classes K and sets Σ as follows:

$$H_M Id K = \{s \approx t \in W_\tau(X)^2 : s \approx t \text{ is an M-hyperidentity of}$$
$$\mathcal{A} \text{ for all } \mathcal{A} \text{ in } K\},$$
$$H_M Mod \Sigma = \{\mathcal{A} \in Alg(\tau) : \text{ all identities in } \Sigma \text{ are}$$
$$\text{hyperidentities of } \mathcal{A}\}.$$

The Galois-closed classes of algebras under this connection are precisely the M-solid varieties of type τ, which then form a complete sublattice of the lattice of all varieties of type τ. Thus studying M-solid and solid varieties gives a way to study complete sublattices of the lattice of all varieties of a given type.

We now introduce some closure operators on the two sets $Alg(\tau)$ and $W_\tau(X)^2$. On the equational side, we can use the extensions of our M-hypersubstitutions to map any terms and identities to new ones. That is, we define an operator χ_M^E by

$$\chi_M^E[u \approx v] = \{\hat{\sigma}[u] \approx \hat{\sigma}[v] : \sigma \in M\}.$$

This extends, additively, to sets of identities, so that for any set Σ of identities we set

$$\chi_M^E[\Sigma] = \{\chi_M^E[u \approx v] : u \approx v \in \Sigma\}.$$

Hypersubstitutions can also be applied to algebras, as follows. Given an algebra $\mathcal{A} = (A; (f_i)_{i \in I})$ and a hypersubstitution σ, we define the algebra $\sigma(\mathcal{A})$: $= (A; (\sigma(f_i)^{\mathcal{A}})_{i \in I})$. This algebra is called the *derived algebra* determined by $\mathcal{A}$ and σ. Notice that by definition it is of the same type as the algebra $\mathcal{A}$. Now we define an operator χ_M^A on the set $Alg(\tau)$, first on individual algebras and then on classes K of algebras, by

$$\chi_M^A[\mathcal{A}] = \{\sigma[\mathcal{A}] : \sigma \in \mathcal{M}\}, \quad \text{and}$$
$$\chi_M^A[K] = \{\chi_M^A[\mathcal{A}] : \mathcal{A} \in \mathcal{K}\}.$$

Proposition 14.3.3 *Let τ be a fixed type and let $\mathcal{M}$ be any submonoid of $Hyp(\tau)$. The two operators χ_M^E and χ_M^A are additive closure operators and are conjugate with respect to the relation R of satisfaction.*

Proof: Since $\mathcal{M}$ is a submonoid of $Hyp(\tau)$, it contains the identity hypersubstitution σ_{id}, and the extension of this hypersubstitution maps any term t to itself. This shows that the operator χ_M^E is extensive. The property of monotonicity is clear from the definition. The idempotency follows from the fact that $\mathcal{M}$ is a monoid: for any σ and ρ in M, the composition $\sigma \circ_h \rho$ is also in M; so for any identity $u \approx v$ we have $\hat{\sigma}[\hat{\rho}[u]] \approx \hat{\sigma}[\hat{\rho}[v]]$ in χ_M^E. Thus χ_M^E is a closure operator. The proof for χ_M^A is similar. Both closure operators are additive by definition, so it remains only to show that they are conjugate with respect to satisfaction. For this we need to show that for any algebra $\mathcal{A}$ and any identity $u \approx v$ of type τ, we have

$$\chi_M^A[\mathcal{A}] \text{ satisfies } u \approx v \text{ iff } \mathcal{A} \text{ satisfies } \chi_M^E[u \approx v].$$

For any $\sigma \in M$, the definition of satisfaction means that $\mathcal{A}$ satisfies $\hat{\sigma}[u] \approx \hat{\sigma}[v]$ iff the induced term operations satisfy $\hat{\sigma}[u]^{\mathcal{A}} = \hat{\sigma}[v]^{\mathcal{A}}$. Similarly, $\sigma[\mathcal{A}]$ satisfies $u \approx v$ iff $u^{\sigma[\mathcal{A}]} = v^{\sigma[\mathcal{A}]}$. But $\hat{\sigma}[u]^{\mathcal{A}} = u^{\sigma[\mathcal{A}]}$, and similarly for v. This shows that $\sigma[\mathcal{A}]$ satisfies $u \approx v$ iff $\mathcal{A}$ satisfies $\hat{\sigma}[u] \approx \hat{\sigma}[v]$, and completes our proof.   ∎

Once we know that our two additive closure operators form a conjugate pair, we can apply our Main Theorem for such conjugate pairs, Theorem 13.1.5. Translating that theorem into the specific case here, we have the following description of the closed objects.

Theorem 14.3.4 *Let M be a monoid of hypersubstitutions of type τ. For any variety V of type τ, the following conditions are equivalent:*

(i) $V = H_M Mod H_M Id V$.

(ii) $\chi_M^A[V] = V$.

(iii) $Id V = H_M Id V$.

(iv) $\chi_M^E[Id V] = Id V$.

And dually, for any equational theory Σ of type τ, the following conditions are equivalent:

(i') $\Sigma = H_M Id H_M Mod \Sigma$.

(ii') $\chi_M^E[\Sigma] = \Sigma$.

(iii') $Mod \Sigma = H_M Mod \Sigma$.

(iv') $\chi_M^A[Mod \Sigma] = Mod \Sigma$.   ∎

Since for any variety V we have $H_M Id V \subseteq Id V$, condition (iii) of this theorem says that V is an M-solid variety, since every identity of V is an M-hyperidentity. In analogy with the (Id, Mod) case, a variety which satisfies condition (i) is called an M-hyperequational class. Thus our M-solid varieties are precisely the M-hyperequational classes. Dual results hold for M-hyperequational theories, using the second half of the theorem. In addition, this tells us that the relation of hypersatisfaction is a Galois-closed subrelation of the satisfaction relation. Moreover, from Theorem 13.1.7 we have the following result.

Theorem 14.3.5 *Let M be a monoid of hypersubstitutions of type τ. Then the class $S_M(\tau)$ of all M-solid varieties of type τ forms a complete sublattice of the lattice $\mathcal{L}(\tau)$ of all varieties of type τ. Dually, the class of all M-hyperequational theories forms a complete sublattice of the lattice of all equational theories of type τ.* ∎

When $\mathcal{M}_1$ and $\mathcal{M}_2$ are both submonoids of $Hyp(\tau)$ and $\mathcal{M}_1$ is a submonoid of $\mathcal{M}_2$, then the corresponding complete lattices satisfy $S_{M_2}(\tau) \subseteq S_{M_1}(\tau)$. As a special case, for any $\mathcal{M} \subseteq Hyp(\tau)$ we see that the lattice $\mathcal{S}(\tau)$ of all solid varieties of type τ is always a sublattice of the lattice $S_M(\tau)$. At the other extreme, for the smallest possible submonoid $\mathcal{M} = \{\sigma_{id}\}$ the corresponding lattice of M-solid varieties is the whole lattice $\mathcal{L}(\tau)$ of all varieties of type τ. Thus we obtain a range of complete sublattices, from all of $\mathcal{L}(\tau)$ to $\mathcal{S}_\tau$. The following definition lists a number of interesting submonoids $\mathcal{M}$ for which the corresponding lattices $S_M(\tau)$ have been studied, both in the general setting and for specific types τ.

Definition 14.3.6 Let τ be a fixed type.
(i) A hypersubstitution $\sigma \in Hyp(\tau)$ is said to be *leftmost* if for every $i \in I$, the first variable in $\hat{\sigma}[f(x_1, \ldots, x_{n_i})]$ is x_1. The set of all leftmost hypersubstitutions of type τ forms a submonoid of $Hyp(\tau)$. The monoid of all rightmost hypersubstitutions is defined dually.

(ii) A hypersubstitution $\sigma \in Hyp(\tau)$ is said to be *outermost* if it is both leftmost and rightmost. The set of all outermost hypersubstitutions of type τ forms a submonoid of $Hyp(\tau)$.

(iii) A hypersubstitution $\sigma \in Hyp(\tau)$ is called *regular* if for every $i \in I$, all the variables $x_1, \ldots, x_{n_i}$ occur in the term $\hat{\sigma}[f_i(x_1, \ldots, x_{n_i})]$. The set $Reg(\tau)$ of all regular hypersubstitutions of type τ forms a submonoid of $Hyp(\tau)$, and a variety which is M-solid for this submonoid $\mathcal{M}$ is called *regular-solid*.

(iv) A hypersubstitution $\sigma \in Hyp(\tau)$ is called a *pre-hypersubstitution* if for every $i \in I$, the term $\sigma(f_i)$ is not a variable. The set of all pre-hypersubstitutions of type τ forms a submonoid of $Hyp(\tau)$, and a variety which is M-solid for this monoid is called *presolid*.

(v) A hypersubstitution $\sigma \in Hyp(\tau)$ is called *symmetrical*, or a *permutation hypersubstitution*, if for every $i \in I$ there is a permutation π on the set of indices $\{1, 2, \ldots, n_i\}$ such that $\hat{\sigma}[f_i(x_1, \ldots, x_{n_i})] = f_i(x_{\pi(1)}, \ldots, x_{\pi(n_i)})$. The set of all symmetrical hypersubstitutions of type τ forms a submonoid of $Hyp(\tau)$, and a variety which is M-solid for this submonoid is called *permutation-solid*.

When V is a variety of type τ, we can form the lattice $\mathcal{L}(V)$ of all subvarieties of V (see Section 6.6). Then the intersection $S_M(V) := S_M(\tau) \cap \mathcal{L}(V)$ is the lattice of all M-solid subvarieties of V. Such lattices have been investigated for a number of choices of V and M, but most work has been done for the case that V is the type (2) variety *Sem* of all semigroups. We give here some examples of results in this direction.

For any variety $W \subseteq Sem$, the associative law is an identity in W. This means that a necessary condition for W to be solid is that it is *hyperassociative*, meaning that it satisfies the associative law as a hyperidentity. It is easy to see that the trivial variety T and the rectangular variety RB of type (2) are both solid, and that any non-trivial solid variety of semigroups must contain the variety RB. At the other extreme, the largest solid variety of semigroups must be the hypermodel class $H_{Hyp}(\tau)Mod\{f(f(x, y), z) \approx f(x, f(y, z))\}$ of the associative law: we know from Theorem 14.3.4 that this hypermodel class is a solid variety in which the associative law is a hyperidentity, and any solid semigroup variety must be contained in this one. This variety was called V_{HS}, the hyperassociative variety, by K. Denecke and J. Koppitz, who first gave a finite (but very large) basis for it in [26]. Another much smaller basis was given by L. Polák in [91].

Theorem 14.3.7 *([91]) The largest solid variety of semigroups is the variety*

$$V_{HS} = Mod\{x(yz) \approx (xy)z, x^2 \approx x^4, xyxzxyx \approx xyzyx, xy^2z^2 \approx xyz^2yz^2, x^2y^2z \approx x^2yx^2yz\}. \quad \blacksquare$$

One direction of this theorem is easy to prove. The variety V_{HS} must of course satisfy the associative law, and by applying the hypersubstitutions taking the binary operation symbol f to the four semigroup terms x^2, xyx, x^2y and xy^2 we get the other four identities in the claimed basis. This shows that V_{HS} is contained in the model class of the set of five identities given in the theorem. The proof of the other direction involves showing that the variety defined by these five identities is indeed hyperassociative, and is too complex for us to give here.

Using this equational basis for the largest solid variety of semigroups, L. Polák also gave a characterization of all solid semigroup varieties.

Theorem 14.3.8 *([92]) Let V be a non-trivial variety of semigroups. Then V is solid iff either $Mod\{x(yz) \approx (xy)z \approx xz\} \subseteq V \subseteq V_{HS}$ and V is permutation-solid, or V is one of the three varieties*
$$RB = Mod\{x(yz) \approx (xy)z, x^2 \approx x, xyz \approx xz\},$$
$$NB = Mod\{x(yz) \approx (xy)z, x^2 \approx x, xyzw \approx xzyw\}, \text{ or}$$
$$RegB = Mod\{x(yz) \approx (xy)z, x^2 \approx x, xyxzxyx \approx xyzyx\}. \quad \blacksquare$$

As a consequence of this characterization theorem it can be shown that there are infinitely many solid semigroup varieties.

M-solidity for semigroup varieties has been investigated for other choices of $\mathcal{M}$ besides the solid case of $\mathcal{M} = Hyp(\tau)$.

Theorem 14.3.9 *([27]) (i) The largest presolid but not solid variety of semigroups is the variety*

$$V_{PS} := Mod\{(xy)z \approx x(yz), xyxzxyx \approx xyzyx, x^2 \approx y^2, x^3 \approx y^3\}.$$

(ii) A non-trivial variety V of semigroups is presolid iff either V is solid, or V is permutation-solid and $Mod\{xy \approx zw\} \subseteq V \subseteq V_{PS}$. $\quad \blacksquare$

Theorem 14.3.10 *([23]) The largest regular-solid variety of semigroups is the variety*

$$V_{RS} := Mod\{(xy)z \approx x(yz), x^2y^2z \approx x^2yx^2yz,$$
$$xy^2z^2 \approx xyz^2yz^2, xyxzxyx \approx xyzyx\}. \quad \blacksquare$$

Similar characterizations have been given for the cases of edge-solid, permutation-solid and other M-solid varieties of semigroups. Hyperidentities and M-solid varieties have also been studied for other kinds of algebras besides semigroups. K. Denecke and P. Jampachon studied regular-solid varieties of commutative and idempotent groupoids in [25], and Denecke and S. Arworn looked at left- and right-edge solid varieties of entropic groupoids in [3]. For type $(2,1)$ algebras, inverse semigroups were studied by D. Cowan and S. L. Wismath in [11] and [12], star-bands by J. Koppitz and S. L. Wismath in [67] and graph algebras by K. Denecke and T. Poomsa-ard in [34].

Quasigroups were investigated by K. Denecke and M. Reichel in [36]. For algebras of type $(2,2)$, D. Schweigert studied lattices in [105] and Denecke and Hounnon looked at solid varieties of semirings in [24].

14.4   Intervals in the Lattice $\mathcal{L}(\tau)$

In Section 13.3 we described a general approach for finding intervals in a complete lattice, based on conjugate pairs of closure operators. Now we apply this general theory to the specific lattice $\mathcal{L}(\tau)$, with the conjugate pair (χ_M^E, χ_M^A) of additive closure operators for a monoid $\mathcal{M}$ of hypersubstitutions. The results of this section were proved by S. Arworn and K. Denecke in [2].

In this setting, Definition 13.3.1 takes the following concrete form.

Definition 14.4.1 Let $K \subseteq Alg(\tau)$ be an arbitrary class of algebras of type τ and let $\mathcal{M} \subseteq Hyp(\tau)$ be an arbitrary monoid of hypersubstitutions of type τ. Then we define

$$\chi_M K := \bigcap \{V' \mid V' \text{ is } M\text{-solid and } V' \supseteq K\}, \text{ and}$$
$$\chi_M K := \bigvee \{V' \mid V' \text{ is M-solid and } V' \subseteq K\}.$$

We also set

$$[\chi_M K, \chi_M K] := \{V' \mid V' \text{ is a variety of type } \tau \text{ and }$$
$$\chi_M K \subseteq V' \subseteq \chi_M K\},$$

and call this set an interval in the lattice $\mathcal{L}(\tau)$.

Applying the theory of Section 13.3 gives the following proposition.

Proposition 14.4.2 *Let V be a variety of type τ and let $\mathcal{M}$ be a monoid of hypersubstitutions. Then*

(i) $\chi^M V = V$ *iff* $\chi_M V = V$ *iff V is M-solid,*
and so $[\chi_M V, \chi^M V] = \{V\}$ *iff V is M-solid.*
(ii) $\chi_M V = H_M Mod IdV = Mod\chi_M^E[IdV]$,
and $\chi^M V = H_M Mod H_M IdV = Mod Id\chi_M^A[V]$.
(iii) If $V_1 \subseteq V_2$ then $\chi_M V_1 \subseteq \chi_M V_2$ *and* $\chi^M V_1 \subseteq \chi^M V_2$;
also $\chi_M V_1 \wedge \chi_M V_2 = \chi_M (V_1 \wedge V_2)$, $\chi^M V_1 \vee \chi^M V_2 = \chi^M (V_1 \vee V_2)$.
(iv) $\chi^M V$ *defines a closure operator satisfying*

$$\chi^M[\bigvee\{V_j \mid j \in J\}] = \bigvee\{\chi^M V_j \mid j \in J\},$$

and $\chi_M V$ *defines a kernel operator satisfying*

$$\chi_M[\bigcap\{V_j \mid j \in J\}] = \bigcap\{\chi_M V_j \mid j \in J\}. \qquad \blacksquare$$

Lemma 14.4.3 *Let $\mathcal{M}_1$ and $\mathcal{M}_2$ be submonoids of $Hyp(\tau)$ and let V be a variety of type τ. If $\mathcal{M}_1 \subseteq \mathcal{M}_2$, then*

(i) $\chi^{M_1} V \subseteq \chi^{M_2} V$,
(ii) $\chi_{M_1} V \supseteq \chi_{M_2} V$,
(iii) $[\chi_{M_1} V, \chi^{M_1} V] \subseteq [\chi_{M_2} V, \chi^{M_2} V]$.

Proof: (i) Additivity of the operator χ_M^A means that if $\mathcal{M}_1 \subseteq \mathcal{M}_2$, then $\chi_{M_1}^A[V] \subseteq \chi_{M_2}^A[V]$. Applying the monotonic closure operator $ModId$ to each side gives $ModId\chi_{M_1}^A[V] \subseteq ModId\chi_{M_2}^A[V]$, and by 14.4.2(ii) we have $\chi^{M_1} V \subseteq \chi^{M_2} V$.

(ii) Additivity of the operator χ_M^E means that if $\mathcal{M}_1 \subseteq \mathcal{M}_2$ then $\chi_{M_1}^E[IdV] \subseteq \chi_{M_2}^E[IdV]$. Application of the anti-isotone operator Mod on both sides gives $Mod\chi_{M_1}^E[IdV] \supseteq Mod\chi_{M_2}^E[IdV]$. Again 14.4.2 (ii) gives $\chi_{M_1} V \supseteq \chi_{M_2} V$.

(iii) It is clear that for intervals, $[\chi_{M_1} V, \chi^{M_1} V] \subseteq [\chi_{M_2} V, \chi^{M_2} V]$ iff $\chi_{M_1} V \supseteq \chi_{M_2} V$ and $\chi^{M_1} V \subseteq \chi^{M_2} V$; and then we apply (i) and (ii). $\blacksquare$

By Lemma 14.4.3, the map φ from the lattice $Sub(Hyp(\tau))$ of all submonoids of $Hyp(\tau)$ to the lattice of all varieties of type τ, which associates to each

$\mathcal{M} \subseteq Hyp(\tau)$ the M-solid variety $^{\chi_M}V$, is order-preserving. Dually, the mapping ψ which associates to every submonoid $\mathcal{M} \subseteq Hyp(\tau)$ the M-solid variety $_{\chi_M}V$ is order-preserving. The behaviour of these mappings with respect to meet and joins is considered next.

Lemma 14.4.4 *Let V be a variety of type τ. Let φ and ψ be mappings from $Sub(Hyp(\tau))$ to $\mathcal{L}(\tau)$ defined by $\varphi : \mathcal{M} \mapsto {}^{\chi_M}V$ and $\psi : \mathcal{M} \mapsto {}_{\chi_M}V$, respectively, for $\mathcal{M} \subseteq Hyp(\tau)$. Then the following are satisfied, for any submonoids $\mathcal{M}_1$ and $\mathcal{M}_2$ of $Hyp(\tau)$:*

(i) $\varphi(\mathcal{M}_1 \cap \mathcal{M}_2) \subseteq \varphi(\mathcal{M}_1) \cap \varphi(\mathcal{M}_2)$,   and
   $\varphi(\mathcal{M}_1 \vee \mathcal{M}_2) \supseteq \varphi(\mathcal{M}_1) \vee \varphi(\mathcal{M}_2)$,
(ii) $\psi(\mathcal{M}_1 \cap \mathcal{M}_2) \supseteq \psi(\mathcal{M}_1) \vee \psi(\mathcal{M}_2)$,   and
   $\psi(\mathcal{M}_1 \vee \mathcal{M}_2) \subseteq \psi(\mathcal{M}_1) \cap \psi(\mathcal{M}_2)$.

Proof: Both claims follow directly from Lemma 14.4.3.   ∎

As a consequence, we have the following inclusions for intervals:

$$[\,_{\chi_{M_1}}V \cap \,_{\chi_{M_2}}V, \, ^{\chi_{M_1}}V \vee \, ^{\chi_{M_2}}V] \subseteq [\,_{\chi_{(M_1 \vee M_2)}}V, \, ^{\chi_{(M_1 \vee M_2)}}V]$$
$$= [\psi(M_1 \vee M_2), \varphi(M_1 \vee M_2)],$$
$$[\,_{\chi_{M_1}}V \vee \,_{\chi_{M_2}}V, \, ^{\chi_{M_1}}V \cap^{\chi_{M_2}}V] \supseteq [\,_{\chi_{(M_1 \cap M_2)}}V, \, ^{\chi_{(M_1 \cap M_2)}}V]$$
$$= [\psi(M_1 \cap M_2), \varphi(M_1 \cap M_2)].$$

We investigate next when we obtain equality for these intervals.

Lemma 14.4.5 *Let $\mathcal{M}_1$ and $\mathcal{M}_2$ be submonoids of $Hyp(\tau)$ and let V be a variety of type τ. If $\mathcal{M}_1 \vee \mathcal{M}_2 = \mathcal{M}_1 \cup \mathcal{M}_2$, then $\varphi(\mathcal{M}_1 \vee \mathcal{M}_2) = \varphi(\mathcal{M}_1) \vee \varphi(\mathcal{M}_2)$ and $\psi(\mathcal{M}_1 \vee \mathcal{M}_2) = \psi(\mathcal{M}_1) \cap \psi(\mathcal{M}_2)$. Therefore*

$$[\,_{\chi_{M_1}}V \cap \,_{\chi_{M_2}}V, \, ^{\chi_{M_1}}V \vee \, ^{\chi_{M_2}}V] = [\,_{\chi_{(M_1 \vee M_2)}}V, \, ^{\chi_{(M_1 \vee M_2)}}V].$$

Proof: Using the assumption, the additivity of the closure operator χ_M^A and the properties of the Galois-connection (Id,Mod), we have

$$\varphi(\mathcal{M}_1 \vee \mathcal{M}_2) = {}^{\chi_{(M_1 \vee M_2)}}V = {}^{\chi_{(M_1 \cup M_2)}}V = ModId\chi_{(M_1 \cup M_2)}^A[V]$$
$$= ModId\chi_{M_1}^A[V] \cup ModId\chi_{M_2}^A[V] = Mod(Id\chi_{M_1}^A[V] \cap Id\chi_{M_2}^A[V])$$
$$= ModId\chi_{M_1}^A[V] \vee ModId\chi_{M_2}^A[V] = {}^{\chi_{M_1}}V \vee {}^{\chi_{M_2}}V$$

$$= \varphi(\mathcal{M}_1) \vee \varphi(\mathcal{M}_2).$$

Similarly, we have

$$\psi(\mathcal{M}_1 \vee \mathcal{M}_2) = \chi_{(\mathcal{M}_1 \vee \mathcal{M}_2)} V = Mod\chi^E_{(\mathcal{M}_1 \cup \mathcal{M}_2)}[IdV]$$
$$= Mod(\chi^E_{\mathcal{M}_1}[IdV] \cup \chi^E_{\mathcal{M}_2}[IdV]) = Mod\chi^E_{\mathcal{M}_1}[IdV] \cap Mod\chi^E_{\mathcal{M}_2}[IdV]$$
$$= \chi_{\mathcal{M}_1} V \cap \chi_{\mathcal{M}_2} V = \psi(\mathcal{M}_1) \cap \psi(\mathcal{M}_2). \qquad \blacksquare$$

14.5 Exercises

14.5.1. Prove Lemma 14.2.1.

14.5.2. Prove that for any two hypersubstitutions σ and ρ of a fixed type τ, we have $(\sigma \circ_h \rho)\hat{} = (\hat{\sigma} \circ \rho)\hat{} = \hat{\sigma} \circ \hat{\rho}$.

14.5.3. Let $\mathcal{M} \subseteq Hyp(\tau)$ be a monoid of hypersubstitutions. A variety V of type τ is called M-hyperequationally simple if V has no solid subvarieties other than the trivial variety T. Prove the following:

a) V is M-hyperequationally simple iff $\chi_M V = T$.
b) If $\mathcal{M}_1$ and $\mathcal{M}_2$ are submonoids of $Hyp(\tau)$ with $\mathcal{M}_1 \subseteq \mathcal{M}_2$, then when V is M_1-hyperequationally simple it is also M_2-hyperequationally simple.

14.5.4. Let V be a variety of semigroups. Show that if V satisfies an identity $u \approx v$ in which the leftmost variables in u and v are different, then V is hyperequationally simple.

14.5.5. A variety V of type τ is called M-hyperidentity-free if the only identities satisfied as M-hyperidentities by V are trivial ones of the form $u \approx u$. Prove that V is M-hyperidentity-free iff $\chi_M V = Alg(\tau)$.

14.5.6. Prove that if $\mathcal{M}_1$ and $\mathcal{M}_2$ are submonoids of $Hyp(\tau)$ with $\mathcal{M}_1 \subseteq \mathcal{M}_2$, then when V is M_1-hyperidentity-free it is also M_2-hyperidentity-free.

14.5.7. Construct some examples of M-hyperidentity-free varieties.

14.5.8. Determine the interval $[\chi_M V, \, \chi_M V]$, for $\mathcal{M} = Hyp(\tau)$, for V equal to the variety of semilattices, the variety of left-zero semigroups, and the variety of right-zero semigroups.

Chapter 15

Hypersubstitutions and Machines

In this chapter we apply the hypersubstitution operation studied in the previous chapter to the Computer Science concepts from Chapters 7 and 8. Our first application is a generalization of the unification problem, which plays an important role in term rewriting systems and logical programming. It is natural to consider the hyperunification problem, which arises when we use hypersubstitutions instead of ordinary substitutions. The second application in this chapter is the generalization of tree-recognizers to hyper-tree-recognizers. Finally, we consider tree transformations and tree transducers generated by hypersubstitutions.

15.1 The Hyperunification Problem

Let V be a variety of type τ. We recall from Chapter 7 that the *word problem* for V is the problem of deciding, given any two terms u and v of type τ, whether $u \approx v$ holds as an identity in V. The concept of a *unifier* is important in solving the word problem. We fix a countably infinite alphabet X, and let $W_\tau(X)$ denote the set of all terms of type τ. Recall that a *substitution* is a mapping $s : X \to W_\tau(X)$, and that any such map has a unique extension $\hat{s} : W_\tau(X) \to W_\tau(X)$. For any term t, the term $\hat{s}(t)$ is obtained by substitution of the term $s(x)$ for each occurrence of a variable x in t. A substitution s for which $\hat{s}(u) \approx \hat{s}(v)$ is an identity in V is called a *unifier* for the terms u and v with respect to the variety V. In the special case that V is the variety $Alg(\tau)$ of all algebras of type τ, we have $\hat{s}(u) \approx \hat{s}(v)$ an identity of V if

347

and only if $\hat{s}(u) = \hat{s}(v)$. In this case we refer to a unifier as a *syntactical unifier*; otherwise, when V is a proper subvariety of $Alg(\tau)$, we refer to a unifier with respect to V as a *semantical unifier*.   To solve the unification problem means to decide, given two terms, whether there exists a unifier for the terms or not. For more information on this topic see J. H. Siekmann, [108].

In this section we describe the work of K. Denecke, J. Koppitz and S. Niwczyk, from [28], regarding the generalization of unifiers and the unification problem to the hyperidentity setting. Instead of looking for substitution mappings s which unify two terms, we look for unifying hypersubstitutions σ.

Definition 15.1.1 Let u and v be two terms of type τ. A hypersubstitution σ of type τ is called a *(syntactical) hyperunifier* for u and v if $\hat{\sigma}[u] = \hat{\sigma}[v]$. When such a hyperunifier exists, we say that the terms u and v are *hyperunifiable*.

By the definition of the kernel of a mapping, a hypersubstitution σ is a hyperunifier for two terms u and v exactly when the pair (u, v) is in the kernel of the extension mapping $\hat{\sigma}$ defined on $W_\tau(X)$. We will refer to this kernel as the kernel of the hypersubstitution σ, and denote it by $ker\sigma$. The first step is to show that any such kernel is a fully invariant congruence relation on the free algebra $\mathcal{F}_\tau(X)$ defined on the set $W_\tau(X)$.

Lemma 15.1.2 *([28]) Let $\tau = (n_i)_{i \in I}$ be a type of algebras with $n_i \geq 1$ for all $i \in I$. Let σ be a hypersubstitution of type τ. Then the relation $ker\sigma$ is a fully invariant congruence on the absolutely free algebra $\mathcal{F}_\tau(X)$.*   ■

In the case of type (n), for $n \geq 1$, Denecke, Koppitz and Niwczyk have completely determined all the congruence relations $ker\sigma$ for any hypersubstitution σ of type (n). To describe their results we need some notation for terms regarded as trees, from Section 5.1. To each node or vertex of a tree representing a term of type (n) we can assign a sequence of integers from the set $\{1, 2, \ldots, n\}$, called the address of the node or vertex. For each value $1 \leq i \leq n$, there is a uniquely determined variable obtained by following the address $ii \cdots i$ in the tree until we have the address of a variable; we shall denote this variable by $var_i(t)$. Let K be any non-empty subset of $\{1, 2, \ldots, n\}$. An address is called a terminating K-sequence for t if it is the address of a variable, it contains only indices from K and no subsequence of the address gives a variable.

Proposition 15.1.3 *([28]) Let* $\tau = (n)$, *with* $n \geq 1$, *with one n-ary operation symbol* f.
(i) If σ *is a regular hypersubstitution of type* τ, *then* $\ker\sigma$ *is the diagonal relation on* $W_\tau(X)$; *that is,* $\hat{\sigma}[u] = \hat{\sigma}[v]$ *iff* $u = v$.
(ii) If σ *is a projection hypersubstitution, so that* $\sigma(f) = x_i$ *for some* $1 \leq i \leq n$, *then* $\ker\sigma = \{(u, v) \in W_\tau(X)^2 \mid var_i(u) = var_i(v)\}$.
(iii) Let $n \geq 2$, *and let* σ *be a non-projection hypersubstitution. Let* K *be the set of variables used in the term* $\sigma(f)$, *with* $1 \leq |K| < n$. *Then a pair* (u, v) *from* $W_\tau(X)^2$ *is in* $\ker\sigma$ *iff any terminating K-sequence of* u *is a terminating K-sequence of* v *and vice versa, and any such sequence addresses the same variable in both* u *and* v. ∎

In the same paper, the authors also extended their results to an arbitrary type τ having no nullary operation symbols, for certain restricted kinds of hypersubstitutions.

In addition to this syntactical hyperunification problem, we may consider the semantical version for any variety V. This is equivalent to studying, for any variety V and hypersubstitution σ of type τ, the relation

$$ker_V\sigma := \{(u, v) \in W_\tau(X)^2 \mid V \text{ satisfies } \hat{\sigma}[u] \approx \hat{\sigma}[v]\}.$$

This concept of the V-kernel of a hypersubstitution has been studied by K. Denecke, J. Koppitz and S. L. Wismath in [29], where it was shown that any such V-kernel is a fully invariant congruence relation on the free algebra defined on $W_\tau(X)$.

15.2 Hyper Tree Recognizers

In Section 8.4 we introduced the concept of a tree-recognizer. A term or tree t of a given type is recognized by a tree-recognizer if there is a finite algebra $\mathcal{A}$ of the type, a subset A' of the universe of $\mathcal{A}$, and an evaluation mapping α which sends the variables occurring in t to elements of A such that $\hat{\alpha}[t]$ belongs to A'. The evaluation mapping carries out the operation of substitution, by replacing the leaves of the tree (corresponding to variables or nullary operation symbols) by terms of the algebra. If in addition we allow the replacement of all vertices of the tree (operation symbols of the term) by term operations of the algebra $\mathcal{A}$ of the same arity, we implement a hypersubstitution. In this way our tree-recognizer becomes a hyperrecognizer. One might think that this kind of "parallel working" might allow us to recognize larger families of languages. But it was shown by K. Denecke and N. Pabhapote in [33] that in the case of a finite alphabet, a language is recognizable iff it is hyperrecognizable. This result means that all vertices in a tree-recognizer can be evaluated at the same time in an appropriate way, giving us additional insight into the power of a tree-recognizer.

We begin by recalling some notation from Section 8.4, where algebras were described by a set Σ of operation symbols rather than a type τ. That is, we let $\Sigma = \Sigma_0 \cup \Sigma_1 \cup \cdots \cup \Sigma_m$ be a set of operation symbols, where the operation symbols in each Σ_n are n-ary. We will also assume here that Σ is finite. As usual we have a countably infinite set X of variables, while for each natural number n we let X_n be the set of variables $x_1, \ldots, x_n$. We will denote by $W_\Sigma(X)$ and $W_\Sigma(X_n)$ the sets of all terms which can be built up from the operation symbols from Σ and the variables from X or from X_n, respectively. We write $\mathcal{A} = (A, \Sigma^{\mathcal{A}})$ for a (finite) algebra whose fundamental operations correspond to the operation symbols from Σ (of the type of Σ). As we saw in Chapter 5, in the case that both Σ and X are finite we can consider terms as trees.

Definition 15.2.1 A $\Sigma - X_n$-tree-hyperrecognizer is a sequence

$$H\mathbf{A} := (clone_n\mathcal{A}, \Sigma, X_n, \sigma^{\mathcal{A}}, C_n^{\mathcal{A}}).$$

Here $clone_n\mathcal{A}$ is the clone of all n-ary term operations of a finite algebra $\mathcal{A}$, and $\sigma^{\mathcal{A}} : \Sigma \cup X_n \to clone_n\mathcal{A}$ is a mapping which is defined in the following way:

$\sigma^{\mathcal{A}}[x_i] := e_i^{n,\mathcal{A}}$ if $x_i \in X_n$ is a variable, and
$\sigma^{\mathcal{A}}(f_i) = t_i^{\mathcal{A}}$ for an n_i-ary operation symbol $f_i \in \Sigma_{n_i}$ and an n_i-ary term operation $t_i^{\mathcal{A}}$ of $\mathcal{A}$.

(Note that n has to be greater than the greatest arity of the operation symbols in Σ.) $C_n^{\mathcal{A}}$ is a subset of $clone_n\mathcal{A}$.

The mapping $\sigma^{\mathcal{A}}$ thus maps each variable and each operation symbol to an n-ary term of the algebra $\mathcal{A}$. We mention that any such mapping can be extended to a mapping $(\sigma^{\mathcal{A}})^{\wedge} : W_{\Sigma}(X_n) \to clone_n\underline{A}$, in the following inductive way:

(i) $(\sigma^{\mathcal{A}})^{\wedge}[x_i] = \sigma^{\mathcal{A}}(x_i)$ if $x_i \in X_n$ is a variable,

(ii) $(\sigma^{\mathcal{A}})^{\wedge}[f_0] = \sigma^{\mathcal{A}}(f_0)$ if $f_0 \in \Sigma_0$ is nullary,

(iii) $(\sigma^{\mathcal{A}})^{\wedge}[f_i(t_1, \ldots, t_{n_i})] = \sigma^{\mathcal{A}}(f_i)((\sigma^{\mathcal{A}})^{\wedge}[t_1], \ldots, (\sigma^{\mathcal{A}})^{\wedge}[t_{n_i}])$ if f_i is an n_i-ary operation symbol and if $(\sigma^{\mathcal{A}})^{\wedge}[t_j]$ is already defined for $1 \le j \le n_i$.

We recall from Section 14.3 that a hypersubstitution (of type Σ) is an arity-preserving mapping $\sigma : \Sigma \to W_{\Sigma}(X_n)$, and that any such hypersubstitution can be uniquely extended to a mapping $\hat{\sigma} : W_{\Sigma}(X) \to W_{\Sigma}(X)$. Then for any term t we have $(\sigma^{\mathcal{A}})^{\wedge}[t] = (\hat{\sigma}[t])^{\mathcal{A}}$, meaning that each operation symbol $f_i \in \Sigma_{n_i}$ is mapped to an n-ary term t such that the term operation induced by the term $\hat{\sigma}[t]$ on $\mathcal{A}$ agrees with $(\sigma^{\mathcal{A}})^{\wedge}[t]$. We remark that hypersubstitutions correspond to the tree-homomorphisms $h : \mathcal{F}_{\Sigma}(X) \to \mathcal{F}_{\Sigma}(X)$ introduced in 8.6 (as in F. Gécseg and M. Steinby, [48]), with the additional restriction that $h(x_i) = x_i$ for all $i = 1, \ldots, n$.

Definition 15.2.2 Let $H\mathbf{A}$ be a $\Sigma - X_n$-tree-hyperrecognizer. The language hyperrecognized by $H\mathbf{A}$ is the $\Sigma - X_n$-language

$$T(H\mathbf{A}) := \{t \in W_{\Sigma}(X_n) \mid (\sigma^{\mathcal{A}})^{\wedge}[t] \in C_n^{\mathcal{A}}\},$$

where $(\sigma^{\mathcal{A}})^{\wedge}$ is the extension of $\sigma^{\mathcal{A}}$. A language $T \subseteq W_{\Sigma}(X_n)$ is called hyperrecognizable if there is a $\Sigma - X_n$-tree-hyperrecognizer $H\mathbf{A}$ such that $T = T(H\mathbf{A})$.

It is not difficult to see that every hyperrecognizable language is recognizable. To prove this, we have to show that given a hyperrecognizer $\mathbf{A}$, we can find an algebra $\mathcal{B}$, an evaluation mapping and a subset of B to use in a recognizer $\mathbf{B}$, in such a way that $T(H\mathbf{A}) = T(\mathbf{B})$. A possible algebra $\mathcal{B}$ satisfying this condition is $\mathcal{B} = (clone_n\mathcal{A}, \sigma(\Sigma)^{\mathcal{B}})$ where $\sigma(\Sigma)$ is the set of all images of the operation symbols from Σ under the hypersubstitution σ. As evaluation mapping we use the restriction of $\sigma^{\mathcal{A}}$ to X. Then it is clear that $H\mathbf{A}$ and $\mathbf{B}$ recognize the same language. Note that the concept of a derived algebra, mentioned in Section 14.3, is involved here.

Example 15.2.3 We set $X_2 = \{x_1, x_2\}$, $\Sigma = \Sigma_2 = \{f\}$ and $\mathcal{A} = (\{0, 1\}; f^A)$ with f^A equal to the conjunction operation $\wedge$. Then $clone_2\mathcal{A}$

$= \{e_1^{2,\mathcal{A}}, e_2^{2,\mathcal{A}}, (x_1 \wedge x_2)^{\mathcal{A}}\}$. We consider the hyperrecognizer $H\mathbf{A} = (clone_2\mathcal{A}, \Sigma, X_2, \sigma^{\mathcal{A}}, C_n^{\mathcal{A}})$ in which $C_n^{\mathcal{A}} = \{(x_1 \wedge x_2)^{\mathcal{A}}\}$ and $\sigma^{\mathcal{A}}$ is given by $\sigma^{\mathcal{A}}(f) = (x_1 \wedge x_2)^{\mathcal{A}}$. Then $T(H\mathbf{A}) = W_\Sigma(X_2)\backslash\{x_1^r, x_2^s \mid r, s \geq 1\}$, where $x_1^r := x_1 \wedge \cdots \wedge x_1$ means that x_1 occurs r times.

Example 15.2.4 We choose X_2, and let Σ contain two operation symbols, one binary and one nullary. Let $\mathcal{A} = (V_4, \cdot, e)$ be the Klein-four group. Let $C_2^{\mathcal{A}} = \{e^{\mathcal{A}}\}$ and $\sigma^{\mathcal{A}}(f) = e^{\mathcal{A}}$ (the identity element $e^{\mathcal{A}}$ considered as binary term operation). Then we have $\sigma^{\mathcal{A}}[x_i] = e_i^{2,\mathcal{A}}$, for $i = 1, 2$. Inductively, when $t = f(t_1, t_2)$, for some $t_1, t_2 \in W_\Sigma(X_2)$, we have $(\sigma^{\mathcal{A}})\hat{}[t] = (\sigma^{\mathcal{A}})\hat{}[f(t_1, t_2)] = \sigma^{\mathcal{A}}(f)((\sigma^{\mathcal{A}})\hat{}[t_1], (\sigma^{\mathcal{A}})\hat{}[t_2]) = e^{\mathcal{A}}((\sigma^{\mathcal{A}})\hat{}[t_1], (\sigma^{\mathcal{A}})\hat{}[t_2]) = e^{\mathcal{A}}$. This shows that $T(H\mathbf{A}) = W_\Sigma(X_2)\backslash\{x_1, x_2\}$.

The following proposition connects tree-hyperrecognizers with identities and varieties. We recall the notation $Id\mathcal{A}$ for the set of identities of the algebra $\mathcal{A}$ and $s^{\mathcal{A}}$ for the term operation on $\mathcal{A}$ induced by a term s.

Proposition 15.2.5 *The language* $T \subseteq W_\Sigma(X_n)$ *is hyperrecognizable iff there is a finite algebra* $\mathcal{A} = (A; \Sigma^{\mathcal{A}})$, *a subset* $C_n^{\mathcal{A}} \subseteq clone_n\mathcal{A}$ *and a hypersubstitution* σ *of type* Σ *such that*

(i) for each $t \in T$ *there exists an element* $s^{\mathcal{A}} \in C_n^{\mathcal{A}}$ *with* $\hat{\sigma}[t] \approx s \in Id\mathcal{A}$

(ii) for each $t \in W_\Sigma(X_n)\backslash T$ *and each* $s^{\mathcal{A}} \in C_n^{\mathcal{A}}$, *we have* $\hat{\sigma}[t] \approx s \notin Id\mathcal{A}$.

Proof: Suppose that T is hyperrecognizable. Then there exists a $\Sigma-X_n$-tree-hyperrecognizer $H\mathbf{A} = (clone_n\mathcal{A}, \Sigma, X_n, \sigma^{\mathcal{A}}, C_n^{\mathcal{A}})$ such that $T(H\mathbf{A}) = T$. If $t \in T$ then there is an element $s^{\mathcal{A}} \in C_n^{\mathcal{A}}$ with $(\sigma^{\mathcal{A}})\hat{}[t] = s^{\mathcal{A}}$. But then there is a hypersubstitution σ with $(\sigma^{\mathcal{A}})\hat{}[t] = (\hat{\sigma}[t])^{\mathcal{A}}$. (Note that σ is not uniquely determined by $\sigma^{\mathcal{A}}$, since every σ' for which $\sigma(f_i) \approx \sigma'(f_i) \in Id\mathcal{A}$ for every operation symbol f_i from Σ satisfies the same equation.) From the last equation we obtain $\hat{\sigma}[t] \approx s \in Id\mathcal{A}$.

If $t \notin T$ then for each $s^{\mathcal{A}} \in C_n^{\mathcal{A}}$ we have $(\sigma^{\mathcal{A}})\hat{}[t] \neq s^{\mathcal{A}}$. This means $\hat{\sigma}[t] \approx s \notin Id\mathcal{A}$ if σ is a hypersubstitution which satisfies $(\sigma^{\mathcal{A}})\hat{}[t] = (\hat{\sigma}[t])^{\mathcal{A}}$.

Conversely, suppose now that there is a finite algebra $\mathcal{A} = (A; \Sigma^{\mathcal{A}})$ with the finite n-clone $clone_n\mathcal{A}$ and a hypersubstitution satisfying (i) and (ii). We will show that the hyperrecognizer $H\mathbf{A} = (clone_n\mathcal{A}, \Sigma, X_n, \sigma^{\mathcal{A}}, C_n^{\mathcal{A}})$ satisfies $T(H\mathbf{A}) = T$. If $t \in T(H\mathbf{A})$ then there is a term operation $s^{\mathcal{A}} \in C_n^{\mathcal{A}}$ such that $(\sigma^{\mathcal{A}})\hat{}[t] = s^{\mathcal{A}}$, i.e., $(\sigma[t])^{\mathcal{A}} = s^{\mathcal{A}}$ and $\hat{\sigma}[t] \approx s \in Id\mathcal{A}$. Because of (ii) we have $t \in T$. Conversely, if $t \in T$ then there exists a term s with $s^{\mathcal{A}} \in C_n^{\mathcal{A}}$ and with $\hat{\sigma}[t] \approx s \in Id\mathcal{A}$ by (i) and thus $\hat{\sigma}[t]^{\mathcal{A}} = (\sigma^{\mathcal{A}})^{\smallfrown}[t] = s^{\mathcal{A}} \in C_n^{\mathcal{A}}$ and t belongs to $T(H\mathbf{A})$.   ∎

The proof of Proposition 15.2.5 (i) shows that the term s which satisfies $\hat{\sigma}[t] \approx s \in Id\mathcal{A}$ for $t \in T$ is not uniquely determined. Therefore we use the following binary relation $\sim_{V(\mathcal{A})}$ defined by J. Płonka in [90] on the set of all hypersubstitutions:

$$\sigma_1 \sim_{V(\mathcal{A})} \sigma_2 :\Leftrightarrow \sigma_1(f_i) \approx \sigma_2(f_i) \in Id\mathcal{A},$$

for all operation symbols $f_i \in \Sigma$. Then we have the following proposition.

Proposition 15.2.6 *(i) Let $H\mathbf{A}_1 := (clone_n\mathcal{A}, \Sigma, X_n, \sigma_1^{\mathcal{A}}, C_n^{\mathcal{A}})$ and $H\mathbf{A}_2 := (clone_n\mathcal{A}, \Sigma, X_n, \sigma_2^{\mathcal{A}}, C_n^{\mathcal{A})})$. If $\sigma_1 \sim_{V(\mathcal{A})} \sigma_2$ then $T(H\mathbf{A}_1) = T(H\mathbf{A}_2)$.*

(ii) Let $\mathcal{A}_1$ and $\mathcal{A}_2$ be Σ-algebras with $Id\mathcal{A}_1 = Id\mathcal{A}_2$ and let $H\mathbf{A}_1 := (clone_n\mathcal{A}_1, \Sigma, X_n, \sigma^{\mathcal{A}_1}, C_n^{\mathcal{A}_1})$ be a tree-hyperrecognizer based on the algebra $\mathcal{A}_1$. Then there exists a tree-hyperrecognizer $T(H\mathbf{A}_2)$ based on $\mathcal{A}_2$ with $T(H\mathbf{A}_1) = T(H\mathbf{A}_2)$.

Proof: (i) If $t \in T(H\mathbf{A}_1)$ then there is an element $s^{\mathcal{A}} \in C_n^{\mathcal{A}}$ and a corresponding term s with $\hat{\sigma}_1[t] \approx s \in Id\mathcal{A}$. From the definition of the relation $\sim_{V(\mathcal{A})}$ it follows that $\hat{\sigma}_1[t] \approx \hat{\sigma}_2[t] \in Id\mathcal{A}$ for all terms t. But then $\hat{\sigma}_2[t] \approx s \in Id\mathcal{A}$.

If $t \in W_\Sigma(X_n) \backslash T(H\mathbf{A}_1)$ then for each $s^{\mathcal{A}} \in C_n^{\mathcal{A}}$ we have $\hat{\sigma}_1[t] \approx s \notin Id\mathcal{A}$. But then also $\hat{\sigma}_2[t] \approx s \notin Id\mathcal{A}$, and by Proposition 15.2.5 we have $t \in T(H\mathbf{A}_2)$. A dual argument shows that $T(H\mathbf{A}_2) \subseteq T(H\mathbf{A}_1)$.

(ii) Let $V(\mathcal{A}_1)$ and $V(\mathcal{A}_2)$ be the varieties generated by $\mathcal{A}_1$ and by $\mathcal{A}_2$, respectively and let $\mathcal{F}_{V(\mathcal{A}_1)}(X_n)$ and $\mathcal{F}_{V(\mathcal{A}_2)}(X_n)$ be the free algebras relative to $V(\mathcal{A}_1)$ and to $V(\mathcal{A}_2)$, respectively. We have seen that the clone of an algebra can be regarded as a multi-based algebra where the m-ary operations for all $0 \leq m \leq n$ are the different sorts and where the operations are the superposition operations. It is also well-known that $clone_n\mathcal{A}_1$ is isomorphic to the clone of the free algebra $\mathcal{F}_{V(\mathcal{A}_1)}(X_n)$. Here we have $Id\mathcal{A}_1 = Id\mathcal{A}_2$, which tells us that $\mathcal{F}_{V(\mathcal{A}_1)}(X_n) = \mathcal{F}_{V(\mathcal{A}_2)}(X_n)$ and therefore the clones are equal. But then $clone_n \mathcal{A}_1$ and $clone_n \mathcal{A}_2$ are isomorphic. Let $C_n'^{\mathcal{A}_2}$ be the image of $C_n^{\mathcal{A}_1}$ under this isomorphism and let $\sigma'^{\mathcal{A}_2}$ be the composition of $\sigma^{\mathcal{A}_1}$ with this isomorphism. Using this mapping it can be shown that $T(H\mathbf{A}_1) = T(H\mathbf{A}_2)$.   ■

Now we are ready to prove that for finite alphabets the concepts of recognizability and hyperrecognizability are equivalent.

Theorem 15.2.7 *When X_n is a finite alphabet, a $\Sigma - X_n$ language is hyperrecognizable iff it is recognizable.*

Proof: Since we have already shown that any hyperrecognizable language is recognizable, we have only to prove the converse. Let T be a recognizable $\Sigma - X_n$ language, with recognizer $\mathbf{A} = (\mathcal{A}, \Sigma, X, \alpha, A')$ such that $T = T(\mathbf{A})$. We denote by $T(\mathbf{A})^{\mathcal{A}}$ the set of all term operations which are induced by the terms from $T(\mathbf{A})$. Note that the term operations from $T(\mathbf{A})^{\mathcal{A}}$ are n-ary. We set $C_n^{\mathcal{A}} = T(\mathbf{A})^{\mathcal{A}}$ and consider the hyperrecognizer $H\mathbf{A} = (clone_n \mathcal{A}, \Sigma, X_n, \sigma_{id}^{\mathcal{A}}, C_n^{\mathcal{A}})$, where $\sigma_{id}^{\mathcal{A}}$ is the mapping which maps each operation symbol f from Σ to the induced fundamental term operation $f^{\mathcal{A}}$ from $\mathcal{A}$. Then we have:

$$t \in T(\mathbf{A}) \Leftrightarrow (\sigma_{id}^{\mathcal{A}})\hat{}[t] = t^{\mathcal{A}} \in T(\mathbf{A})^{\mathcal{A}} = C_n^{\mathcal{A}},$$

and this means that $T(\mathbf{A}) = T(H\mathbf{A})$.   ■

Next we describe another way to construct a tree-recognizer equivalent to a given hyperrecognizable language, which illustrates the interconnections between hyperrecognizers and equivalent tree-recognizers. Assume that T is hyperrecognizable. Then there exists a hyperrecognizer

$$H\mathbf{A} = (clone_n \mathcal{A}, \Sigma, X_n, \sigma^{\mathcal{A}}, C_n^{\mathcal{A}})$$

such that $T = T(H\mathbf{A})$. We want to show that there is a tree-recognizer $\mathbf{A}$ such that $T(H\mathbf{A}) = T(\mathbf{A})$.

Since $C_n^{\mathcal{A}}$ is a finite subset of $clone_n \mathcal{A}$ we can write $C_n^{\mathcal{A}} = \{s_1^{\mathcal{A}}, \ldots, s_m^{\mathcal{A}}\}$. Consider the tree-hyperrecognizers

$$H\mathbf{A}_i = (clone_n \mathcal{A}, \Sigma, X_n, \sigma^{\mathcal{A}}, \{s_i^{\mathcal{A}}\}),$$

for $1 \leq i \leq m$. Then we have

$$
\begin{aligned}
t \in T(H\mathbf{A}) \quad &\Leftrightarrow \quad (\sigma^{\mathcal{A}})\widehat{\ }[t] \in C_n^{\mathcal{A}} \\
&\Leftrightarrow \quad (\sigma^{\mathcal{A}})\widehat{\ }[t] = s_i^{\mathcal{A}} \ \text{for some} \ i \in \{1, \dots, m\} \\
&\Leftrightarrow \quad t \in T(H\mathbf{A}_i) \ \text{for some} \ i \in \{1, \dots, m\} \\
&\Leftrightarrow \quad t \in \bigcup_{i=1}^{m} T(H\mathbf{A}_i).
\end{aligned}
$$

Therefore $T(H\mathbf{A}) = \bigcup_{i=1}^{m} T(H\mathbf{A}_i)$.

Consider now an n-ary term $t \in T(H\mathbf{A}_i)$. Then we have

$$
\begin{aligned}
t \in T(H\mathbf{A}_i) \quad &\Rightarrow \quad (\sigma^{\mathcal{A}})\widehat{\ }[t] = s_i^{\mathcal{A}} \\
&\Rightarrow \quad (\hat{\sigma}[t])^{\mathcal{A}} = s_i^{\mathcal{A}} \ \text{if} \ \sigma \ \text{is a hypersubstitution} \\
&\qquad \text{satisfying} \ (\sigma^{\mathcal{A}})\widehat{\ }[t] = (\hat{\sigma}[t])^{\mathcal{A}} \\
&\qquad \text{for all} \ t \in W_\Sigma(X_n) \\
&\Rightarrow \quad (\hat{\sigma}[t])^{\mathcal{A}}(a_1, \dots, a_n) = s_i^{\mathcal{A}}(a_1, \dots, a_n) \\
&\qquad \text{for all} \ a_1, \dots, a_n \in A.
\end{aligned}
$$

Since the image of $s_i^{\mathcal{A}}$ is finite we can write $Im s_i^{\mathcal{A}} = \{c_{i_1}, \dots, c_{i_{k_i}}\}$ where $c_{i_1}, \dots, c_{i_{k_i}} \in A$. Consider the tree-recognizers $\mathbf{A}_{i_l} = (A, \Sigma, X_n, \alpha'_{i_l}, \{c_{i_l}\})$, for $1 \le l \le k_i$. Here the evaluation mapping α'_{i_l} is defined by $\hat{\alpha}'_{i_l} = \hat{\alpha}_{i_l} \circ \hat{\sigma}$ where α_{i_l} maps $(x_1, \dots, x_n)$ to the n-tuple $(s_i^{\mathcal{A}})^{-1}(c_{i_l})$ for all $l = 1, \dots, k_i$ and σ is a hypersubstitution with $(\hat{\sigma}[t])^{\mathcal{A}} = (\sigma^{\mathcal{A}})\widehat{\ }[t]$. Then we have

$$
\begin{aligned}
t \in T(H\mathbf{A}_i) \quad &\Leftrightarrow \quad (\sigma^{\mathcal{A}})\widehat{\ }[t] = s_i^{\mathcal{A}} \\
&\Leftrightarrow \quad (\hat{\sigma}[t])^{\mathcal{A}} = s_i^{\mathcal{A}} \\
&\Leftrightarrow \quad \hat{\alpha}_{i_l}[\hat{\sigma}[t]] = \hat{\alpha}_{i_l}[s_i] \ \text{for all} \ l = 1, \dots, k_i \\
&\Leftrightarrow \quad (\hat{\alpha}_{i_l} \circ \hat{\sigma})[t] = c_{i_l} \ \text{for all} \ l = 1, \dots, k_i \\
&\Leftrightarrow \quad t \in T(\mathbf{A}_{i_l}) \ \text{for all} \ l = 1, \dots, k_i \\
&\Leftrightarrow \quad t \in \bigcap_{l=1}^{k_i} T(\mathbf{A}_{i_l}),
\end{aligned}
$$

showing that $T(H\mathbf{A}_i) = \bigcap_{l=1}^{k_i} T(\mathbf{A}_{i_l})$.

As we saw in Chapter 8, it is well known that the intersection of recognizable languages is recognizable (see for instance [48]). Therefore there exists a tree-recognizer $\mathbf{A}_i$ such that $T(\mathbf{A}_i) = \bigcap_{l=1}^{k_i} T(\mathbf{A}_{i_l})$ for each $i = 1, \dots, m$ and then also $T(H\mathbf{A}) = \bigcup_{i=1}^{m} T(\mathbf{A}_i)$.

Since the join of recognizable languages is also recognizable, there is a tree-recognizer $\mathbf{B}$ such that $T(\mathbf{B}) = \bigcup_{i=1}^{m} T(\mathbf{A}_i)$ and then $T(H\mathbf{A}) = T(\mathbf{B})$.

Tree-recognizers were introduced as generalizations of finite automata independently by J. E. Doner ([40], [41]) and by J. W. Thatcher and J. B. Wright ([114], [115]). They can be defined for both finite and infinite alphabets. Our definition and results so far have been for hyperrecognizers with a finite alphabet. But if we extend our definition to the case of a countably infinite alphabet X, it turns out that recognizability and hyperrecognizability are no longer equivalent.

Proposition 15.2.8 *If $T \subseteq W_\Sigma(X)$ is a recognizable language for which there is no $n \in I\!\!N$ such that $T \subseteq W_\Sigma(X_n)$, then T is not hyperrecognizable.*

Proof: Assume that T is hyperrecognizable. Then there is a hyperrecognizer $H\mathbf{A} = (clone_n\mathcal{A}, \Sigma, X_n, \sigma^\mathcal{A}, C_n^\mathcal{A})$ such that $T = T(H\mathbf{A})$. Since $T \nsubseteq W_\Sigma(X_n)$, there is a tree $t \in T$ with $t \notin W_\Sigma(X_n)$. Therefore $t^\mathcal{A} \notin clone_n\mathcal{A}$, $(\sigma^\mathcal{A})\hat{}[t] \notin clone_n\mathcal{A}$ and $(\sigma^\mathcal{A})\hat{}[t] \notin C_n^\mathcal{A}$. This contradicts $t \in T = T(H\mathbf{A})$. ∎

Example 15.2.9 Here is an example of a language which is recognizable but not hyperrecognizable. Let us take $\Sigma = \Sigma_2 = \{f\}$ and $T := \{f(x_1, x_j) \mid 2 \leq j \in I\!\!N\}$. For our tree-recognizer, we let $\mathcal{A} = (\{a, b, c, d\}; f^\mathcal{A})$, with $f^\mathcal{A}(a, b) = c$ and $f^\mathcal{A}(x, y) = d$ on all other inputs. We choose $A' = \{c\}$, and let α be defined by $\alpha(x_1) = a$ and $\alpha(x_j) = b$ for all $2 \leq j \in I\!\!N$. We will show that the tree-recognizer $\mathbf{A} = (\mathcal{A}, \Sigma, X, \alpha, A')$ recognizes our language T. If $t \in T(\mathbf{A})$, then $\hat{\alpha}[t] = c$ and t cannot be a variable. This means there exist terms $r, s \in W_\Sigma(X)$ such that $t = f(r, s)$. Then $\hat{\alpha}[t] = f^\mathcal{A}(\hat{\alpha}[r], \hat{\alpha}[s]) = c$ and $\hat{\alpha}[r] = a$, $\hat{\alpha}[s] = b$. If $r, s \in W_\Sigma(X)\backslash X$ we have $\hat{\alpha}[r], \hat{\alpha}[s] \in \{c, d\}$. Therefore $r = x_1$ and $s \in \{x_i \mid 2 \leq i \in I\!\!N\}$. Thus $t = f(x_1, x_j)$ for some natural number $j \geq 2$, and we have shown that $T(\mathbf{A}) \subseteq T$. Conversely, if $t \in T$, then $\hat{\alpha}[t] = f^\mathcal{A}(a, b) = c$, so that $t \in T(\mathbf{A})$.

Thus T is recognizable. However, since there is no $n \in I\!\!N$ such that $T \subseteq W_\Sigma(X_n)$, we see by Proposition 15.2.8 that T is not hyperrecognizable.

We remark that our approach to hyperrecognizers uses the concept of a clone. Instead of clones one could also consider algebraic theories in the sense of Lawvere, as was done by Z. Ésik in [43].

15.3 Tree Transformations

The tree transducers of Section 8.8 are generalizations of automata: just as automata transform words (terms of a particular type) into words, transducers transform terms of any one fixed type into terms of a second fixed type. In doing so they produce transformations, which are sets consisting of pairs of trees, where the first components are trees from the first type or language and the second components are trees of the second type.

Our definition of a hypersubstitution, from Section 14.3, gave us a mapping which took operation symbols of one type or language to terms of that same type. Now we generalize this definition too, to include mappings from operation symbols of one language into terms of a second language. We also consider the corresponding tree transformations. We shall prove that the set of all tree transformations which are defined by hypersubstitutions of a given type forms a monoid with respect to the composition of binary relations, and that this monoid is isomorphic to the monoid of all hypersubstitutions of this type. We characterize transitivity, reflexivity and symmetry of tree transformations by properties of the corresponding hypersubstitutions. The results will be illustrated for type (2), with $\Sigma = \Sigma_2 = \{f\}$.

In general, let $\Sigma := \{f_i \mid i \in I\}$ be a set of operation symbols of type $\tau_1 = (n_i)_{i \in I}$, where f_i is n_i-ary, $n_i \in \mathbb{N}$ and let $\Omega = \{g_j \mid j \in J\}$ be a set of operation symbols of type $\tau_2 = (n_j)_{j \in J}$ where g_j is n_j-ary. As usual, we denote by $W_{\tau_1}(X)$ and by $W_{\tau_2}(X)$ the sets of all terms of types τ_1 and τ_2, respectively.

Definition 15.3.1 A $(\tau_1 - \tau_2)$-*hypersubstitution* is a mapping

$$\sigma : \{f_i \mid i \in I\} \to W_{\tau_2}(X)$$

which maps each operation symbol f_i of type τ_1 to a term $\sigma(f_i)$ of type τ_2 of the same arity as f_i.

As before, every $(\tau_1 - \tau_2)$-hypersubstitution σ can be extended to a mapping

$$\hat{\sigma} : W_{\tau_1}(X) \to W_{\tau_2}(X)$$

in the following inductive way:

(i) $\hat{\sigma}[x] := x$,

(ii) $\hat{\sigma}[f_i(t_i, \ldots, t_{n_i})] := \sigma(f_i)(\hat{\sigma}[t_i], \ldots, \hat{\sigma}[t_{n_i}])$.

Definition 15.3.2 Let σ be a $(\tau_1 - \tau_2)$-hypersubstitution. Then

$$T_\sigma := \{(t, \hat{\sigma}[t]) \mid t \in W_{\tau_1}(X)\}$$

is called the *tree transformation defined by* σ.

When the types τ_1 and τ_2 are the same, this definition reduces to the usual definition of a (type τ_1) hypersubstitution. But there is another way to consider these new mixed-type hypersubstitutions as hypersubstitutions of one type. Given the two different types τ_1 and τ_2, we form a new type τ by taking the union of the types; that is, we form the union $\Sigma \cup \Omega$ of the sets of operation symbols of the two types. As we saw in Section 14.3, the set of all type τ hypersubstitutions forms a monoid $Hyp(\tau)$, under the composition operation

$$\sigma_1 \circ_h \sigma_2 := \hat{\sigma}_1 \circ \sigma_2.$$

Now every $(\tau_1 - \tau_2)$-hypersubstitution can be considered as a hypersubstitution of type τ which fixes the symbols of type τ_2, in the sense that each operation symbol g_j of type τ_2 is mapped to the fundamental term $g_j(x_1, \ldots, x_{n_j})$. Since the composition of two hypersubstitutions which fix the operation symbols from Ω is again a hypersubstitution which fixes the operation symbols from Ω, the set of all $(\tau_1 - \tau_2)$-hypersubstitution regarded as hypersubstitutions of the type τ forms a submonoid of the monoid $Hyp(\tau)$. This allows us to consider tree transformations T_σ where $\sigma \in Hyp(\tau)$, $t \in W_\tau(X)$ and $\hat{\sigma}[t] \in W_\tau(X)$.

Tree transducers were defined in Section 8.8. It turns out that tree transformations T_σ for a hypersubstitution σ are induced by tree transducers. The following proposition is a special case of 8.8.5, since the extensions of hypersubstitutions are a particular kind of tree homomorphisms as introduced in Section 8.6.

Proposition 15.3.3 *If σ is a $(\tau_1 - \tau_2)$-hypersubstitution and if $\underline{A} = (\Sigma, X, A, \Omega, P, A')$ is the tree transducer with $A = A' = \{a\}$ and $P = \{x \to_A ax \mid x \in X\} \cup \{f_i(a(\xi_1), \cdots, a(\xi_{n_i})) \to_A a\sigma(f_i)(\xi_1, \cdots, \xi_{n_i}) \mid i \in I\}$, then $T_{\underline{A}} = T_\sigma$.* ∎

Note that tree transducers considered in the previous proposition are examples of the *H-transducers* introduced in Section 8.8.

In the remainder of this section we assume that we have only one type τ. We denote by $T_{\sigma_1} \circ T_{\sigma_2}$ the composition of the tree transformations T_{σ_1} and T_{σ_2}. We can also consider inverses, domains and ranges of tree transformations. We define $T_{Hyp(\tau)} := \{T_\sigma \mid \sigma \in Hyp(\tau)\}$.

Theorem 15.3.4 $(T_{Hyp(\tau)}; \circ, T_{\sigma_{id}})$ *is a monoid which is isomorphic to the monoid* $Hyp(\tau)$ *of all hypersubstitutions of type* τ.

Proof: We define a mapping $\varphi : Hyp(\tau) \to T_{Hyp(\tau)}$ by $\sigma \mapsto T_\sigma$. Clearly, φ is well defined. To show that φ is a homomorphism, we will show that $T_{\sigma_1} \circ T_{\sigma_2} = T_{\sigma_1 \circ_h \sigma_2}$, so that $\varphi(\sigma_1 \circ_h \sigma_2) = \varphi(\sigma_1) \circ \varphi(\sigma_2)$. We have

$$
\begin{aligned}
& (t, t'') \in T_{\sigma_1} \circ T_{\sigma_2} \\
\Leftrightarrow\ & \exists t'((t, t') \in T_{\sigma_2} \text{ and } (t', t'') \in T_{\sigma_1}) \\
\Leftrightarrow\ & t' = \hat{\sigma}_2[t] \text{ and } t'' = \hat{\sigma}_1[t'] \\
\Leftrightarrow\ & t'' = \hat{\sigma}_1[\hat{\sigma}_2[t]] \\
\Leftrightarrow\ & t'' = (\sigma_1 \circ_h \sigma_2)\hat{\ }[t] \\
\Leftrightarrow\ & (t, t'') \in T_{\sigma_1 \circ_h \sigma_2}.
\end{aligned}
$$

To see that φ is one-to-one, let $T_{\sigma_1} = T_{\sigma_2}$. Then for all $t \in W_\tau(X)$ we have $\hat{\sigma}_1[t] = \hat{\sigma}_2[t]$. But this means that for all operation symbols f_i we also have

$$
\hat{\sigma}_1[f_i(x_i, \cdots, x_{n_i})] = \sigma_1(f_i) = \sigma_2(f_i) = \hat{\sigma}_2[f_i(x_i, \cdots, x_{n_i})],
$$

and therefore $\sigma_1 = \sigma_2$. Finally, since $T_{\sigma_1} \circ T_{\sigma_2} = T_{\sigma_1 \circ_h \sigma_2}$, the tree transformation $T_{\sigma_{id}}$ is an identity element with respect to the composition $\circ$. ∎

The previous theorem now allows us to describe properties of the relation T_σ by properties of the hypersubstitution σ, and vice versa.

Theorem 15.3.5 *Let* $\sigma \in Hyp(\tau)$ *be a hypersubstitution of type* τ *and let* T_σ *be the corresponding tree transformation. Then*

(i) T_σ *is transitive iff* σ *is idempotent,*

(ii) T_σ *is reflexive iff* $\sigma = \sigma_{id}$,

(iii) T_σ *is symmetric iff* $\sigma \circ_h \sigma = \sigma_{id}$.

Proof: (i) When σ is idempotent, we have $T_{\sigma \circ_h \sigma} = T_\sigma \circ T_\sigma = T_\sigma$ by Theorem 15.3.4, and T_σ is transitive. Conversely, when T_σ is transitive, we have $T_\sigma \circ$

$T_\sigma \subseteq T_\sigma$, so that $T_{\sigma \circ_h \sigma} \subseteq T_\sigma$. Then

$$(t_1, (\sigma \circ_h \sigma)\,\hat{}[t]) \in T_{\sigma \circ_h \sigma} \Rightarrow (t_1, (\sigma \circ_h \sigma)\,\hat{}[t]) \in T_\sigma \Rightarrow (\sigma \circ_h \sigma)[t] = \hat{\sigma}[t],$$

for all $t \in W_\tau(X)$, and σ is idempotent.

(ii) Assume that T_σ is reflexive, so that $T_{\sigma_{id}} = \triangle_{W_\tau(X)} \subseteq T_\sigma$. Therefore $(t, t) \in T_\sigma$ for all $t \in W_\tau(X)$ and then $\hat{\sigma}[t] = t$ for all $t \in W_\tau(X)$, making $\sigma = \sigma_{id}$.

If conversely $\sigma = \sigma_{id}$, then $T_{\sigma_{id}} = \{(t, \hat{\sigma}_{id}[t]) \mid t \in W_\tau(X)\} = \{(t, t) \mid t \in W_\tau(X)\} = \triangle_{W_\tau(X)}$ and T_σ is reflexive.

(iii) If T_σ is symmetric, then for all $t \in W_\tau(X)$ we have

$$(t, \hat{\sigma}[t]) \in T_\sigma \Rightarrow (\hat{\sigma}[t], t) \in T_\sigma.$$

Therefore $t = \hat{\sigma}[\hat{\sigma}[t]]$ and $\hat{\sigma}_{id}[t] = (\sigma \circ_h \sigma)\,\hat{}[t]$ for all $t \in W_\tau(X)$, and we have $\sigma \circ_h \sigma = \sigma_{id}$.

If conversely $\sigma \circ_h \sigma = \sigma_{id}$ then we have $T_{\sigma \circ_h \sigma} = T_\sigma \circ T_\sigma = T_{\sigma_{id}}$. But this means $T_\sigma = (T_\sigma)^{-1}$, and T_σ is symmetric. ∎

In general, the range of a tree transformation σ, which is the set

$$\hat{\sigma}(W_\tau(X)) = \{t' \mid \exists t \in W_\tau(X)(t' = \hat{\sigma}[t])\},$$

is a subset of $W_\tau(X)$. Therefore, we consider T_σ as a relation between $W_\tau(X)$ and $\sigma(W_\tau(X))$, so that $T_\sigma \subseteq W_\tau(X) \times \sigma(W_\tau(X))$. We notice that $T_\sigma \circ (T_\sigma)^{-1} = T_{\sigma_{id}} = \triangle_{W_\tau(X)}$ and that $(T_\sigma)^{-1} \circ T_\sigma = \{(t, t') \mid \hat{\sigma}[t] = \hat{\sigma}[t']\} = \ker\sigma$ (the kernel of σ). This gives the following result.

Proposition 15.3.6 *Let $\sigma \in Hyp(\tau)$ be a hypersubstitution of type τ and let $T_\sigma = W_\tau(X) \times \hat{\sigma}(W_\tau(X))$ be the corresponding tree transformation. Then T_σ is bijective iff $\ker\sigma = \triangle_{W_\tau(X)} = T_{\sigma_{id}}$.*

Proof: T_σ is bijective iff $T_\sigma \circ (T_\sigma)^{-1} = (T_\sigma)^{-1} \circ T_\sigma = T_{\sigma_{id}} = \triangle_{W_\tau(X)}$. Now we use the previous remark. ∎

In Theorem 15.1.3 it was shown that for the type $\tau = (n)$ with $n \geq 2$, any regular hypersubstitution σ has the property that $\ker\sigma$ is equal to the diagonal relation $\triangle_{W_\tau(X)}$.

Corollary 15.3.7 *Let σ be a regular hypersubstitution of type $\tau = (n)$, for $n \geq 2$, and let T_σ be the tree transformation defined by σ. Then T_σ is bijective.* ∎

15.4 Exercises

15.4.1. Let V be a variety and σ be a hypersubstitution of type τ. The set

$$T_\sigma^V := \{(t,t') \mid t,t' \in W_\tau(X) \text{ and } \hat{\sigma}[t] \approx t' \in IdV\}$$

is called the V-tree-transformation defined by σ. Prove that when $\sigma_1 \sim_V \sigma_2$, under the relation $\sim_V$ defined just after Proposition 15.2.5, then $T_{\sigma_1}^V = T_{\sigma_2}^V$.

15.4.2. For which hypersubstitutions σ is it the case that

$$T_{\sigma_1}^V \circ T_{\sigma_2}^V = T_{\sigma_1 \circ \sigma_2}^V?$$

15.4.3. Prove that T_σ^V is surjective iff $T_\sigma^V \circ (T_\sigma^V)^{-1} = IdV$.

15.4.4. Prove that T_σ^V is injective iff $(T_\sigma^V)^{-1} \circ (T_\sigma^V) = ker_V \sigma = \Delta_{W_\tau(X)}$.

15.4.5. A mapping $\sigma : \{f_i \mid i \in I\} \to W_\tau(X)$, from the set of operation symbols of a type τ to the set of terms of that type, which does not necessarily preserve arities, is called a *generalized hypersubstitution*. Prove that the kernel of a generalized hypersubstitution σ is a fully invariant congruence relation on the free algebra $\mathcal{F}_\tau(X)$ iff σ does not map any n_i-ary operation symbol f_i to a variable different from $x_1, \ldots, x_{n_i}$.

Bibliography

[1] Arworn, S., *Groupoids of Hypersubstitutions and G-solid varieties*, Shaker-Verlag, Aachen, 2000.

[2] Arworn, S. and K. Denecke, *Intervals defined by M-solid varieties*, in: General Algebra and Applications, Proc. 59th Workshop on General Algebra and 15th Conference for Young Algebraists, Potsdam 2000, Shaker-Verlag, Aachen, 2000, 1 - 18.

[3] Arworn, S. and K. Denecke, *Left- and right-edge solid varieties of entropic groupoids*, Demonstratio Mathematica, Vol. XXXII No. 1 (1999), 1 - 11.

[4] Arworn, S., K. Denecke and R. Pöschel, *Closure Operators on Complete Lattices*, to appear in Proc. International Conference on Ordered Algebraic Structures, Nanjing, 1998.

[5] Baker, K. A., *Finite equational bases for finite algebras in congruence-distributive equational classes*, Adv. in Math. **24** (1977), 201 - 243.

[6] Baker, K. and A. Pixley, *Polynomial interpolation and the Chinese Remainder Theorem for algebraic systems*, Math. Z. **143** (1975), 165 - 174.

[7] Berman, J., *A proof of Lyndon's finite basis theorem*, Discrete Math. **29** (1980), 229 - 233.

[8] Birkhoff, G., *The structure of abstract algebras*, Proc. Cambridge Philosophical Society **31** (1935), 433 - 454.

[9] Birkhoff, G. and J. D. Lipson, *Heterogeneous algebras*, J. Combinat. Theory **8**(1970), 115 - 133.

[10] Cohn, P. M., *Universal Algebra*, Harper & Row, New York, 1965.

[11] Cowan, D. and S. L. Wismath, *Unary iterative hyperidentities for semigroups and inverse semigroups*, Semigroup Forum **55** (1997), 221 - 231.

[12] Cowan, D. and S. L. Wismath, *Unary hyperidentities for varieties of inverse semigroups*, Semigroup Forum, **58** (1999), 106 - 125.

[13] Csákány, B. and T. Gavalcová, *Finite homogeneous algebras I.*, Acta Sci. Math. (Szeged) **42** (1980), 57 - 65.

[14] Czedli, G., *A characterization for congruence semi-distributivity*, Universal Algebra and Lattice Theory, Proceedings Puebla 1982, Springer-Verlag, Berlin, Heidelberg, New York, Tokyo, 1983, 104 - 110.

[15] Davis, M. Computability and Unsolvability, New York, 1958.

[16] Day, A., *A characterization of modularity for congruence lattices of algebras*, Canad. Math. Bull. **12** (1969), 167 - 173.

[17] Demetrovics, J. and L. Hannák, *On the cardinality of self-dual closed classes in k-valued logics*, Közl.-MTA Számitástech. Aut. Kutató Int. Budapest, **23** (1979), 7 - 17.

[18] Demetrovics, J., L. Hannák and S.S. Marchenkov, *On closed classes of self-dual functions in P_3*, Colloq. Math. Soc. Janos Bolyai **28**, Finite Algebra and Multiple-valued logic Szeged (Hungary) 1979, 183 - 189.

[19] Demetrovics, J., L. Hannák and S.S. Marchenko, *On closed classes of self-dual functions in P_3* (Russian), Metodi diskretnogo analiza v reshenii kombinatornych zadach, Sbornik trudov Instituta Matematiki SO Akademii Nauk SSSR **34** (1980), 38 - 73.

[20] Denecke, K., *Preprimal Algebras*, Akademie-Verlag Berlin 1982.

[21] Denecke, K., *Eine Charakterisierung der funktionalen Vollständigkeit in kongruenzvertauschbaren Varietäten durch Hyperidentitäten*, Rostocker Mathematisches Kolloquium **36** (1989), 73 - 80.

[22] Denecke, K., *Clones closed with respect to closure operators*, Multi. Val. Logic, Vol. **4** (1999), 229 - 247.

[23] Denecke, K., L. Freiberg and J. Koppitz, *Algorithmic problems in M-solid varieties of semigroups*, in: Semigroups, Proceedings of the International Conference in Semigroup and its Related Topics, Kunming 1995, Springer-Verlag, Singapore, 1998, 104 - 117.

[24] Denecke, K. and H. Hounnon, *All solid varieties of semirings*, preprint, 2000.

[25] Denecke, K. and P. Jampachon, *Regular-solid varieties of commutative and idempotent groupoids*, Algebras and Combinatorics, Proc. of the International Congress, ICAC 97, Hong Kong, Springer-Verlag, Singapore, 1999, 177 - 188.

[26] Denecke, K. and J. Koppitz, *Hyperassociative varieties of semigroups*, Semigroup Forum **49** (1994), 41 - 48.

[27] Denecke, K. and J. Koppitz, *Pre-solid varieties of semigroups*, Archivum Mathematicum (Brno) **31** (1995), 171 - 181.

[28] Denecke, K., J. Koppitz and S. Niwczyk, *Equational Theories Generated by Hypersubstitutions of Type (n)*, preprint, 2001.

[29] Denecke, K. J. Koppitz and S. L. Wismath, *The Semantical Hyperunification Problem*, preprint, 2001.

[30] Denecke, K. and S. Leeratanavalee, *Weak hypersubstitutions and weakly derived algebras*, Contributions to General Algebra 11, Verlag Johannes Heyn, Klagenfurth 1999, 59 - 75.

[31] Denecke, K. and S. Leeratanavalee, *Solid polynomial varieties of semigroups which are definable by identities*, Contributions to General Algebra 12, Verlag Johannes Heyn, Klagenfurth 2000, 155 - 164.

[32] Denecke, K., D. Lau, R. Pöschel, and D. Schweigert, *Hyperidentities, Hyperequational classes and clone congruences*, Contributions to General Algebra 7, Verlag Hölder-Pichler-Tempsky, Wien, 1991, 97 - 118.

[33] Denecke, K. and N. Pabhapote, *Tree-recognizers and tree-hyperrecognizers*, Contributions to General Algebra **13**, Verlag Johannes Heyn, Klagenfurth, 2001, 107 - 114.

[34] Denecke, K. and T. Poomsa-ard, *Hyperidentities in graph algebras*, General Algebra and Applications in Discrete Mathematics, Shaker-Verlag, Aachen 1997, 59 - 68.

[35] Denecke, K. and M. Reichel, *Monoids of hypersubstitutions and M-solid varieties*, Contributions to General Algebra 9, Verlag Hölder-Pichler-Tempsky, Wien 1995 - Verlag B.G. Teubner, Stuttgart, 117 - 125.

[36] Denecke, K. and M. Reichel, *Hyperidentities in Quasigroups*, Beiträge zur Jahrestagung Algebra und Grenzgebiete, Güstrow, 1990, 67 - 75.

[37] Denecke, K. and S. L. Wismath, *Hyperidentities and Clones*, Gordon and Breach Science Publishers, 2000.

[38] Denecke, K. and S. L. Wismath, *Galois connections and complete sublattices*, preprint, 2001.

[39] Dikranjan, D. and E. Giuli, *Closure Operators I*, Topology Appl. **27**(1987), 129 - 143.

[40] Doner, J. E., *Decidability of the weak second-order theory of two successors, Generalized finite automata*, Notices Amer. Math. Soc. **12** (1965), abstract No. 65T-468, 819.

[41] Doner, J. E., *Tree acceptors and some of their applications*, J. CSS **4** (1970), 406 - 451.

[42] Eder, E., *Properties of substitutions and unifications*, J. Symbolic Computation, Vol.1 (1985), 31 - 46.

[43] Ésik, Z., *A variety theorem for trees and theories*, Publicationes Mathematicae, Debrecen, Tomus **54** Supplement (1999), 711 - 762.

[44] Fichtner, K., *Distributivity and modularity in varieties of algebras*, Acta Sci. Math.(Szeged) **33** (1972), 343 - 346.

[45] Foster, A. L., *Generalized Boolean theory of universal algebras, Part I*, Math. Zeitschr. **58** (1953), 306-336; Part II, Math. Zeitschr. **59** (1953), 191 - 199.

[46] Foster, A. L., *Functional completeness in the small. Algebraic structure theorems and identities*, Math. Ann. **143**(1961), 127 - 146.

[47] Ganter, B. and R. Wille, *Formale Begriffsanalyse*, Springer-Verlag, 1996.

[48] Gécseg, F. and M. Steinby, *Tree Automata*, Akademiai Kiado, Budapest, 1984.

[49] Gluschkow, W. M., G. J. Zeitlin and J. L. Justschenko, *Algebra, Sprachen, Programmierung*, Akademie-Verlag, Berlin, 1980.

[50] Gorlov, V. V. and R. Pöschel, *Clones closed with respect to permutation groups or transformation semigroups*, Beiträge zur Algebra und Geometrie **39** (1998), no. 1, 181 - 204.

[51] Grätzer, G., *Universal Algebra*, 2nd edition, Springer-Verlag, Berlin, Heidelberg, New York, 1979.

[52] Gumm, H. P., *Congruence modularity is permutability composed with distributivity*, Arch. Math. **36** (1981), 569 - 576.

[53] Hagemann, J. A. and A. Mitschke, *On n-permutable congruences*, Algebra Universalis **3** (1973), 8 - 12.

[54] *Handbook of Formal Languages*, Vol. 3, Springer, 1997.

[55] Higgins, P. J., *Algebras with a scheme of operators*, Math. Nachr. **27** (1963), 115 - 132.

[56] van Hoa, N., *On the structure of self-dual closed classes of three-valued logic*, Diskr. Mathematika **4** (1992), 82 - 95.

[57] Hobby, D. and R. McKenzie, *The Structure of Finite Algebras* (Tame Congruence Theory), AMS Contemporary Mathematics Series, Providence, Rhode Island, 1988.

[58] Ihringer, Th., *Allgemeine Algebra*, Verlag B.G. Teubner, Stuttgart, 1993.

[59] Jablonskij, S.V., *Functional constructions in multivalued logics*, (Russian), Trudy Inst. Mat. Steklov **51** (1958), 5 - 142.

[60] Jablonskij, S.V., G. P. Gawrilow and W. B. Kudrjawzew, *Boolesche Funktionen und Postsche Klassen*, Akademie-Verlag, Berlin, 1970.

[61] Jonsson, B., *Algebras whose congruence lattices are distributive*, Math. Scand. **21** (1967), 110 - 121.

[62] Kiss, E. W. and E. Pröhle, *Problems and results in tame congruence theory*, Mathematical Institute of the Hungarian Academy of Sciences, Preprint No. 60 (1988).

[63] Kleene, S. C., *Introduction to Metamathematics*, D. Van Nostrand Co., Inc., 1950.

[64] Klein, F., *Vergleichende Betrachtungen über neuere geometrische Forschungen*, Math. Ann. **43** (1893), 63 - 100.

[65] Knoebel, A., *The equational classes generated by single functionally precomplete algebras*, Memoirs of the Amer. Math. Soc. **57**, 332, Providence, Rhode Island, 1985.

[66] Knuth, D. E. and P.E. Bendix, *Simple Word Problem in Universal algebra*, in: Computational problems in abstract algebra, Pergamon Press, Oxford, 1970, 263 - 297.

[67] Koppitz, J. and S. L. Wismath, *Hyperidentities for varieties of star bands*, Sci. Math. **3** no. 3 (2000), 299 - 307.

[68] Kruse, R. L., *Identities satisfied by a finite ring*, J. Algebra **26** (1973), 298 - 318.

[69] Lau, D., *On closed subsets of Boolean functions, (A new proof for Post's theorem)*, J. Inform. Process. Cybernet. **EIK 27** (1991), 167 - 178.

[70] Lau, D., *Ein neuer Beweis für Rosenberg's Vollständigkeitskriterium*, J. Inform. Process. Cybernet. **EIK 28**, 4 (1992), 149 - 195.

[71] Lawvere, F. W., *Functorial semantics of algebraic theories*, Proc. Nat. Acad. Sci. **50** (1963), 869 - 872.

[72] Lyndon, R. C., *Identities in two-valued calculi*, Trans. Amer. Math. Soc. **71** (1954), 457 - 465.

[73] Lyndon, R. C., *Identities in finite algebras*, Proc. Amer. Math. Soc. **5** (1954), 8 - 9.

[74] Mal'cev, A.I., *On the general theory of algebraic systems*, (Russian), Mat. Sbornik **35**, 77 (1954), 3 - 20.

[75] Mal'cev, A.I., *Algebraic Systems*, Akademie-Verlag, Berlin 1973.

[76] Marchenkov, S.S., *On closed classes of self-dual functions of multiple-valued logic* (Russian), Problemy Kibernetiki **36**, 5 - 22.

[77] R. McKenzie, *Para-primal varieties: A study of finite axiomatizability and definable principal congruences in locally finite varieties*, Alg. Universalis **8** (1978), 336 - 348.

[78] McKenzie, R., *On minimal, locally finite varieties with permuting congruence relations*, preprint, 1976.

[79] McKenzie, R., *The residual bound of finite algebra is not computable*, J. of Algebra and Computation **6** No. 1 (1996), 29 - 48.

[80] McKenzie, R., *Tarski's finite basis problem is undecidable*, J. of Algebra and Computation **6** No. 1 (1996), 49 - 104.

[81] McKenzie, R., *The residual bounds of finite algebras*, J. of Algebra and Computation **6** No. 1 (1996), 1 - 28.

[82] McNulty, G., *Residual finiteness and finite equational bases: undecidable properties of finite algebras*, Lectures 2000.

[83] Murskij, V. L., *The existence in the three-valued logic of a closed class with a finite basis not having a finite complete system of identities*, Soviet Math. Dokl. **6** (1965), 1020 - 1024.

[84] Oates, S. and M. B. Powell, *Identical relations in finite groups*, J. Algebra **1** (1965), 11 - 39.

[85] Papert, D., *Congruence relations in semilattices*, London Math. Soc. **39** (1964), 723 - 729.

[86] Perkins, P., *Bases for equational theories of semigroups*, J. Algebra **11** (1969), 298 - 314.

[87] Pixley, A.F., *The ternary discriminator function in Universal Algebra*, Math. Ann. **191** (1971), 167 - 180.

[88] Pixley, A. F., *A note on hemi-primal algebras*, Math. Zeitschr. **124**(1972), 213 - 214.

[89] Pixley, A. F., *Functional and affine completeness and arithmetical varieties*, in: Algebras and orders, NATO Adv. Sci. Inst. Ser. C Math. Phys. Sci., 389, Kluwer Acad. Publ., Dordrecht, 1993, 317 - 357.

[90] Płonka, J. *Proper and inner hypersubstitutions of varieties*, in: Proceedings of the International Conference: Summer school on General algebra and ordered sets 1994, Palacky University Olomouc, 1994, 106 - 115.

[91] Polák, L., *On hyperassociativity*, Algebra Universalis **36** No. 3 (1996), 363 - 378.

[92] Polák, L. *All solid varieties of semigroups*, J. Algebra **219** no. 2 (1999), 421 - 436.

[93] Polin, S. V., *Identities of finite algebras*, Siberian Math. J. **17** (1976), 992 - 999.

[94] Post, E.L., *Introduction to a general theory of elementary propositions*, Amer. J. Math. **43** (1921), 163 - 185.

[95] Post, E.L., *The two-valued iterative systems of mathematical logic*, Ann. Math. Studies **5**, Princeton Univ. Press, 1941.

[96] Pöschel, R. and L. A. Kalužnin, *Funktionen- und Relationenalgebren*, VEB Deutscher Verlag der Wissenschaften, Berlin; 1979.

[97] Quackenbush, R.W., *A new proof of Rosenberg's primal algebra characterization theorem*, in: Finite Algebra and Multiple-valued Logic, (Proc. Conf. Szeged, 1979), Colloq. Math. Soc. J. Bolyai, North-Holland Amsterdam, vol. 28, 603 - 634.

[98] Reichel, M., Bi-Homomorphismen und Hyperidentitäten, Dissertation, Universität Potsdam, 1994.

[99] Reschke, M. and K. Denecke, *Ein neuer Beweis für die Ergebnisse von E.L. Post über abgeschlossene Klassen Boolescher Funktionen*, J. Inform. Process. Cybernet. **EIK 25** (1981), 361 - 380.

[100] Reschke, M., O. Lüders and K. Denecke, *Kongruenzdistributivität, Kongruenzvertauschbarkeit und Kongruenzmodularität zweielementiger Algebren*, J. Inform. Process. Cybern. EIK **24** (1988) 1/2, 65 - 78.

[101] Robinson, J. A., *A machine oriented logic based on the Resolution Principle*; Journal of the ACM, Vol. 12 (1965), 23 - 41.

[102] Rosenberg, I.G., *La structure des functions de plusieurs variables sur un ensemble fini*, C.R. Acad. Sci. Paris Ser. A-B **260** (1965), 3817 - 3819.

[103] Rosenberg, I.G., *Über die funktionale Vollständigkeit in den mehrwertigen Logiken*, Rozpr. ČSAV, Řada Mat. Přír. Věd. Praha **80**, 1 (1970), 3 - 93.

[104] Rousseau, G., *Completeness in finite algebras with a single operation*, Proc. Amer. Math. Soc. **18** (1967), 1009 - 1013.

[105] Schweigert, D., *Hyperidentities*, in: Algebras and Orders, Kluwer Academic Publishers, Dordrecht, Boston, London, 1993, 405-506.

[106] Šešelja, B. and A. Tepavčević, *On a partial closure operator*, preprint, 2000.

[107] Shum, K. P. and A. Yang, *Interior operators on complete lattices*, Pu. M. A. Ser. A, Vol. **3** (1992), No. 1-2, 73 - 80.

[108] Siekmann, J. H., *Universal Unification*, in: Lecture Notes in Computer Science, eds. G. Goos and J. Hartmanis, 7th International Conference on Automated Deductions, Napa, California, May 1984, Springer-Verlag, Berlin, Heidelberg, New York.

[109] Słupecki, J., *Completeness criterion for systems of many-valued propositional calculus*, Studia Logica **30** (1972), 153 - 157.

[110] Tardos, G., *A maximal clone of monotone operations which is not finitely generated*, Order **3** (1986), 211 - 218.

[111] Tarski, A., *A lattice theoretical fix point theorem and its application*, Pacific. J. Math. **5** (1955), 285 - 310.

[112] Taylor, W., *Characterizing Mal'cev conditions*, Algebra Universalis **3** (1973), 351 - 397.

[113] Thatcher, J. W., *Tree Automata: an informal survey - Currents in the theory of computing*, ed. A. V. Ano, Prentice-Hall, Englewood Cliffs, N J (1973), 143 - 172.

[114] Thatcher, J. W. and J. B. Wright, *Generalized finite automata*, Notices Amer. Math. Soc. **12.** (1965), abstract no. 65T-649, 820.

[115] Thatcher, J. W. and J. B. Wright, *Generalized finite automata theory with an application to a decision problem of second order logic*, MST **2** (1968), 57 - 81.

[116] Traczyk, T., *An equational definition of a class of Post algebras*, Bull. Acad.Pol. Sc. **12** (1964), 147 - 149.

[117] Webb, D. L., *Definition of Post's generalized negative and maximum in terms of one binary operation*, Amer. J. Math **58** (1936), 193 - 194.

[118] Werner, H., *Discriminator Algebras*, Akademie-Verlag, Berlin 1970.

[119] Willard, R., *Extending Baker's theorem*, Conference on Lattices and Universal Algebra (Szeged, 1998), Algebra Universalis **45** (2001), no. 2-3, 335 - 344.

[120] R. Wille, *Restructuring lattice theory: an approach based on hierarchies of concepts*, in Ordered Sets, ed. I. Rival, Reidel, Dordrecht-Boston, 1982, 445 - 470.

[121] Wille, R., Kongruenzklassengeometrien, Lecture Notes in Mathematics **113**, Springer, Berlin, 1970.

Index

abelian, 5
abelian algebra, 280, 292
abelian group, 272, 280
absolutely free algebra, 80
absorption law, 7
acceptor, 11
additive closure operator, 304
address of a node, 79, 350
affine algebra, 239
affine complete algebra, 241
$\text{Alg}(\tau)$, 109
algebra, 4
    abelian, 280, 292
    absolutely free, 80
    affine, 239
    Boolean, 9
    congruence modular, 203
    congruence regular, 205
    constantive, 246
    demiprimal, 241
    derived, 339, 353
    directly irreducible, 68
    equivalent, 220
    factor, 27
    finite, 251
    finitely axiomatizable, 110
    finitely based, 110
    finitely generated, 19
    free, 115
    full iterative, 10, 215
    functionally complete, 231

hemiprimal, 241
heterogeneous, 11, 159, 216
homogeneous, 11, 215
indexed, 4
induced on a set, 252
infinitely generated, 19
join-semidistributive, 281
locally finite, 108
meet-semidistributive, 281
minimal, 256, 258
multi-based, 11, 159
multi-sorted, 11
non-deterministic, 168
non-indexed, 4
one-based, 11
paraprimal, 241
permutation, 257, 269
polynomially equivalent, 298
Post, 233
preprimal, 245
primal, 231, 278, 286
quasiprimal, 241
quotient, 27, 115
regular subprimal, 241
relatively free, 98
semidistributive, 281
simple, 23, 71
singular subprimal, 241
solvable, 298
strongly abelian, 280
subdirectly irreducible, 71

subprimal, 241
tame, 262
term, 80
two-element, 206
affine complete, 241
cryptoprimal, 241
semiprimal, 240
algebraic lattice, 37
algebraic theories, 216
alphabet, 76
ranked, 165
arithmetical variety, 201
arity of a function, 3
Arworn, S., 310, 311, 344
associative law, 5
attribute, 42
automaton, 11, 147
deterministic, 158
equivalent, 162
finite, 12
initial, 157
non-deterministic, 158
quotient, 161
reduced, 163
weak initial, 157
with output, 158
automorphism, 49, 52
relative, 53
automorphism group, 52

Baker, K., 110, 199, 223, 224
Baker's Theorem, 112
Baker-Pixley Theorem, 199, 234
base set, 4
basis, 28
basis of identities, 110
Bendix, P.E., 130, 136, 139
Berman, J., 219
binary operation, 4

Birkhoff's Theorem, 106
Birkhoff, G., 72, 216
Boolean algebra, 110
Boolean clone lattice, 219
Boolean clones, 274
Boolean function, 38
dual, 218
linear, 219
monotone, 219
self-dual, 218
Boolean operation, 3, 206, 218
linear, 222
Boolean type, 276

canonical normal form, 123
categories, 216
center of a relation, 231
central relation, 231
centroid element, 246
centroid of an algebra, 246
Church-Rosser property, 121
Church-Rosser Theorem, 122
class
equational, 93
locally finite, 108
clone, 9, 38, 83, 87, 215
generated by a set of operations, 83
maximal, 245
relational, 218
polynomial, 231
term, 231
closed set, 32, 93
closure, 33
closure operator, 16, 32, 102, 303
additive, 304
inductive, 36
closure properties, 16, 25
closure system, 33

coatom, 262
Cohn, P., 37
commutative, 5
commutator, 291
    i-th iterated, 297
commutator congruence, 296
commutator group, 291
compact element, 37
compatibility, 22, 217
complement, 9
complement laws, 9
complete lattice, 93
complete reduction system, 123
completeness, 98
Completeness Criterion, 21
completion of a reduction system, 126
complexity of a term, 77
composition of operations, 9
confluent relation, 121
congruence, 161, 177
    fully invariant, 95
    generated by set, 25
    modulo m, 22
    n-permutability, 204
    permutable, 204
    regular, 205
    relation, 23
    tame, 263
    permutable, 193
congruence distributive algebra, 196
congruence distributive variety, 112, 196, 220, 274
congruence lattice of an algebra, 26
congruence modular algebra, 203
congruence modular variety, 203, 219

congruence permutable algebra, 204
congruence permutable variety, 274
congruence regular, 205
congruence regular algebra, 205
congruence regular variety, 205
conjugate pair of closure operators, 304
connected tree-recognizer, 179
consistency, 98
Consistency and Completeness Theorem, 98
constant, 86
constantive algebra, 246
Cowan, D., 344
critical pair, 126, 138
cryptoprimal algebra, 241
Csákány, B., 329
Czedli, G., 283

Davis, M., 116
Day, A., 203
decidable problem, 116
decision procedure, 116
deduction rules, 97, 130
Demetrovics, J., 329
demiprimal algebra, 241
Denecke, K., 86, 206, 224, 246, 248, 285, 311, 325, 332, 337, 342–344, 350–352
depth of a term, 77
derivation rules, 97, 169, 183
derived algebra, 339, 353
derived language, 154
deterministic automaton, 158
Dikranjan, D., 304
direct product, 63
directly irreducible algebra, 68

disjunctive normal form, 229
distributive laws, 6
Doner, J.E., 358
dual Boolean functions, 218

Eder, E., 134
effectively solvable problem, 115
embedding, 49
empty word, 148
endomorphism, 49, 52
    join-, 260
    meet-, 260
equation, 91
equational class, 93
equational logic, 98
equational theory, 93
    Main Theorem, 106
equationally complete variety, 109
equationally definable class of al-
        gebras, 93
equivalence relation, 115
equivalent algebras, 220
equivalent automata, 162
equivalent grammars, 170
equivalent reduction systems, 126
equivalent states, 162
Ésik, Z., 358
essentially binary operation, 269
evaluation mapping, 165
extended regular grammar, 172
extension of a function, 53
extensivity, 16, 25, 32, 260
extent, 42

factor algebra, 27
Fichtner, K., 198
field, 6, 32
finite algebra, 251
finite automaton, 12, 147

finite state machine, 147
finitely axiomatizable
    algebra, 110
    variety, 110
finitely based algebra, 110
finitely based variety, 110
finitely generated, 19
First Isomorphism Theorem, 58
fixed point, 52, 93
formal concept, 42
formal context, 42
formal language, 76
Foster, A.L., 233, 246
free algebra, 115
free monoid, 101
free semigroup, 101
Freiberg, L., 343
full clone, 9, 83
full iterative algebra, 10, 215
fully invariant congruence, 95
functional completeness, 215
functionally complete algebra, 231
fundamental operation, 4

G-clone, 329
G-relational clone, 329
Galois-closed subrelation, 310
Galois-connection, 40, 92, 217, 303
    $(H_M Id, H_M Mod)$, 338
    (GPol,GInv), 330
    (HPol, HInv), 334
    (Id,Mod), 92, 110, 337
    (Pol,Inv), 39, 217, 327
Ganter, B., 44, 311, 312
Gavalcová, T., 329
Gawrilow, G.P., 332
Gécseg, F., 176, 353, 357
generalized hypersubstitution, 363
generated, finitely, 19

generated, infinitely, 19
generating system, 16, 19
Gluschkow, W.M., 21
Gorlov, V.V., 329, 333
grammar
    extended regular, 172
    normal form, 171
    tree, 169
Grätzer, G., 37
graph of a relation, 3
graph of an operation, 14
graph-algebra, 29
group, 5
    abelian, 280
groupoid, 5
    abelian, 5
    commutative, 5
Guili, E., 304
Gumm, H.P., 203, 298

H-clones, 335
Hagemann, J.A., 204
Halting Problem, 187
Hannák, L., 329
hemiprimal algebra, 241
heterogeneous algebra, 11, 159, 216
Higgins, P.J., 216
Hoa, N. van, 329
Hobby, D., 251, 277, 281
homogeneous algebra, 11, 215
Homomorphic Image Theorem, 57
Homomorphic Image Theorem for Tree-Recognizers, 178
homomorphism, 49, 176
    natural, 50
Homomorphism Theorem for Automata, 161
Hounnon, H., 344

hyperassociative variety, 342
hyperidentity, 338
hypersatisfaction, 338
hypersubstitution, 175, 359
    leftmost, 341
    outermost, 341
    permutation, 342
    regular, 341
    symmetrical, 342
hyperunifiable terms, 350
hyperunifier, 350

idempotency, 16, 25, 32
idempotent law, 7
idempotent mapping, 252
identity, 91
Ihringer, Th., 36, 67
individual variable, 165, 169
induced algebra, 252
induction, principle of structural, 18
infinitely generated, 19
initial automaton, 157
initial symbol, 169
input alphabet, 185
intent, 42
interpolation property, 232
intersection of algebras, 15
invariant, 217
invertibility, 5
irreducible element, 119
isomorphism, 49
    relative, 53
Isomorphism Theorem
    First, 58
    Second, 60
iteration, 149

Jablonskij, S.V., 229, 332

Jampachon, P., 344
join operation, 7
join-endomorphism, 260
join-semidistributive, 281
Jonsson, B., 197, 201
Jonsson's Condition, 197
Justschenko, J.L., 21

K-free algebra, 98
Kalužnin, L.A., 39, 218
kernel, 161
    of a function, 22
    of a homomorphism, 54
kernel operator, 45, 321
kernel system, 45
Kiss, E.W., 282
Kleene, S.C., 187
Kleene's Theorem, 151, 157
Klein, F., 39
Knoebel, A., 248
Knuth, D.E., 130, 136, 139
Knuth-Bendix completion proce-
        dure, 130, 140
Knuth-Bendix ordering, 142
Knuth-Bendix Theorem, 139
Koppitz, J., 342–344, 350, 351
Kruse, R.L., 110
Kudrjawzew, W.B., 332

Lüders, O., 206, 285
language, 166
    derived, 154
    recognizable, 151, 166
    regular, 149
lattice, 6
    0-1-simple, 267
    algebraic, 37
    bounded, 8
    complete, 8, 36

distributive, 7, 110
modular, 7
of Boolean clones, 219
Post's, 219
tight, 267, 277
lattice homomorphism
    0-1-separating, 267
    0-separating, 267
    1-separating, 267
lattice type, 276
Lau, D., 219, 229, 337
Lawvere, F.W., 216
Leeratanavalee, S., 86
leftmost hypersubstitution, 341
length of a word, 148
linear Boolean function, 219
Lipson, J.D., 216
locally confluent reduction system,
        123
locally finite
    algebra, 108
    class, 108
    variety, 282, 286
loop, 271
Lyndon, R.C., 110

M-hyperidentity, 338
M-solid variety, 338
Main Theorem for Conjugate
        Pairs of Closure Opera-
        tors, 306
Main Theorem of Equational The-
        ory, 106
majority term, 196, 223, 234
Mal'cev, A.I., 193, 215, 229
Mal'cev term, 193, 194
Mal'cev-type condition, 193
Marchenkov, S.S., 329
match of terms, 133

maximal clone, 245
maximal subalgebra, 20
McKenzie, R., 112, 187, 222, 239,
      251, 277, 281
meet operation, 7
meet-endomorphism, 260
meet-semidistributive, 281
meet-semidistributive law, 112
minimal algebra, 256, 258
      type of, 276
minimal set, 258
minimal set of an algebra, 256
minimal tree-recognizer, 179
minimal variety, 109
Mitschke, A., 204
modular law, 203
module, 10
monoid, 5
monotone Boolean function, 219
monotone operation, 221
monotonicity, 16, 25, 32
multi-based algebra, 159
Murskij, V.L., 111

near-unanimity function, 224
Niwczyk, S., 350
noetherian relation, 120
non-deterministic
      algebra, 168
      automaton, 158
      operation, 167
      tree-recognizer, 168
non-terminating symbol, 169
normal form, 101, 119, 130
      canonical, 123
normal form grammar, 171
nullary operation, 4

Oates, S., 110

Omitting Type Theorem, 281
operation
      Boolean, 3, 206, 218
      depending on i-th variable,
            269
      essentially binary, 269
      fundamental, 4
      join, 7
      linear, 38
      meet, 7
      monotone, 221
      non-deterministic, 167
      nullary, 3
      on a set, 3
      polynomial, 87
      term, 82
operators H, S, P, 102
outermost hypersubstitution, 341

Pöschel, R., 39, 218, 311, 329, 333,
      337
Płonka, J., 355
Pabhapote, N., 352
Papert, D., 285
paraprimal algebra, 241
partial closure operator, 326
partial order, 7
Perkins, P., 111
permutable congruences, 193
permutable relations, 65
permutation algebra, 257, 269
permutation group, 257
permutation hypersubstitution,
      342
permutation-solid variety, 342
Pixley, A.F., 199, 223, 224, 235,
      244
Polák, L., 342
Polin, S.V., 111

polymorphism, 217
polynomial, 86
polynomial clone, 231
polynomial equivalence, 270, 298
polynomial isomorphism, 265
polynomial operations, 87
Poomsa-ard, T., 344
Post, E.L., 207, 218, 219, 228, 229,
        232, 233, 245, 332
Post Algebra, 233
Post's lattice, 219
Powell, M.B., 110
Pröhle, E., 282
pre-hypersubstitution, 342
prefix order, 132
preprimal algebra, 245
preservation relation, 38, 217, 327
presolid variety, 342
primal algebra, 231, 278, 286
prime interval, 277
Principle of Noetherian Induction,
        124
product
    direct, 63
    subdirect, 69
production, 169, 183
projection, 9, 64
proper order relation, 141

Quackenbush, R.W., 229
quasigroup, 6, 270
quasilinear function, 230
quasiprimal algebra, 241
quotient algebra, 27, 115
quotient automaton, 161

R-module, 10
ranked alphabet, 165
recognition relation, 147

recognizable
    language, 151, 166
    word, 151
recognizer, 11
reduced automaton, 163
reduced tree-recognizer, 179
reducible element, 119
reduct element, 119
reduction
    non-terminating, 120
    terminating, 119
reduction order, 141
reduction rule, 119, 129, 132
reduction system, 123
    complete, 123
    completion, 126
    equivalent, 126
    locally confluent, 123
reflexive closure, 117
reflexive relation, 246
regular expressions, 149
regular hypersubstitution, 341
regular language, 149
regular subprimal algebra, 241
regular-solid variety, 341
Reichel, M., 304, 332, 344
relation
    central, 231
    Church-Rosser, 121
    confluent, 121
    invariant, 39
    noetherian, 120
    proper order, 141
    reduction, 141
    reflexive, 246
    reflexive closure, 117
    symmetric closure, 117
    t-universal, 231
    terminating, 120

totally reflexive, 230, 246
totally symmetric, 230
transitive closure, 117
trivial, 23
relational clones, 218
relative automorphism, 53
relative isomorphism, 53
relatively free algebra, 98
replacement rule, 97, 130
Reschke, M., 206, 224, 285
residual bound, 112
residually finite variety, 112, 187
residually small variety, 112, 286
restriction, 53
    of a mapping, 251
    of an algebra to a set, 252
    of an equivalence relation, 252
    of an operation, 252
ring, 6
Robinson, J.A., 134
root of a tree, 78
Rosenberg, I.G., 229, 329, 330
Rousseau, G., 240
rules of consequence, 97

satisfaction, 91
scalar, 11
Schweigert, D., 337, 344
Second Isomorphism Theorem, 60
self-dual, 218
semantic tree, 78
semantical unifier, 350
semidistributive algebra, 281
semidistributive variety, 281
semigroup, 5
    free, 101, 148
    left-zero, 110
    right-zero, 110
    zero, 110

semilattice, 8, 110
semilattice type, 276
semiprimal algebra, 240
Sešelja, B., 326
Shum, K.P., 325
Siekmann, J.H., 350
simple algebra, 23, 71
singular subprimal algebra, 241
skew field, 6
Słupecki, J., 231
Słupecki criterion, 231
snag, 282
solid variety, 338
solvable algebra, 298
Steinby, M., 176, 353, 357
strong term condition, 279
strongly abelian algebra, 280
strongly extensive mapping, 260
structural induction, 18
subalgebra, 14
    maximal, 20
Subalgebra Criterion, 14
subalgebra lattice, 19
subautomaton, 159
subclone, 83
subdirect product, 69
subdirectly irreducible algebra, 71,
    104, 112
subprimal algebra, 241
substitution, 130, 349
substitution rule, 97, 130
subvariety, 109
subvariety lattice, 110
superposition, 9, 216
symbol
    initial, 169
    non-terminating, 169
symmetric closure, 117

symmetrical hypersubstitution, 342
syntactical hyperunifier, 350
syntactical unifier, 350
system of sets
  inductive, 35
  upward directed, 35

t-universal relation, 231
tame algebra, 262
tame congruence, 263
tape symbol, 185
Tardos, G., 248
Tarski, A., 322
Tarski's Finite Basis Problem, 187
Taylor, W., 216
Tepavčević, A., 326
term
  inductive definition, 76
  majority, 196
  Mal'cev, 193
  nullary, 77
term algebra, 80
term clone, 84, 220, 231
term condition, 280, 292
  strong, 279
term condition for congruences, 294
term operation, 82, 91
term reduction system, 129, 132
term rewriting system, 129, 132
terminating reduction, 119
terminating relation, 120
ternary discriminator, 242
Thatcher, J.W., 182, 358
tight lattice, 267, 277
totally reflexive relation, 230, 246
totally symmetric relation, 230
trace of a minimal algebra, 259

trace of an interval, 277
Traczyk, T., 234
transformation, 333
transitive closure, 117
transitive hull operator, 26
translation, 24
tree, 78
tree grammar, 169
tree homomorphism, 175
tree transducer, 183
tree transformation, 183, 360
tree-recognizer, 165
  connected, 179
  deterministic, 168
  minimal, 179
  non-deterministic, 168
  reduced, 179
trivial relation, 23
trivial variety, 109
Turing, A., 185
Turing machine, 185
type, 76
  of a minimal algebra, 276
  of a prime interval, 277
  of a tame algebra, 276
  of a tame interval, 277
  of a variety, 277
  of an algebra, 4
  of terms, 76

unar, 5
unary operation, 4
unary type, 276
unifiable terms, 133
unifier, 349
  semantical, 350
  syntactical, 350
universe of an algebra, 4

variable, 76
    individual, 169
variety, 102
    arithmetical, 201
    congruence distributive, 112, 196, 274
    congruence modular, 203
    congruence permutable, 274
    congruence regular, 205
    finitely based, 110
    generated by class K, 104
    hyperassociative, 342
    locally finite, 282, 286
    M-solid, 338
    minimal, 109
    permutation-solid, 342
    presolid, 342
    regular-solid, 341
    residually finite, 112, 187
    residually small, 112, 286
    semidistributive, 281
    solid, 338
vector, 11
vector space, 11, 32, 270
vector space type, 276

weak initial automaton, 157
Webb, D.L., 233
weight function, 142
Werner, H., 235
Willard, R., 113
Wille, R., 42, 44, 205, 311, 312
Wismath, S.L., 325, 344, 351
word, 148
word problem, 116, 349
words in the free semigroup, 101
Wright, J.B., 358

Yang, A., 325

yields, 97

Zeitlin, G.J., 21

Printed in the United States
by Baker & Taylor Publisher Services